Landmarks in Mechanical Engineering

LANDMARKS *in* MECHANICAL ENGINEERING

ASME International
History and Heritage

PURDUE UNIVERSITY PRESS
WEST LAFAYETTE INDIANA

01 00 99 98 97 5 4 3 2 1

The paper used in this book meets the minimum requirements of American National Standard for Information Sciences—Permanence of Paper for Printed Library Materials, ANSI Z39.48-1992.

Printed in the United States of America
Design by inari
Cover photo credits

Front: Icing Research Tunnel, NASA Lewis Research Center; *top inset,* Saturn V rocket; *bottom inset,* Wyman-Gordon 50,000-ton hydraulic forging press (Courtesy Jet Lowe, Library of Congress Collections

Back: top, Kaplan turbine; *middle,* Thomas Edison and his phonograph; *bottom,* "Big Brutus" mine shovel

Unless otherwise indicated, all photographs and illustrations were provided from the ASME landmarks archive.

Library of Congress Cataloging-in-Publication Data

Landmarks in mechanical engineering / ASME International History and Heritage.

p. cm.

Includes bibliographical references and index.
ISBN 1-55753-093-9 (cloth : alk. paper).—
ISBN 1-55753-094-7 (pbk. : alk. paper)
1. Mechanical engineering—United States—History.
2. Mechanical engineering—History.
I. American Society of Mechanical Engineers. History and Heritage Committee.

TJ23.L35 1996
621'.0973—dc20 96-31573
CIP

CONTENTS

Water Transportation

Rail Transportation

PREFACE

Historic mechanical engineering landmarks are machines, systems, or devices that help shape our civilization, either in industry or the personal lives we live. The variety of those presented on these pages is remarkable, ranging from marine steam engines to food-processing equipment to manufacturing plants to postage meters to medical devices to nuclear power plants to the collection of a specialized technical museum. They are found in every region of the United States as well as in other countries. The spectrum of significance is equally broad, stretching from the steam engine of Thomas Newcomen (1712), which was a major element in the advent of the Industrial Revolution, up to the Saturn V rocket (1967). While by no means a comprehensive list, these landmarks represent what the mechanical engineering profession considers to be both unusual and significant achievements through the eyes of ASME International (the American Society of Mechanical Engineers).

As a worldwide engineering society focused on technical, educational, and research issues, ASME International conducts one of the world's largest technical publishing operations, holds some thirty technical conferences and two hundred professional development courses each year, and sets many industrial and manufacturing standards. Since ASME's founding in 1880, engineers engaged in the mechanical arts and sciences have found a professional home in the American Society of Mechanical Engineers. Here they could meet and share ideas, plans, discoveries, and the results of their research. The History and Heritage program of ASME, in its present-day activities, began when a committee was formed in 1971 to administer its recognition program through a grassroots nomination process.

This book describes the 135 historic mechanical engineering landmarks designated by ASME International between 1973 and 1989. This publication is actually a successor to an earlier volume entitled *National Historic Mechanical Engineering Landmarks*, which was prepared by Richard S. Hartenberg, P.E., a distinguished founding member of ASME's History and Heritage Committee. Professor Hartenberg's work described the first twenty-eight landmarks, designated between 1973 and 1977.

In 1987 the committee decided that it was appropriate to produce a new publication that would bring the story up to date. A search was conducted for an author who could prepare the manuscript under the general supervision of the committee. Carol Poh Miller, a historical consultant based in Cleveland who has written widely in the areas of industrial and technological history, was selected.

Ms. Miller prepared the individual landmark entries and sidebars, organized the text, and selected the majority of the illustrations. She also conceived the idea of providing information on the location and accessibility of the landmarks, together with suggestions for further reading, for those who may wish to visit or learn more about them.

The committee has been closely involved with the preparation of the book. In addition to lending occasional technical expertise, the members of the committee have prepared the introductory essays that open each chapter. These are intended to set the material of the chapter in the broader context of the history of mechanical engineering. In reading these introductory essays, as well as the main body of the chapter, the committee hopes that the user will obtain an understanding of the topic in both a wider sense and, as far as particular landmarks are concerned, in greater depth. Illustrating these landmarks in ways that are useful to portraying their mechanical aspects has been challenging. There is a mix of those that reflect the time period in which the machines or businesses functioned alongside those that give the reader an idea of what to expect when visiting one today.

The landmarks are chosen by a careful procedure intended to ensure that they are truly outstanding examples of the art and science of mechanical engineering. They are nominated by the local sections of the society in whose geographical area the artifacts reside or by its technical divisions under which the technology can be categorized. A carefully documented statement of the credentials of the potential landmark is submitted to the History and Heritage Committee for review. The committee's membership consists of mechanical engineers with a good understanding of engineering history and historians with a sound background in the history of technology. If the committee members agree with the nominator's assessment of the significance of the nominated artifact, then it is designated as a Historic Mechanical Engineering Landmark. Sometimes the committee's deliberations involve a request for further information and research before the committee can be satisfied that the nominee is truly worthy of the landmark designation. Once an item has been so approved, the nominating section is responsible for arranging a ceremony at which a senior officer of the society, usually the president, presents to the owners of the landmark a bronze plaque, attesting to the status of the landmark and the reasons for its being so distinguished. The plaque can then be displayed on or near the landmark. The landmarks program is a continuing activity of the society and, at the time of publication, nearly two hundred landmarks have been identified.

The artifacts described here were chosen for a variety of reasons. In some cases, a landmark represents the beginning of a particular new technology. Another might be chosen because it was an outstanding representative of the mechanical engineer's art. Some were selected because they operated in the most efficient manner and were thereby examples to the profession of what could be achieved. Often the profession responded to the challenge and, subsequently, the performance of a landmark was surpassed. Occasionally size has played a part in the choice of a landmark in recognition of the achievement in designing, constructing, and operating a machine of unusually large dimensions. Very small dimensions could also be a qualifying factor, as exemplified by a landmark that was designated after completion of the manuscript of this book. This is the Texas Instruments ABACUS II, which is used to manipulate and solder connections to electronic microchips. Survival has always been an important criterion in the selection of landmarks, on the premise that this would be indicative of a sound original design and because it would give current and future generations an opportunity to study the work of earlier engineers.

The attention of the reader should be drawn to some practical points regarding visits to the landmarks. Directions for reaching each of the landmarks are provided wherever possible, but readers should note that in some cases prior arrangements will have to be made with the owners of the artifacts. Also, the directions must be treated with caution. In spite of the committee's best efforts to keep track of the landmarks, some may have been moved from the locations given in this book. Furthermore, in other cases the plaque has been separated from the artifact.

An additional point in visiting landmarks concerns the identification of landmarks by name. In certain cases, the designation used to identify a landmark does not correspond to the title on the plaque. This situation has arisen because of the evolution over time of the ownership or due to refinements by the History and Heritage Committee. We hope this will not confuse visitors to the various landmarks.

Unfortunately, historically important machines have not typically been easy to preserve. After years of service, obsolete machines are replaced and often scrapped or abandoned. Museums and corporate archives occasionally are able to store an artifact until suitable display can be arranged. For example, the Corning ribbon machine was rescued from warehouse storage for its display in the Henry Ford Museum in Dearborn, Michigan. Some industrial systems are simply too big for storage and therefore too costly to display. ASME will not remove a landmark from its roster if it is altered or destroyed in hopes that any remaining documentation will not be lost or forgotten as well.

The range of technologies represented and the chronological depth of the period covered by the landmarks gives the interested observer an outstanding opportunity to view, appreciate, and understand the work of the mechanical engineer. In a sense, the collection of landmarks represents a giant museum, and this book is a guide to that museum. It is a museum assembled by the energy and interest of mechanical engineers, with the intention of showing their fellow professionals and the public at large what mechanical engineers have wrought. ASME

International and its History and Heritage Committee hope very much that readers will obtain a greater understanding of mechanical engineering, particularly in its historical aspects, and in consequence be led to an enhanced appreciation of the world in which we currently live.

History and Heritage Committee
American Society of Mechanical Engineers

ACKNOWLEDGMENTS

The History and Heritage Committee would like to express its special appreciation to Carron Garvin-Donohue, formerly assistant director of the Public Information Department of the American Society of Mechanical Engineers. She was present at the committee's creation in 1971 and served with outstanding dedication as its staff support member until 1993. All the activities of the History and Heritage Committee and of the landmarks program in particular benefited greatly from her administrative and political skills.

The committee would also like to acknowledge the strong support of Patricia Smith, former director of the Public Information Department of ASME, as well as June Scangarello, her successor. Diane Kaylor, manager of special projects in that department, has also been of great help in the final stages of book preparation, including with indexing. Special thanks goes to Carolyn A. McGrew, the editor at Purdue University Press who led us through production.

Oscar Fisher, P.E., deserves special mention for his review of the manuscript to ensure that the units employed in the specifications of the various landmarks conformed to the best usage of the Système International d'Unités (SI).

The committee would also like to take this opportunity to acknowledge the support throughout the time of this book's preparation of the ASME Board on Public Information, the ASME Council on Public Affairs, and the society's Board of Governors. Their interest, and that of their respective chairs and of past and current presidents of the society, did much to ensure that the idea behind this book became a reality.

During the period 1987 to 1994 when the manuscript was being produced, the members of the History and Heritage Committee of the American Society of Mechanical Engineers were Robert B. Gaither; Richard S. Hartenberg, P.E.; J. Paul Hartman, P.E.; J. Lawrence Lee, P.E.; John H. Lienhard; Euan F. C. Somerscales; Joseph P. van Overveen, P.E.; Robert M. Vogel; and William J. Warren, P.E. Each of these individuals volunteered substantial effort to reviewing the manuscript, preparing essays, and providing general advice and assistance.

Too numerous to mention—but perhaps the most significant individuals in the History and Heritage landmark process—are the nominators and organizers for the landmarks themselves. These ASME members championed their projects throughout the documentation and ceremonial processes, providing us with the opportunity to recognize these examples of engineering excellence. Their public service to the society and the profession finds its reward in the legacy they leave for the future.

Landmarks in Mechanical Engineering

Pumping

INTRODUCTION by William J. Warren

Water. The life stream of our planet. Since time began, people have tried to move water. First by hand, then by animal power, and finally by using mechanical devices, we have tried to supply water as a necessity for life itself. At the same time, we have struggled to control the flow of water when it was convenient for our needs. Some of our earliest mechanical devices were designed to raise water from a stream for the irrigation of crops. The Archimedean screw, first developed in Hellenistic Greece, was one such device. This same principle was adapted nearly two thousand years later to transfer brine for salt production in southern San Francisco Bay.

One of our most famous mechanical devices was the Newcomen steam engine designed to dewater British coal mines in the early eighteenth century. The Industrial Revolution can be traced to our ability to harness steam to our use, and this pioneer pumping system allowed greatly expanded coal production, fueling the boilers of infant industries.

The distribution of water allowed the development of cities as we know them. Philadelphia's Fairmount Waterworks harnessed the energy of a river to fill reservoirs in the early nineteenth century. Later systems, such as the Leavitt-Riedler Pumping Engine in Boston, enabled the occupation of elevated sections of growing cities by supplying pressurized water systems. Still larger systems, such as the Chestnut Street Pumping Engine in Erie, Pennsylvania, and the Reynolds-Corliss Pumping Engine in Jacksonville, Florida, illustrate the development of steam-driven pumping systems to meet the ever expanding need of growing cities. One of the last of these is the Worthington Horizontal Cross-compound Pumping Engine in York, Pennsylvania, a complex design of the early twentieth century that soon was to be eclipsed by the centrifugal pump.

But drinking water was not the only municipal concern. In the 1860s, Birdsill Holly's unique water system was installed in Lockport, New York. This was the forerunner of today's pressurized fire protection system, which is found in every city in the country. Similarly, the Chesapeake & Delaware Canal Scoop Wheel and Steam Engines allowed uninterrupted commerce on a nineteenth-century canal linking Philadelphia and Baltimore. Without them, the canal could not function in times of low water.

The great mines of northern Michigan suffered from too much water. The Chapin Mine Pumping Engine, installed in the 1890s, proved to be the largest pumping engine ever built in the United States. It functioned well but would also be replaced by centrifugal pumps within a short time. It remains a crowning monument to the principle of Newcomen's first pumping engine.

And too much water can be a vexing problem for cities built at sea level or, in the case of New Orleans, below the level of surrounding rivers and lakes. The Wood Low-lift Screw Pump, an early axial-flow pump design, has allowed New Orleans to survive since the installation of the first units in 1915. These same pumps are still in service today.

Without mechanical engineers to control and channel the flow of water, we might still be limited to living and working within walking distance of the nearest stream, and our lives would certainly be vastly different from those we enjoy today.

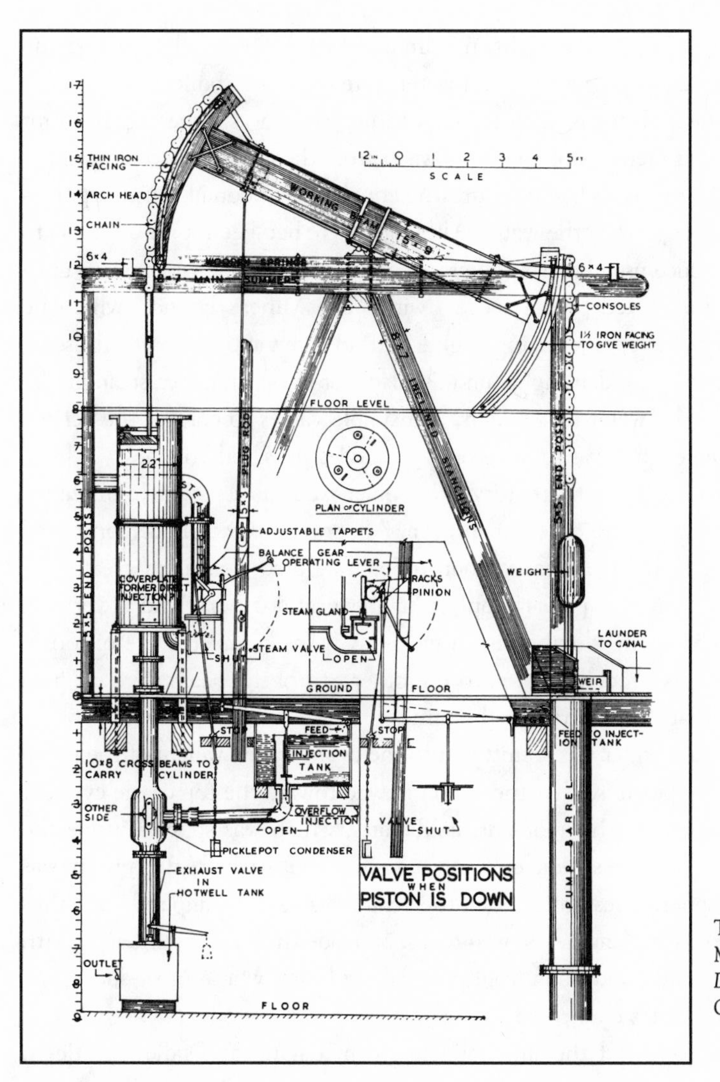

The Newcomen Memorial Engine. *Drawing by Dr. C. T. G. Boucher.*

Newcomen Memorial Engine

Dartmouth, Devon, England

The atmospheric steam engine developed in 1712 by Thomas Newcomen (1663–1729) of Dartmouth, England, marked the beginning of commercially practical thermal prime movers. With its combination of boiler, cylinder, piston, and self-acting valve gear, it was the forerunner of all the steam engines that were to follow. Indeed, Newcomen's engine would prove to be one of the most momentous inventions in world history.

Before the Newcomen engine, there were only three ways to produce mechanical power: through the muscular effort of people or animals, by waterwheels, or by windmills. The steam engine was something entirely new—something that could work tirelessly day and night, not limited by the flow of the river or the vagaries of weather, as long as fuel and water were fed to the boiler.

Toward the end of the seventeenth century, there was a pressing need for better and cheaper methods of removing water from the deepening coal mines of Great Britain. Many mines had been drowned out and abandoned; existing pumps simply could not cope with the water. Although there had been some attempts to use steam to produce useful power, no practical pumping engine was devised until Thomas Savery achieved partial success with "The Miner's Friend," which he patented in 1698. Savery's pumping engine had no heavy moving parts; it used a vacuum produced by condensing steam to suck water into a chamber, steam under pressure to force the water to a height, and simple valves to control the action. But the pumping engine was suitable only for modest lifts and volumes.

In the early eighteenth century, Newcomen and his assistant John Calley developed an engine quite different in form. Adopting a principle demonstrated by Denis Papin in his laboratory about 1690, they used a vacuum created by condensing steam from a pressure only just above atmospheric and a vertical, open-topped cylinder in which a piston moved. Chains connected the piston to one end of a massive rocking beam; to the other end of the beam were chained the pump rods that descended into the mine.

Steam from a boiler was admitted into the cylinder; the weight of the pump rods activated the beam, so that the piston moved toward the top of the cylinder and drew in steam. At this moment, cold water, which was sprayed inside the cylinder, condensed the steam, creating a vacuum into which the piston was forced by atmospheric pressure, rocking the beam, raising the pump rods, and thus creating a stroke of the engine. Newcomen soon made the engine self-acting with the addition of a plug rod hung from the rocking beam, whose pegs, during the stroke, activated levers connected to the valves.

Newcomen erected the first reliable steam engine at Dudley Castle in Staffordshire in 1712. The atmospheric beam engine, as L. T. C. Rolt and J. S. Allen quote, "vibrates 12 times in a Minute & each stroke lifts up 10 Gall [38 l] of water 51 yards [47 m] p'pender [perpendicular]"—the equivalent of a power output of about 5½ horsepower (4 kW). This first engine gradually was followed by others in the coal districts of England and Wales. In addition to mining use, the engine was adopted as a water-supply pump.

Newcomen operated within the Savery patent, which had been granted in very broad terms and further extended to 1733. But Savery died in 1715, and the "Proprietors of the Invention for raising Water by Fire" was organized to exploit his invention. Although, by the time he died in 1729, hundreds of Newcomen engines were at work across Britain and in Hungary, Belgium, France, Germany, Sweden, Austria, and possibly Spain, Newcomen's achievement went largely unrecognized during his lifetime.

In 1963, on the tercentenary of Newcomen's birth, the Newcomen Society for the Study of the History of Engineering and Technology sought to create a suitable memorial. At Hawkesbury, Warwickshire, was a direct descendant of Newcomen's first machine, which the Coventry Canal Company had purchased secondhand from a colliery at Measham in 1821. The small engine was similar to the first at Dudley Castle, although the small 22-inch (559-mm) cylinder was made of iron, not brass. The engine had a simple, untrussed wooden beam with arch-heads and chain connections, and wooden spring-beams. Such details, together with its small size, pointed to an early origin. The Newcomen Society moved the engine, which had lain idle for half a century, to Dartmouth and reerected it as a permanent memorial to the man who first showed the world how useful power could be harnessed by a cylinder and a piston.

Location/Access

The Newcomen Engine House, Mayors Avenue, is operated by Friends of Dartmouth Museum Association, Dartmouth, Devon TQ6 9PZ, England. Admission fee.

FURTHER READING

Richard L. Hills, *Power from Steam: A History of the Stationary Steam Engine* (Cambridge: Cambridge University Press, 1989).

L. T. C. Rolt, *Thomas Newcomen: The Prehistory of the Steam Engine* (Dawlish, Devon: David and Charles, 1963).

L. T. C. Rolt and J. S. Allen, *The Steam Engine of Thomas Newcomen* (New York: Science History Publications/USA, Neale Watson Academic Publications, Inc., 1977).

Fairmount Waterworks

Philadelphia, Pennsylvania

"Philadelphia is most bountifully provided with fresh water," Charles Dickens observed in American Notes (1842), "which is showered and jerked about, and turned on, and poured off, everywhere. The Waterworks, which are on a height near the city, are no less ornamental than useful, being tastefully laid out as a public garden, and kept in the best and neatest order." Housed in a succession of neoclassical temples along the Schuykill River and set on a small plot of landscaped ground that eventually grew to the 8,700-acre (3,520-ha) Fairmount Park, Fairmount Waterworks included the first large-scale, steam pumping station in the United States.

The city's first waterworks, consisting of a combined steam pumping station and water tower at Centre Square (now the site of City Hall) and conduits of hollowed logs, was put into service in 1799, but owing to the rapid growth of the

Force pump and expansion tanks at Fairmount Waterworks in 1876.

city, a larger works was soon needed. In 1811 the Philadelphia Watering Committee directed Frederick Graff (1774–1847) to examine the best methods of procuring water for the city. Graff, who had assisted Benjamin Henry Latrobe in designing the waterworks at Centre Square and afterward served there as engineer, proposed that a steam pumping works be erected near Morris Hill (Faire Mount) to pump water from the Schuykill River into reservoirs constructed on the hill.

In 1812 the Watering Committee purchased 5 acres (2 ha) for a new waterworks and awarded contracts for two steam engines. One was for a Boulton & Watt–type low-pressure condensing beam engine of 44-inch (1,118-mm) bore and 6-foot (1,829-mm) stroke. The second contract, awarded to the distinguished Philadelphia engine builder Oliver Evans, was for a high-pressure noncondensing engine of 20-inch (508-mm) bore and 5-foot (1,524-mm) stroke; Evans called it, patriotically, a "Columbian" steam engine. The engines did periodic duty until 1822, with disappointing results: operating expenses, especially for fuel, were high, and there were two boiler explosions, in 1818 and 1821, killing three.

In 1819 Watering Committee chairman Joseph S. Lewis proposed constructing a dam at Fairmount and substituting river-driven waterwheels for steam power. The dam was finished in 1821. A race that was 419 feet (128 m) long, 90 feet (27 m) wide, and 16 to 60 feet (5 m to 18 m) deep channeled water to three breast wheels driving pumps inside a monumental neoclassical mill house, another Graff design. The first wheel, made of wood, went into operation on July 1, 1822.

Rimmed by public gardens embellished with romantic sculptures, the Fairmount Waterworks became a popular recreational attraction and a symbol of the city. By 1843, the plant had been expanded to eight wheels and four hilltop reservoirs, with a combined capacity of 22 million gallons (83.27 million liters). From specially built galleries in the mill house, visitors watched the wheels turn, almost noiselessly, at 11 to 14 rpm. The wheels, 15 feet (4,572 mm) wide and 15 to 18 feet (4,572 to 5,486 mm) in diameter, drove 16-inch (406-mm) double-acting pumps with strokes of 4.5 to 6 feet (1,370 to 1,830 mm). Each pump was capable, in a twenty-four-hour period, of raising 1.5 million gallons (5.68 million liters) of water a perpendicular height of 92 feet (28 m) into the reservoirs, from which water was distributed by gravity through a system of cast-iron mains and pipes.

The waterwheels remained in service until 1866, although beginning in 1851 they were gradually replaced by seven Jonval turbines. These large vertical-shaft, axial-flow machines, suited to low heads, were each geared to the crankshafts of two pumps 16 inches (406 mm) in diameter, similar to those driven by the breast wheels. The turbines continued to supply portions of the city until March 1911, when river pollution caused the historic waterworks to be closed. The waterworks housed the city aquarium until 1962, after which the complex was abandoned and left to decay.

Today the waterworks is undergoing restoration, thanks to the combined efforts of the Fairmount Park Commission, the Philadelphia Water Department, and the Junior League of Philadelphia. The complex consists of the engine house, old and new mill houses, the caretaker's house, the Watering Committee building, and a neoclassical pavilion (added in the 1870s). A single (incomplete) Jonval turbine and pump of 1851 are the only extant machinery.

Location/Access

The Fairmount Waterworks is located in Fairmount Park behind the Philadelphia Museum of Art (which occupies the site of the former Fairmount reservoirs), 26th Street and Benjamin Franklin Parkway. Tours are available. Contact: Director, Fairmount Waterworks Interpretive Center, Philadelphia Water Department, 1101 Market Street, 3d Floor, Philadelphia, PA 19107; phone (215) 592-4908.

FURTHER READING

Jane Mork Gibson, "The Fairmount Waterworks," *Philadelphia Museum of Art Bulletin* 84 (Summer 1988): 1–40.

C. S. Keyser, *Fairmount Park* (Philadelphia: Claxton, Remsen, and Haffelfinger, 1872).

Chesapeake & Delaware Canal Scoop Wheel and Steam Engines

Chesapeake City, Maryland

The Chesapeake & Delaware Canal forms a shortcut across the narrow neck of the 180-mile- (290-km-) long Delmarva Peninsula, connecting Chesapeake Bay on the west with Delaware Bay and the Atlantic Ocean on the east and shortening the route between Baltimore and Philadelphia by 316 miles (508 km). Opened for navigation in 1829, the C & D was short—just 13.6 miles (22 km) long—and required a maximum total lift of just 16 feet (4,876 mm), employing a tide lock at each end and one other lift lock. Difficult terrain, including more than a mile of tidal marsh and a 3-mile (4.8-km) cut through a low ridge running down the middle of the peninsula, made it the most expensive canal of its time. The final cost, borne by private stock subscription and the states of Pennsylvania, Maryland, and Delaware, was $2.2 million.

Initially, the canal used natural watercourses along the route for its water supply. But the supply of water at the summit was deficient during dry months and deep-draft vessels often had to be turned away. In 1837 a steam pump (about which little is known) was installed at Chesapeake City but soon proved inadequate. In 1848 the Chesapeake & Delaware Canal Company announced a contest for the best design of a steam pump capable of lifting 200,000 cubic feet (5,660 m^3) of water per hour to a height of 16 feet (4,876 mm). Two Philadelphia engineers, Samuel V. Merrick and John H. Towne of Merrick & Son, submitted plans for a steam-operated lifting or scoop wheel. In 1851 Merrick & Son was engaged to install the scoop wheel and a 175-horsepower (130-kW) condensing beam engine.

One of two pumping engines that supplied water to the Chesapeake & Delaware Canal.

The Merrick scoop wheel was described in the Journal of the Franklin Institute in 1853. Made of wood and iron, the wheel was 39 feet (11.9 m) in diameter and 10 feet (3 m) wide, with twelve buckets. Water was channeled from Back Creek into a deep well under the scoop wheel. As the wheel revolved, water that was scooped into the buckets flowed out of lateral discharge openings located near the center of the wheel into an upper race, which carried it into the canal at a point about 960 feet (293 m) east of the Chesapeake City lock. The scoop wheel was geared to the crankshaft of a condensing beam engine with a cylinder of 36 inches (914 mm) diameter and 7-foot (2,133 mm) stroke. At 24 rpm (the usual speed), the wheel made 2.46 revolutions per minute, delivering the contents of 29½ buckets. Merrick & Son added a second steam engine in 1854.

The scoop wheel was put to the test during 1855, when, because of dry weather, it ran continuously from February until December. "It is believed that the machinery for producing the necessary supply of water is as economical, efficient, and simple, both as regards its principle and construction, as can be devised," President A. C. Gray stated in the thirty-seventh annual report of the Chesapeake & Delaware Canal Company in 1856. The huge waterwheel and steam engines remained in continuous use for the duration of the C & D's life as a lock canal, until the plant shut down in 1927.

During its long life, the Chesapeake & Delaware Canal stimulated waterborne commerce. The year ending June 1872 was the peak year for tonnage, with 1.3 million tons (1.18 million t) of cargo hauled during the nine-month shipping season. In addition, the C & D proved a vital lifeline in wartime. In 1861 troops and supplies from Philadelphia were brought through the canal to protect Washington from threatened invasion. In 1919 the United States government purchased the canal and turned over its operation to the Army Corps of Engineers, which deepened and widened it, and removed the locks in 1927. Today, as part of the Intracoastal Waterway system, the Chesapeake & Delaware Canal is more than 400 feet (122 m) wide and 27 feet (8.2 m) deep, the result of subsequent enlargements. It can accommodate all but the largest ocean liners and battleships.

Location/Access

The scoop wheel and steam engines, as well as exhibits about the canal, may be seen at the Old Lock Pump House, State Route 286, Chesapeake City, MD 21915; phone (410) 885-5621 for hours and information.

FURTHER READING

"Chesapeake and Delaware Canal Wheel for Raising Water," *Journal of the Franklin Institute* 55 (February 1853): 93–95.

Greville Bathe, *An Engineer's Miscellany* (Philadelphia: Patterson & White Company, 1938).

Holly System of Fire Protection and Water Supply

Lockport, New York

Birdsill Holly (1822–94), inventor and manufacturer, installed the nation's first pressurized fire-hydrant system in Lockport, New York, in 1863. In 1866 the system was expanded to provide water to businesses and residences. While municipal water and fire-protection systems were nothing new, Holly's achievement was to develop a single system that would both furnish water and extinguish fires.

Holly's water-supply system maintained pressure in the mains solely by the pump that was controlled by a pressure governor, rather than by the gravity head of an elevated reservoir or standpipe. Because of its simplicity, the system rapidly came into wide use. By the time of Holly's death, the Holly Manufacturing Company had placed his system in more than two thousand cities and towns in the United States and Canada.

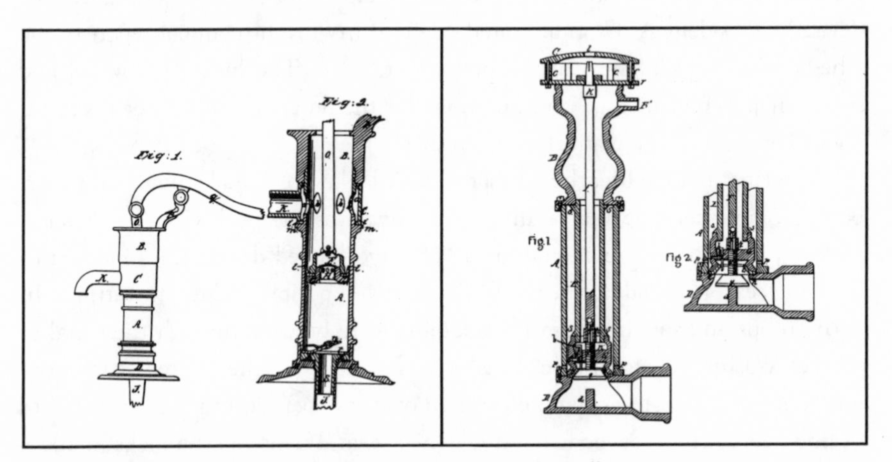

Patent drawings of typical hand pump and fire hydrant in the Holly system.

In 1987 the American Society of Mechanical Engineers designated the Holly System of Fire Protection and Water Supply and the Holly System of District Heating (see p. 10 and p. 201) as Mechanical Engineering Heritage Sites. The designations, the first of their kind, recognize important developments in the history of mechanical engineering, even though a structure or object is no longer extant.

Location/Access

The American Society of Mechanical Engineers plaque is located at the Erie Canal Museum, New York State Canal Corporation, 80 Richmond Avenue, Lockport, NY 14094; phone (716) 434-3140.

FURTHER READING

Morris A. Pierce, "The Introduction of Direct Pressure Water Supply, Cogeneration, and District Heating in Urban and Institutional Communities, 1863–1882" (Ph.D. diss., University of Rochester, 1993).

Archimedean Screw Pump

Newark, California

The gold rush of 1849 sharply increased the demand for salt in California. By 1868, eighteen companies had set up plants on the eastern shore of San Francisco Bay, where they produced salt from seawater by using heat from the sun to evaporate brine in open ponds. Wind-powered pumps transferred the brine from one concentration pond to another.

One of the earliest of California's solar producers, the Oliver Salt Company (founded in 1872), used screw pumps to move the concentrated brine. The screw pump has been attributed to Archimedes, a Greek who lived in Sicily from 287 to 212 b.c. It consists of a deep screw thread encased in an inclined, watertight wood cylinder, with its lower end immersed in the water. As the screw is turned, it carries water up the thread and discharges it at the top. It was originally foot-powered, but the power of the wind was applied in seventeenth-century Holland, where such pumps were widely employed to reclaim land from the sea.

In 1978 Don Holmquist, pond superintendent for Leslie Salt Company, decided to restore an Archimedean screw pump to working order. Using O. E. Oliver's drawings of 1891, Holmquist and his colleagues rebuilt the pump and

Wind-powered Archimedean screw pumps once served San Francisco Bay salt producers by moving brine from one concentration pond to another.

placed it in a pond on Leslie's property. The faithful replica consists of a continuous spiral formed around an inclined redwood shaft 22 feet (6.7 m) long. The four blades of the 20-foot- (6-m-) diameter fan—in essence, a windmill—rotate the screw, raising the water. With full sail and a wind of 25 miles per hour (40 km/hr), the pump, turning at 60 rpm, can raise 1,500 to 2,000 gallons (5,700 to 7,600 l) of brine per minute. Assuming a 4-foot (1.2-m) lift, this represents a power output of some 1.5 to 2 horsepower (1.1 to 1.5 kW).

Although the wind-powered pumps were efficient, wind was not always available. In the early twentieth century, the San Francisco Bay salt industry turned to electric pumps, which could be turned on and off at will and were able to pump against higher heads.

Location/Access

The restored Archimedean pump is displayed outside at the Cargill Salt Company within the San Francisco Bay National Wildlife Refuge. Contact the Cargill Salt Company, 7220 Central Avenue, Newark, CA 94560; phone (510) 797-8157.

FURTHER READING

Garnett Laidlaw Eskew, *Salt, the Fifth Element: The Story of a Basic American Industry* (Chicago: J. G. Ferguson and Associates, 1948).

Robert P. Multhauf, *Neptune's Gift: A History of Common Salt,* Johns Hopkins Studies in the History of Technology (Baltimore: The Johns Hopkins University Press, 1978).

Chapin Mine Pumping Engine

Iron Mountain, Michigan

Founded in 1879, the Chapin Mine, in the heart of Michigan's Menominee Range, was one of the greatest iron mines in the Lake Superior district. It was also one of the wettest. The sloping ore body, a half-mile (0.8-km) long and almost 1,500 feet (457 m) deep, was located almost entirely beneath a swamp that defied every effort to remove its treasure of ore. In 1891 the Chapin Mining Company contracted with Milwaukee's Edward P. Allis Company to build a pumping engine capable of removing all water from the mine for years to come. It would prove to be the largest steam-driven pumping engine ever built in the United States.

Designed by Edwin Reynolds (1831–1909), Allis chief engineer, the Chapin Mine pumping engine was a vertical steeple-compound engine with a high-pressure cylinder 50 inches (1,270 mm) in diameter, a low-pressure cylinder 100 inches (2,540 mm) in diameter, and a piston stroke of 10 feet (3,050 mm). Standing 54 feet (16.4 m) tall, with a flywheel 40 feet (12 m) in diameter, it weighed 600 tons (544 t).

The Chapin Mine pumping engine, the largest steam-driven pumping engine ever built in the United States, was the workhorse of the Chapin Mine from 1892 until 1914.

Through a massive bell-crank walking beam and connecting rod, the engine drove a series of single-acting plunger ("Cornish") pumps, which were arrayed down the shaft. Each pump stood in a wrought-iron tank, or "sump," that discharged mine water through a 28-inch (711-mm) rising main into the tank of the pump above. The engine was designed to run on compressed air, supplied by the company's water-powered plant at nearby Quinnesec Falls, as well as steam.

The pumping engine worked well at the Chapin Mine, pumping from a depth of 600 feet (183 m) until the shaft shifted out of alignment and was closed. The engine was dismantled in 1899, put into storage, and later moved to its present location at the Ludington Mine "C" shaft, which Chapin acquired in 1894), where it continued to operate until it was replaced by electric pumps in 1914.

Certainly the engine was an impressive one. In 1915 *Power* noted that during a one-year period, the pump operated 99.5 percent of the time at a rate of 6.63 rpm, pumping 1,922 gallons (7,275 l) per minute against a head of 1,513 feet (461 m), with an average horsepower of 736 (549 kW), giving the engine a mechanical efficiency of 88.6 percent.

Celebrated as it was, however, the Chapin pumping engine had been planned during a period of rapid technological advancement that saw the classic reciprocating engine fall from favor as a prime mover. In fact, in reporting the engine's installation in 1893, *Engineering News* editorialized against it, pronouncing, "The work could be done far more cheaply and perfectly . . . by electricity."

Location/Access

The Chapin Mine pumping engine is the focal point of the Cornish Pump and Mining Museum, operated by the Menominee Range Historical Foundation, 300 Kent Street at Carpenter Avenue, P.O. Box 237, Iron Mountain, MI 49801; phone (906) 774-1086.

FURTHER READING

"The Chapin Mine Pumping Engine," *Engineering News* 30 (19 October 1893): 310–11. (See also pp. 315–16 for *Engineering News*'s editorial against the Chapin pumping engine and in favor of electric pumps.)

Louis C. Hunter, *A History of Industrial Power in the United States*, vol. 2, *Steam Power* (Charlottesville, Va.: University Press of Virginia, 1985).

C. Ziemke, "Old Pumping Engine Preserved for Posterity," *Compressed Air Magazine*, November 1947, 276–77.

Leavitt-Riedler Pumping Engine

Boston, Massachusetts

In 1894 a new high-service pumping engine was installed in Boston's Chestnut Hill Pumping Station (1887) to augment water supply to elevated sections of a growing city. Designed by Erasmus D. Leavitt, Jr. (1836–1916), of Cambridge, Massachusetts, and built by the Quintard Iron Works of New York, Pumping Engine No. 3 attracted national attention as "the most efficient pumping engine in the world" (according to *Power*), and because its novel design represented "an advance on previous practice" (according to *Scientific American*).

The Leavitt engine is a triple-expansion, three-crank "rocker" engine with pistons 13.70, 24.38, and 39 inches (348, 619, and 991 mm) in diameter and 6-foot (1,829-mm) stroke. The cylinders are vertical and inverted, and are carried, together with the valve gear, on an entablature supported by six vertical and six diagonal columns. From each rocker run two connecting rods: one to the crankshaft carrying a 15-foot (4,570-mm) flywheel, the other to one of the three pump plunger-rods.

Each pump contains two suction and two delivery valves, each about 3 feet (914 mm) in diameter. The pumping engine owed its great efficiency to the use of these large valves and to the novel design of the pump-valve mechanism, which Leavitt based on the invention of Professor Alois Riedler of the Royal Polytechnic University in Berlin. This invention consisted of closing each valve positively at just the moment of reversal of stroke by means of levers and rods not unlike those of a Corliss engine. After closing the valves, the mechanism released, leaving the valves free to open by the suction pressure.

Pumping against a head of 128 feet (39 m), or about 55 psig (379 kPa), the

The Leavitt-Riedler pumping engine was illustrated in *Scientific American* on September 14, 1895.

Boston engine was designed to run easily at 60 rpm, a speed made possible by the Riedler valve gear. At the normal speed of 50 rpm, the pumping engine had a capacity of 20 million gallons (75.7 million liters) in twenty-four hours. Steam was supplied by a single Belpaire-firebox boiler of Leavitt's design (no longer extant) with two separate furnaces and a common combustion chamber. Pumping Engine No. 3 served for thirty-four years before it was relegated to standby duty in 1928.

Location/Access

Open upon application to Chestnut Hill Pumping Station, 2436 Beacon Street, Chestnut Hill, Boston, MA 02167; phone (617) 734-9194.

FURTHER READING

F. W. Dean, "An Account of the Engineering Work of E. D. Leavitt," *Transactions of the American Society of Mechanical Engineers* 39 (December 1915): 993–1036.

Edward F. Miller, "Description and Computation of a Twenty-Four Hour Duty Test on the Twenty Million Gallon Leavitt Pumping Engine at Chestnut Hill," *Technology Quarterly* 9 (June–September 1896): 72–115.

"New High Service Pump, Boston Water Works," *Scientific American* 73 (14 September 1895): 166.

"Record Making Pumping Engine, Chestnut Hill Pumping Station, Boston, Mass.," *Power* 16 (April 1896): 1–6.

Erasmus D. Leavitt, Jr.

Erasmus D. Leavitt, Jr. (1836–1916).

Without formal technical training, Erasmus Darwin Leavitt, Jr., achieved the highest distinction in the ranks of mechanical engineering. According to ASME Transactions *in 1916, as a machinery designer "he did more than any other engineer in this country to establish sound principles and propriety of design" and was "among the very first engineers . . . to appreciate the importance of weight in machinery." He was born in Lowell, Massachusetts, on October 27, 1836, the son of Erasmus Darwin and Almira (Fay) Leavitt, and was educated in the local schools. At sixteen he began a three-year apprenticeship in the machine shop of the Lowell Manufacturing Company. He was employed for a year with Corliss & Nightingale, Providence, Rhode Island, before returning to Boston as assistant foreman of the City Point Works of Harrison Loring; there, he had charge of building the engine for the USS* Hartford.

From 1859 to 1861, Leavitt was chief draftsman for Thurston, Gardner & Company, steam-engine builders of Providence. He left to join the U.S. Navy at the start of the Civil War. He was assigned to the gunboat Sagamore, *then to construction duty at Baltimore, Boston, and Brooklyn. In 1865 he was detailed to the U.S. Naval Academy at Annapolis, where he was an instructor in steam engineering. He resigned that position in 1867 to enter private practice as a consulting engineer.*

*A beam compound pumping engine that he designed for Lynn, Massachusetts, in 1873 "marked an era in the economy of pumping engines throughout the world" (*ASME Transactions, *1916) and brought Leavitt to the attention of the engineering profession. He became acquainted with the leading engineers of Europe, including Alois Riedler, from whom he acquired the right to use the Riedler pump and valve gear in the United States. The Leavitt-Riedler Pumping Engine, installed in Boston's Chestnut Hill Station in 1894, was among his best-known and most successful designs.*

From 1874 to 1904, Leavitt served as consulting mechanical engineer for the Calumet & Hecla Mining Company, designing more than forty engines for pumping, air-compression, hoisting, stamping, and powering for the company's extensive mines in Michigan. As a consulting engineer, Leavitt worked on an array of important projects. He designed an engine for hydraulic forging at the Bethlehem Steel Company; engines, boilers, and other machinery for the El Callao Mining Company, Venezuela; and pumping engines for the waterworks of Louisville, Kentucky, as well as Boston, Cambridge, Lawrence, and New Bedford, Massachusetts. An admirer of Krupp forgings, Leavitt for a time kept an inspector at the Krupp works at Essen, Germany.

Leavitt received the first honorary Doctor of Engineering degree from the Stevens Institute of Technology in 1884. He was a founding member of the American Society of Mechanical Engineers and served as a vice-president from 1881 to 1882 and as president in 1883. He was also a fellow of the American Academy of Arts and Sciences. He died on March 11, 1916, in Cambridge, Massachusetts, his longtime home.

Sources: Obituary, ASME *Transactions* 38 (1916): 1347–51; F. W. Dean, "An Account of the Engineering Work of E. D. Leavitt," ASME *Transactions* 39 (1918): 993–1036.

Chestnut Street Pumping Engine

Erie, Pennsylvania

The concept of using steam first at high pressure in a small cylinder and then at low pressure in a larger cylinder was patented by Jonathan Hornblower in England in 1781. Another English engineer, Arthur Woolf, added a high-pressure cylinder to an existing engine at London's Meux brewery in 1803. By the mid-nineteenth century, "compounding," as it was called, was well developed as a means of obtaining greater efficiencies (i.e., greater energy extraction from each unit of fuel) from steam engines.

By the early twentieth century, the concept of triple-compounding, or "triple-expansion," was well established for certain applications, especially water-pumping and marine engines. The first triple-expansion pumping engine, built in 1886, was designed by Edwin Reynolds of the Edward P. Allis Company for the City of Milwaukee. In both size and efficiency, massive triple- and sometimes even quadruple-expansion steam engines represented the zenith of reciprocating steam-engine design.

The Chestnut Street Pumping Station contains one of the last and largest examples of a reciprocating steam engine built to drive water pumps. Built by Bethlehem Steel Company in 1913, this triple-expansion steam engine has three cylinders—high-, intermediate-, and low-pressure—directly coupled to large plunger-type pumps. The massive unit, with two flywheels 20 feet (6,096 mm) in diameter each, fills a building almost 60 feet (18 m) high.

The Chestnut Street pumping engine had a capacity of 20 million gallons (75.7 million liters) per day. It operated until 1951, when it was replaced by four vertical-turbine and three horizontal-centrifugal pumps, all electrically powered.

SPECIFICATIONS

Chestnut Street Pumping Engine

Cylinders: three; 33 inches (838 mm), 66 inches (1,676 mm), 98 inches (2,489 mm) in diameter
Stroke: 5½ feet (1,676 mm)
Speed: 25 rpm
Flywheels: two; each 20 feet (6,096 mm) in diameter
Horsepower: 600 (447 kW)

Location/Access

Contact the Erie City Water Authority, Administration Building, 340 West Bay Front Parkway, Erie, PA 16507; phone (814) 870-8000.

FURTHER READING

Louis C. Hunter, *A History of Industrial Power in the United States*, vol. 2, *Steam Power* (Charlottesville, Va.: University Press of Virginia, 1985).

A. B. Wood Low-lift Screw Pump

New Orleans, Louisiana

Because of the low elevation of New Orleans and the fact that it is entirely surrounded by levees and dikes, its drainage system differs radically from that of other American cities. Rainwater must be disposed of mechanically. In the late nineteenth century, drainage was handled by a wholly inadequate system of deep gutters intercepted by open canals, from which wastewater was pumped by steam-powered lift wheels into canals leading to Lake Pontchartrain. At most the system could pump 1½ inches (38 mm) of rain a day. The city suffered from cholera, yellow fever, malaria, and other diseases, and was reputed to be one of the unhealthiest places in America.

In 1893 the New Orleans Advisory Board on Drainage was established to oversee a topographical and hydrographical survey and recommend a drainage system for the city. The advisory board proposed a gravity system of canals and pumping stations that would discharge rainwater into Lake Borgne via Bayou Bienvenu, but they faced an apathetic public and limited funding until an outbreak of yellow fever aroused support and led to passage of a bond issue and creation of the New Orleans Sewerage and Water Board in 1899.

The city installed a system of vertical-shaft screw pumps, 8 feet (2,438 mm) in diameter with submerged screws, the best then available. The pumps were

Fourteen-foot (4,267-mm) A. B. Wood screw pump during construction, ca. 1930.

inefficient, difficult to service, and often overloaded their motors. (A screw pump is a rotary machine having an impeller with a row of twisted blades that are, in essence, short sections of a thin helix or screw thread.) With the need for better drainage pumps pressing, Albert Baldwin Wood (1879–1956), a young mechanical engineer with the Sewerage and Water Board, designed the first of a series of horizontal-shaft screw pumps, for which he would win international fame. In 1913 Wood presented plans for a 12-foot (3,657-mm) screw pump and gave the board perpetual rights to his invention. Known as "Wood screw pumps," the pumps were designed to discharge great quantities of water against widely varying lifts. The first four were installed in 1915.

Rainwater flows by gravity from buildings and streets into underground canals sloped in the direction of the nearest pumping station. The screw pump lifts the water to a higher level and sends it flowing to the next lift station and eventual disposal. Manufactured and installed by the Nordberg Manufacturing Company of Milwaukee, the first Wood screw pump consisted of a cylindrical casing 12 feet (3,657 mm) in diameter and 13 feet, 9 inches (4,190 mm) long, lying with its axis horizontal and containing the impeller (or moving) blades and the stationary (or diffusion) blades. The diffusion blades were mounted in a watertight, cone-shaped housing that was 8 feet (2,438 mm) in diameter at the widest point, within which were located a self-aligning main bearing and a marine-type thrust bearing. These could be reached through a watertight manhole at the top of the pump; thus, the inner cone was readily entered for inspection or adjustment of the bearings, even while the pump was operating. The placement of the 12-foot (3,657 mm) screw at the top of a siphon, instead of submerging it, was a notable advantage of the new pump, which was driven by a 600-horsepower (447-kW), three-phase, 6,000-volt synchronous motor built by Allis-Chalmers.

Tests conducted by Professor W. H. P. Creighton, dean of the Department of Technology of Tulane University, proved the high efficiency of the Wood screw pump—that is, the relatively high discharge per horsepower at low working heads, an essential requirement in times of flood. "While the pump surpasses in efficiency, under normal conditions, those of previous installations," Creighton concluded, "the superiority is much greater just when the greatest service is required. Emergency service is probably the weak point of the old pumps. It is the forte of the new. Results show that the pumps . . . are the largest and most efficient low-lift pumps in the world."

By 1925, eleven units had been installed in six different drainage stations throughout the city. By 1932, with the installation of larger, 14-foot (4,267-mm) screw pumps, the city's drainage system could remove 14 inches (355 mm) of rain per day. It was put to the test in 1978, when some 11 inches (279 mm) of rain fell in seven hours. The pumping system drained 11 billion gallons (41.6 billion liters) in twenty-four hours from some 55,000 acres (22,257 ha)—roughly equivalent to a lake 10 square miles (26 km²) in area and 5½ feet (1,676 mm) deep. Following their successful performance in New Orleans, Wood screw pumps were installed in Holland, Egypt, China, and India.

Wood, a native of New Orleans and a graduate of Tulane University, was a lifelong employee of the New Orleans Sewerage and Water Board. He refused more lucrative offers of employment that would have taken him to other cities and countries, although he served as a consulting engineer to Chicago, Memphis, Baltimore, and other cities. Thirty-eight patents attest to his inventive mind. In 1939 Tulane University awarded Wood an honorary Doctor of Engineering degree, citing him as an "engineer, designer, and inventor whose genius has contributed much to the comfort, safety, and livelihood of multitudes of human beings."

Location/Access

Wood screw pumps are located at Melpomene Pumping Station No. 1, 2501 South Broad Avenue, New Orleans, Louisiana. Permission to view the pumps must be obtained from the New Orleans Sewerage and Water Board, Community and Intergovernmental Relations, 625 St. Joseph, New Orleans, LA 70165; phone (504) 585-2169.

FURTHER READING

O. J. Abell, "Making Unusual Pumps for New Orleans," *Iron Age* 94 (5 November 1914): 1060–63.

"The Drainage of New Orleans," *Engineering Record* 31 (25 May 1895): 454–56.

"Mammoth Screw Pumps of New Design Develop High Efficiencies for Low Lifts," *Engineering Record* 73 (8 January 1916): 54–56.

Reynolds-Corliss Pumping Engine

Jacksonville, Florida

Jacksonville's water supply improvement program of 1914–17 saw the installation of two 5-million-gallon-per-day (22-million-liter-per-day) pumps driven by reciprocating steam engines in the city's new Main Street Pumping Station. These provided the city's sole water supply until 1930, when the present electrically driven peripheral pumping stations came on line. Steam operation ceased in 1956, and the first pump, of 1915, built by the Epping-Carpenter Company, was scrapped. The second, built by the Allis-Chalmers Company of Milwaukee and installed in 1917, remains on standby, coupled to a Reynolds-Corliss engine. The engine is of particular interest as a surviving example of the Corliss type patented in 1849 and improved by later engine builders, in this case the chief engineer of Allis-Chalmers, Edwin Reynolds (1831–1909).

Corliss engines are distinguished by having four semirotary valves per cylinder, two for steam and two for exhaust, set at right angles to the center line of the cylinder. The valve ports are short, and entering steam does not pass through ports previously cooled by the exhaust. An oscillating wrist-plate enables the fine tuning of the steam inlet valve for precise cutoff and of the exhaust valve for the ideal point of release. The Corliss engine was some 35 percent more efficient than the older slide-valve engines. The design became popular in the United States and was widely copied by European engineers. When Corliss's patent expired in 1873, anyone was free to use the idea—and many did.

Valve gear on Reynolds-Corliss pumping engine.

After ten years as superintendent of the Corliss works in Providence, Rhode Island, Reynolds joined the Edward P. Allis Company of Milwaukee as superintendent in 1877. By 1878, the Reynolds-Corliss engine, with an improved releasing valve mechanism—it was quieter and could run at much higher speeds—went into production. By 1885, Allis had sold more than five hundred such engines for driving pumps, mine hoists, air compressors, blowing engines, and electrical generators. With the formation of the Allis-Chalmers Company in 1901, Reynolds became chief engineer of that firm. In all, he held more than forty patents, including that for the first cross-compound mine hoisting engine.

Location/Access

Main Street Pumping Station, 182 North Main Street at Hogan Creek, Jacksonville, FL 32206. Open upon application to City of Jacksonville Water Division, Public Education, phone (904) 630-0730.

FURTHER READING

Walter F. Peterson, *An Industrial Heritage: Allis-Chalmers Corporation* (Milwaukee: Milwaukee County Historical Society, 1978).

Worthington Horizontal Cross-compound Pumping Engine

York, Pennsylvania

Manufactured by the Worthington Pump & Machinery Corporation's Snow-Holly Works in Buffalo, New York, this small but efficient Corliss pumping engine served at the Brillhart Station of the York Water Company from 1925 until 1956, when it was relegated to standby duty in favor of electrically powered pumps. Between the 1890s and World War I, many water companies nationwide installed similar pumping engines, which could supply between 5 million and 12 million gallons (18.9 million and 45.4 million liters) per day and were considerably smaller and cheaper than the triple-expansion vertical pumping engines typically chosen for larger stations. Most of these were subsequently scrapped, making the York engine—the only known operable engine of its type in Pennsylvania, Maryland, New Jersey, and Delaware—a rare survivor.

The York engine's rated capacity is 5 million gallons (18.9 million liters) per day. A coal-fired, 277-horsepower (206-kW) Stirling boiler supplied steam to the engine's high- and low-pressure cylinders, each of which was connected to a water pump (hence the description "cross-compound").

The York engine was put back to work during hurricanes Agnes (1972) and

Small units such as this Worthington-built Corliss cross-compound pumping engine were the popular choice of small water stations in the United States during the early twentieth century. *Courtesy Stephen Heaver, Jr.*

Eloise (1975), when those storms knocked the station's electric pumps out of service. The pumping engine was removed from service in 1982 but remains in place.

Location/Access

Brillhart Pumping Station is on Codorus Creek, 3 miles (4.8 km) south of York, Pennsylvania. Open upon application to the York Water Company, 130 East Market, York, PA 17401; phone (717) 845-3601.

FURTHER READING

Arthur M. Greene, Jr., *Pumping Machinery: A Treatise on the History, Design, Construction, and Operation of Various Forms of Pumps*, 2d ed., rev. (New York: John Wiley & Sons, 1919).

SPECIFICATIONS

Pumping Engine No. 2, York Water Company

High-pressure cylinder: 18¼ inches (463 mm) in diameter
Low-pressure cylinder: 44 inches (1,117 mm) in diameter
Water cylinders: two; 13½ inches (342 mm) in diameter
Stroke: 36 inches (914 mm)
Steam pressure: 165 psig (1,138 kPa)
Speed: 40 rpm
Horsepower: 225 (168 kW)

Mechanical Power Production

Water

INTRODUCTION by Robert M. Vogel

Until the introduction of Newcomen's steam engine early in the eighteenth century, the principal means for supplementing the power of muscles was harnessing the energy of moving water. Centuries before, the Roman Empire mill stones were turned by a primitive form of turbine or "Norse mill." This consisted of a vertical shaft into which were set wooden blades that were struck by a small stream of water causing them, the shaft, and the movable stone to revolve. Roman engineers employed a form of overshot waterwheel to grind grain, and throughout the East, current wheels—"norias"—were used to raise water for irrigation. The lower part of the noria was set in the flowing stream to be turned by the current. Waterwheels of various types, to a far greater extent than windmills, were a vital element of civilization's spread and growth for grain milling, water raising, sawing, oil pressing, cloth fulling, the working of metals, and other labor-intensive tasks that could be performed on one spot (as opposed to the tilling of land, for example).

With the advent of the Industrial Revolution and a burgeoning need for power, there arose a proportionate need for the effective exploitation of the available water-power sites. This led to an increase in the scale of water-power machinery and an improvement in the efficiency of both the wheels and the transmission systems conveying the wheels' power to the driven machinery. While this need for mechanical power had inspired the invention and application of the steam engine, in terms of the world's overall production of power it played a relatively minor role. It was not until the mid-nineteenth century that more horsepower was produced by steam than by water.

While the Romans were, of course, accomplished hydraulic engineers, water-power machinery never was a major element of their undertakings. Not until the end of the sixteenth century did the true profession of hydraulic engineer–millwright emerge, with the production of such epochal works as supplying London with water, raised from the Thames by a series of pumps driven by large current wheels set within some of the arches of London Bridge. Other works on this scale followed, invariably for water supply, with a notable example being the great pumping works erected in 1682 and powered by the River Seine to supply the gardens at Versailles.

By the end of the eighteenth century, the appearance of large factories, chiefly textile mills in Britain and the United States, provided a powerful impetus to advance the field of water-power engineering. Iron replaced wood in wheel construction and transmission systems; the overshot and breast wheel replaced the inefficient undershot or current wheel; and governing devices were introduced to provide the close speed regulation required by increasingly refined manufacturing machinery. But most importantly, the design of the waterwheel itself and its many adjuncts—both waterways and mechanical devices—were transformed from a largely empirical craft in the hands of the millwright to a nearly exact science in the hands of the mechanical engineer. Again, the result was a leap in both plant capacity and the efficient utilization of the available energy in the moving water.

In the United States, the real birth of water-power engineering, however, sprang from a uniquely American concept: a newly built industrial complex based on a major water-power site, with a single corporation purchasing the surrounding land and water-power privileges, constructing a dam and system of power canals, selling mill sites along the canals, then making the water power available to these sites on an annual lease. The scheme was first attempted on a large scale with the landmark Great Falls Raceway and Power Canal System, organized in 1791 to tap the enormous power potential in the falls of the Passaic River at what became Paterson, New Jersey. The success of the Paterson project duly encouraged large blocks of New England capital in similar undertakings. The first of these was in 1813 at Waltham, Massachusetts, organized largely as a trial by a group of Boston investors. The cotton mills they erected on the Charles River did well and emboldened the backers to embrace what must be seen as one of the most ambitious industrial schemes of that time. The vast textile complex of Lowell, established at the Falls of the Merrimack starting in 1822, became the prototype not only for similar ventures in the United States—the majority of them also powered by New England's mighty rivers as they crossed the fall line—but the world. The great industrial cities of Lewiston and Saco-Biddeford in Maine; Manchester and Nashua in New Hampshire; Lawrence, Fall River, and Holyoke in Massachusetts; Pawtucket, Rhode Island; and Cohoes, New York, all followed more or less closely the Lowell model.

The demands of scale and efficiency at these massive concentrations of water power provided a powerful incentive to improve the performance of the hydraulic prime mover. Although the ponderous iron breast wheels—as much as 20 feet in diameter and width—that drove the first mills at Lowell and its early successors were at the cutting edge of the technology of their eras, their shortcomings were recognized. They were inherently slow, typically turning at 1 or 2 rpm, requiring trains of speed-increase gearing to accommodate the speed of the fast-running textile machinery, which introduced additional friction into the transmission system with resultant loss of power. Being large, these wheels were costly and consumed a good deal of the mills' real estate. Even though invariably they were housed in basements, they were subject to being slowed, stopped, and damaged by the ice of northern winters. And although a well-designed wheel might operate at 65-percent efficiency, that could be achieved only when the wheel was just free of the tail water. If the level fell, the wheel could not take advantage of the increased head. Worse, if the level rose, the drag of the water would impede the wheel with a considerable loss of power.

Although engineers and millwrights were well aware of the many advantages to be found in a hydraulic prime mover that was smaller and faster than the various types of waterwheels, it was clear—either through tentative experimentation or perhaps intuition—that the design of a turbine capable of improving on the performance of contemporary wheels required a degree of sophistication not readily available until the nineteenth century. By the 1840s, practical, fairly efficient turbines had been developed to a commercial level by several French engineer-inventors and introduced to this country on a small scale. The turbine's advantages were recognized by, among others, Uriah Boyden (1804–79), a consultant to the Lowell water-power corporation. In 1844 Boyden undertook to improve the French Fourneyron turbine, building a large machine to replace the breast wheel in one of the Lowell mills. This turbine, operating at nearly 80-percent efficiency, doomed the breast wheel throughout the industry. The fact that, in addition to its other advantages, the turbine could operate effectively at heads both higher and lower than the waterwheel and could be built for vastly greater capacities signaled a revolution in the exploitation of water power, immediately for the direct drive of machinery and later for the large-scale production of hydroelectricity.

Great Falls Raceway and Power System

Paterson, New Jersey

With the incorporation of the Society for Establishing Useful Manufactures, familiarly known as "the S.U.M.," in 1791, Secretary of the Treasury Alexander Hamilton laid the foundation for America's first planned industrial city. The New Jersey legislature granted the S.U.M. perpetual exemption from county and township taxes and the rights to hold property, improve rivers, build canals, and raise $100,000 by lottery. From a number of sites offered, the S.U.M. selected the Great Falls of the Passaic River. There, the city of Paterson, named after New Jersey Governor William Paterson, grew out of the society's 700 acres (283 ha) above and below the falls. It became an incubator for countless engineering and industrial innovations, including the Colt revolver, the Rogers steam locomotive, the Holland submarine, the Curtiss-Wright aircraft engine, and textile manufacturing that made Paterson famous as the "Silk City."

Hamilton had visited the Great Falls of the Passaic River during the American Revolution. The ceaseless flow and power of the waterfall—77 feet (23.5 m) high and 280 feet (85.3 m) wide—inspired Hamilton's dream of American industrial strength and economic independence from foreign markets to assure the hard-won gains of the revolution. The S.U.M. hired Major Pierre Charles L'Enfant, architect and planner of Washington, D.C., to design a system of raceways. L'Enfant's plan was modified by Peter Colt, treasurer of the state of Connecticut and an associate of the Hartford woolen mill (the first woolen mill in the country), who moved to Paterson in 1793 to take charge of the cotton mill of the S.U.M. Both men envisioned a multitiered raceway system that would channel water to provide power to mills. The original raceway system, built and operated

The upper raceway of the Great Falls power system, ca. 1850. The Rogers Locomotive Shops are on the right.

by the S.U.M. from 1794 to 1797, drew water from the Passaic by use of a wooden diversion dam above the falls. The water then entered a reservoir, passing through the raceway to a flume and waterwheel. After providing power for the first S.U.M. factory, a cotton mill, the water was channeled back to the Passaic below the falls.

After 1800, following financial difficulty operating as a manufacturing corporation, the S.U.M. became a power and real estate developer. It was evident that the raceway would have to be extended to provide power for more mills. Between 1800 and 1827, the S.U.M. built two additional raceways and sold new lots and water rights to manufacturers. Mill activity expanded rapidly, and in the late 1820s, the S.U.M. undertook a major realignment of the raceway and power system in order to provide water for a new upper tier of mill sites. The last modification to the system occurred in 1838, when a new channel and dam were built to divert the river into three raceways; the diverted water served three tiers of factories before it was returned to the river. By 1840, the Great Falls provided power for four fulling mills, nineteen cotton factories, a woolen factory, two dyeing and printing establishments, two paper factories, a tannery, and a sawmill.

Throughout the nineteenth century, the Great Falls raceway and power system was the primary power source for manufacturing in Paterson. The abundant, inexpensive energy attracted countless creative enterprises, including, in 1840, silk manufacturing, which surpassed cotton a decade later. At its peak, more than forty thousand workers were employed in Paterson's industries, manufacturing textiles and textile machinery, clothing, revolvers, steam locomotives, and aircraft engines. As other power sources gained favor, the S.U.M. adapted and supplied them to its customers. From 1912 to 1914 the company built a hydroelectric generating station at the foot of the Great Falls that remained in service until 1969. In 1915 it added a steam generating plant to supply power during periods when the river was low. The S.U.M. continued to operate until 1945, when its assets were sold to the city of Paterson.

In the mid-1960s, most of the raceway and power system, as well as many of the more than forty old mills adjacent to it, were threatened with demolition for construction of a highway. A citizens' group successfully fought to preserve the area, which is now designated the Great Falls Historic District. The Great Falls Development Corporation was organized in 1971 to oversee preservation of the district, which today is a mixed-use development that includes offices, housing, a museum–cultural arts center, and some manufacturing plants. On June 6, 1976, President Gerald R. Ford came to Paterson to designate the district a National Historic Landmark.

Location/Access

The 119-acre (48-ha) Great Falls Historic District, on the Passaic River between Grand Street and Ryle Avenue, preserves sections of raceways and a number of early factories, including Peter Colt's Gun Mill (1836) at Mill and Van Houten streets. For group tours and maps for self-guided tours, contact the Great Falls Visitor Center, 65 McBride Avenue, Paterson, NJ 07505; phone (201) 279-9587. The Paterson Museum, 2 Market Street, Paterson, NJ 07505, occupies the first floor of the Thomas Rogers locomotive erecting shop (1871). It contains the hull of a submarine built by Paterson schoolteacher John Philip Holland in 1878. Hours: Tuesday–Friday, 10 A.M. to 4 P.M.; Saturday and Sunday, 12:30–4:30 P.M. Phone (201) 881-3874.

FURTHER READING

Russel I. Fries, "European vs. American Engineering: Pierre Charles L'Enfant and the Water Power System of Paterson, N.J.," *Northeast Historical Archaeology* 4, nos. 1 & 2 (Spring 1975): 69–96.

Alexander Hamilton, *Industrial and Commercial Correspondence of Alexander Hamilton,* edited by Arthur Harrison Cole, 2 vols. (Chicago: Business Historical Society, Inc./A. W. Shaw Company, 1928).

Christopher Norwood, *About Paterson: The Making and Unmaking of an American City* (New York: Saturday Review Press/E. P. Dutton & Co., Inc., 1974).

Lowell Power Canal System and Pawtucket Gatehouse Turbine

Lowell, Massachusetts

In 1821 Boston capitalists who were previously successful with manufacturing cotton cloth on the Charles River at Waltham decided to build a complex of textile mills to take advantage of the vast water power of the Merrimack River at Pawtucket Falls near Chelmsford. They purchased the controlling stock of the Proprietors of the Locks and Canals on Merrimack River (chartered in 1792 for the purpose of improving navigation of the Merrimack River), and, tapping the power of the broad and fast-flowing river, built and administered an extensive system of power canals between the Pawtucket (navigational bypass) Canal and the confluence of the Merrimack and Concord rivers. It was one of the earliest and most successful efforts to tap the water power of North America.

Lowell boomed. By 1840, the city had a population of twenty thousand and was home to eight major textile firms employing almost eight thousand workers. The bulk of Lowell's unskilled work force were young, unmarried women recruited throughout New England. Living in company boardinghouses, they worked twelve hours a day, six days a week, winning fame for the Lowell factory system.

The Lowell mills used water power on a scale unprecedented in America. As chief engineer of the Proprietors of Locks and Canals from 1837 to 1884, James B. Francis (see sidebar) was responsible for meeting manufacturers' demands for ever-increasing amounts of power. Since 1826, engineers had been able to increase

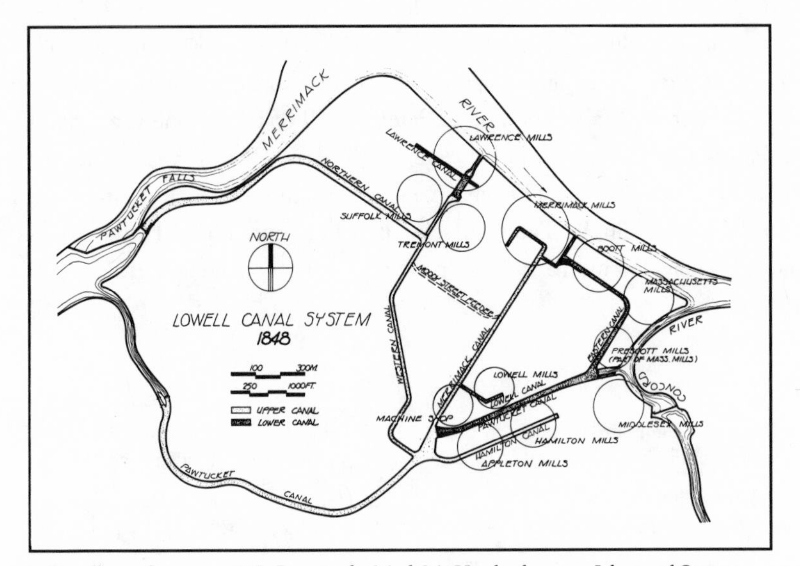

Lowell canal system, 1848. *Drawing by Mark M. Howland, 1975, Library of Congress Collections.*

the flow into the Lowell power canal system by constructing an enlarged dam at Pawtucket Falls. The dam did not satisfy water needs for long, however, and by 1840 shortages were commonplace. To alleviate the problem, Francis purchased control of a number of the Merrimack River's water sources in central New Hampshire and from 1846 to 1847 supervised construction of a new feeder canal. The Northern Canal set a new standard in civil and hydraulic engineering and introduced the famous Francis turbine to the world.

Built at a cost of just over $500,000, the Northern Canal was Lowell's largest and most complex waterway. More than 4,000 feet (1,200 m) long, 100 feet (30 m) wide, and 16 to 21 feet (4,876 to 6,400 mm) deep, it ran from the head of Pawtucket Falls to the upper level of the Western Canal. Francis had to cut through difficult, rocky terrain and place a major section of the canal in the bed of the Merrimack River.

To hold the canal above the Merrimack rapids, Francis built a great river wall, 2,300 feet (700 m) long, of random coursed granite rubble and concrete. To pond the river, he rebuilt the 1,093-foot (333-m) Pawtucket Dam. To control the flow of water into the new canal, he equipped the Pawtucket Gatehouse with sluice gates raised by a small Francis turbine—i.e., a modern, mixed-flow reaction turbine based on a design patented in 1838 by Samuel B. Howd.

Francis's studies of turbine operation, meanwhile, which he is said to have conducted in special testing chambers of the Pawtucket Gatehouse, persuaded manufacturers to switch from the large breast wheels then generally in use to more efficient turbines. As chief engineer for the Proprietors of Locks and Canals, Francis designed and supervised the widespread installation of turbines at Lowell after 1849.

Pawtucket Dam and Gatehouse, looking east from the north side of the Merrimack River, 1976. *Photograph by Jack Boucher, Library of Congress Collections.*

By 1880 water and steam, almost equally, powered the textile machinery of Lowell's ten large cotton and woolen manufacturers, which furnished employment to more than sixteen thousand men, women, and children. Lowell was now polyglot, its workforce comprised of large numbers of Greeks, Eastern Europeans, and French-Canadians. Lowell lost its utopian image.

Beginning in the 1880s, the textile industry began to move to the South, seeking lower labor costs. The trend gathered momentum in the twentieth century. First cotton, then the woolen-worsted industry departed. In recent years, Lowell's handsome brick mill buildings have provided homes for new service industries and the impetus for creation of the Lowell National Historical Park to interpret the history of the nation's first major industrial center and the contributions of James Francis.

Location/Access

Lowell is 33 miles (53 km) northwest of Boston via U.S. Route 3. The Dutton Street parking lot provides visitor parking for the Lowell National Historical Park and State Heritage Park. The Northern Canal begins above Pawtucket Dam and ends at the Western Canal, Francis and Suffolk streets; a section of Francis's great river wall has been replaced with concrete. The Pawtucket Gatehouse is located at the Pawtucket Dam, Merrimack River at School Street; electric motors replaced the Francis turbine early in the twentieth century, but most of the original equipment, including the turbine, is still intact. Obtain information, including a map for a self-guided tour of Lowell and its power canal system, from: Lowell National Historical Park, P.O. Box 1098, Lowell, MA 01853; phone (508) 970-5000.

FURTHER READING

Nathan Appleton, *Introduction of the Power Loom and Origin of Lowell* (Lowell, Mass.: B. H. Penhallow, 1858).

Robert F. Dalzell, Jr., *Enterprising Elite: The Boston Associates and the World They Made*, Harvard Studies in Business History (Cambridge: Harvard University Press, 1987).

James B. Francis, *Lowell Hydraulic Experiments, Being a Selection from Experiments on Hydraulic Motors, on the Flow of Water over Weirs, in Open Canals of Uniform Rectangular Section, and through Submerged Orifices and Diverging Tubes, Made at Lowell, Massachusetts*, 3d ed. (New York: D. Van Nostrand, 1871).

"Memoirs of Deceased Members: James Bicheno Francis," *American Society of Civil Engineers Proceedings* 19 (April 1893): 74–88.

Louis C. Hunter, *A History of Industrial Power in the United States, 1780–1930*, vol. 1, *Waterpower in the Century of the Steam Engine* (Charlottesville, Va.: University of Virginia Press, 1979).

Larry D. Lankton and Patrick M. Malone, *The Power Canals of Lowell, Massachusetts* (Lowell, Mass.: Human Services Corporation, 1973).

Patrick M. Malone, *Canals and Industry: Engineering in Lowell, 1825–1880* (Lowell, Mass.: Lowell Museum, 1983).

———, *The Lowell Canal System* (Lowell, Mass.: Lowell Museum, 1976).

James B. Francis, the Maker of Lowell

James Bicheno Francis was born at Southleigh, Oxfordshire, England, on May 18, 1815, the son of John and Eliza Frith (Bicheno) Francis. John Francis was superintendent of one of the early short-line railroads in Wales, and James was trained to follow in his footsteps.

Following a brief formal education, he became an assistant to his father on the construction of a canal and harbor works connected with the railroad. After performing construction work for the Great Western Canal Company for two years with his father and two others, Francis emigrated to the United States, arriving in New York City in 1833. He found employment with Major George W. Whistler on the construction of the Stonington Railroad in Connecticut. When Whistler became chief engineer of the Proprietors of Locks and Canals in Lowell, he recruited Francis as a draftsman. Francis, then only eighteen, joined the Proprietors in 1834. One of his first jobs was to disassemble, measure, and make detailed working drawings of a new locomotive built by English engineer Robert Stephenson, purchased to serve as a model for the engines of the Boston & Lowell Railroad.

When Whistler resigned to oversee railroad construction in Russia, Francis, at age twenty-two, replaced him as chief engineer. For almost forty years, Francis not only looked after Lowell's water power but also served as consultant to the consortium of Lowell manufacturers that owned and used it, contributing materially to the city's industrial preeminence. "He was the maker of Lowell," stated one biographer in 1907 in The National Cyclopaedia of American Biography.

About 1849, Francis designed the first scientifically designed turbine to be manufactured in the United States in any quantity. The Francis turbine was a mixed-flow type; water flowed radially into the guide vanes and on into the runner, from which it emerged axially. It is still the most common turbine type because of the wide range of heads with which it can be used. Francis's studies of the flow of water through turbines, over weirs, and in canals were disseminated in his acclaimed Lowell Hydraulic Experiments *(1855, revised 1868 and 1871).*

Following his retirement in 1884, Francis was employed as a consulting engineer on the construction of the Quaker Bridge Dam on the Croton River, New York, and the retaining dam at St. Anthony Falls on the Mississippi River at Minneapolis. He joined the American Society of Civil Engineers at its first meeting in 1852 and in 1880 served as its president. He died at Lowell on September 18, 1892.

Sources: *The Dictionary of American Biography* (New York: Charles Scribner's Sons, 1964);*The National Cyclopaedia of American Biography*, vol. 9 (Clifton, N.J.: J. T. White and Co., 1907).

Holyoke Water Power System

Holyoke, Massachusetts

By September 1847, for a total cost of $300,000 for real estate and water rights, Boston capitalists were in possession of the greatest potential mill development in New England. The Hadley Falls Company proceeded to develop the company town of Holyoke just above Hadley Falls on the Connecticut River. There, James K. Mills and George W. Lyman, engineers in charge of construction, planned a water-power system so perfect that, almost ninety years later, no fundamental changes had been made (although the first dam, completed in 1848, failed and was replaced a year later).

The system was designed to use the Connecticut River twice—that is, by sets of mills on two different levels. Water was first received from the dammed Connecticut into a main canal; the main canal then branched to form an upper canal and a lower canal. A raceway parallel with the upper canal received the water as it came through the wheels of the factories and carried it back to the head of the lower canal to be used over again. The plan was modified in 1854 by converting the raceway into a middle canal to create additional mill sites.

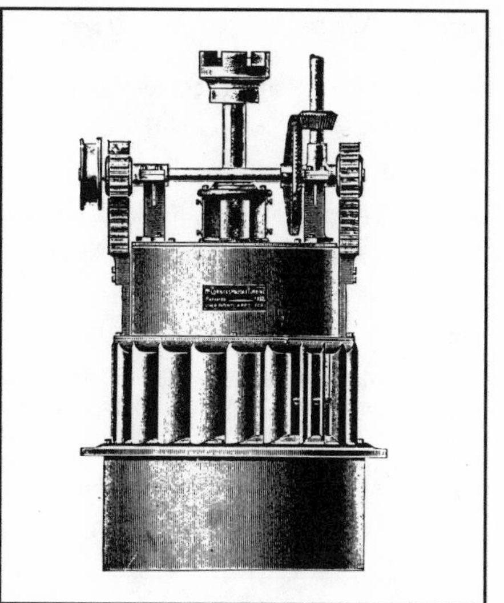

McCormick's Holyoke turbine.

But Holyoke grew slowly, and the textile manufacturing city envisioned by the Hadley Falls Company never took hold. Holyoke instead became home to diverse industries: cotton, woolen, thread, and wire-drawing mills; foundries and machine shops; and, after the Civil War, a vast papermaking industry. In the meantime, poor management and the Panic of 1857 led to the dissolution of the Hadley Falls Company in 1860 and subsequent control by the Holyoke Water Power Company. The latter company continued developing Holyoke as a manufacturing center, completing the city's 4½-mile (7.2-km) canal system in 1892.

By 1880, having reached the limit of available power, the Holyoke Water Power Company was urging lessees to install the most efficient water turbines available and had built a flume where accurate hydraulic power tests could be performed. Designed by hydraulic engineer Clemens Herschel (1842–1930), the flume was equipped for testing turbines of up to 300 horsepower (224 kW). The flume tests gave well-earned publicity to the efficient turbines developed by John B. McCormick (1834–1924) and the Holyoke Machine Company. The achievements of Herschel and McCormick reached far beyond Holyoke. McCormick's turbines gained international fame, while Herschel's venturi flowmeter,* first tested at the flume in 1886, not only allowed the Holyoke Water Power Company to maintain closer control over each mill's water use but became a standard means of measuring the flow rate of liquids.

After 1900, the mill machinery of Holyoke gradually was adapted to electrical power; power could now, with comparative ease, be brought to the manufacturer instead of bringing the manufacturer to the power. While the city's industrial base declined after 1920, its water-power system remains substantially intact, and the Holyoke Water Power Company continues to sell water power to a number of mills whose wheels, now driving generators, produce electricity.

Location/Access

The Holyoke Heritage State Park, 221 Appleton Street, Holyoke, MA 01040, includes a visitors' center with exhibits of the city's engineering achievements and industrial history, and walking tours of the mills and workers' housing; phone (413) 534-1723. Nearby, a McCormick turbine manufactured by Holyoke's J. & W. Jolly, Inc., may be seen outside Holyoke City Hall.

FURTHER READING

Constance McLaughlin Green, *Holyoke, Massachusetts: A Case History of the Industrial Revolution in America* (New Haven, Conn.: Yale University Press, 1939).

Robert Thurston, "The Systematic Testing of Turbine Water Wheels in the United States," *Transactions of the American Society of Mechanical Engineers* 8 (1886–87): 366–74.

* Named after G. B. Venturi (1746–1822), the Italian physicist who first studied the effects of constricted channels on flow.

Morris Canal Scotch (Reaction) Turbine

Greenwich Township (Warren County), New Jersey

In 1972 James Lee uncovered what is believed to be the only Scotch (reaction) turbine in the United States surviving *in situ.* (Only three such turbines are known to exist.) Lee found the turbine, which once powered the winding gear of inclined plane No. 9 West of the long-abandoned Morris Canal, at the bottom of its 30-foot (9,144-mm) supply shaft.

Built between 1825 and 1832 to connect the coal-rich Lehigh Valley of Pennsylvania with the manufacturing centers of New Jersey and New York, the Morris Canal was the highest climber of all the nation's towpath canals. Boats traveling westward from tidewater at Newark Bay to the summit at the tip of Lake Hopatcong, a distance of 51 miles (82 km), climbed 914 feet (279 m), then dropped 760 feet (232 m) to the Delaware River at Phillipsburg (opposite Easton, Pennsylvania), for a total rise and fall of 1,674 feet (510 m) in just over 90 miles (145 km). Had locks with their limited lift been used, between two hundred and three hundred of them would have been required, at a prohibitive cost of both money and travel time. Instead, consulting engineer James Renwick (1790–1863) overcame the steep grades by designing a system of twenty-three inclined planes to supplement the canal's twenty-three lift locks.

The inclined plane was, in essence, a boat railway. A boat was floated onto a wheeled frame running on rails, called a plane car. The plane car was then attached to the wire ropes of a powered winding drum, which hauled it along the incline to a higher (or lower) level of the canal. The plane required two operators: one on the plane car, the other in the plane house sheltering the winding machinery. The trip up or down the incline took approximately eight minutes. Plane No. 9 West, one of three with a double set of tracks, was the highest and longest

Morris Canal Plane No. 9 West, ca. 1900, with Scotch turbine located in the building at the top of the hill. *Library of Congress Collections*.

James Lee and Plane No. 9 West

James Lee came to Plane No. 9 West in 1947. He bought the old plane tender's house and set about restoring it and collecting and preserving memorabilia of the Morris Canal. The cast-iron penstock was still there—albeit filled in—and he reasoned that, if the penstock hadn't gone for scrap during the war, the turbine must still be there, too.

For years, Lee was unable to test his theory, for heavy equipment and many strong backs would be required to unearth the turbine. But in July 1972, with the help of neighbor Scott Hamlen and other volunteers, Lee began hauling rock and dirt out of the rectangular stone chamber. On the night of August 6, the work crew reached the turbine. It was somewhat damaged but still intact. Further digging revealed that the rectangular stone supply shaft led to a vaulted chamber and the discharge tunnel.

Lee's fascination with the Morris Canal began in childhood, when he built a raft and used it in the canal basin at Port Delaware. "My raft," Lee later recalled in The Morris Canal: A Photographic History, *"could hold two small boys quite well; and a friend and I poled it back and forth over a half mile section of the canal. . . . I remember listening to men, many older than my father, tell me stories about life on the Morris Canal. . . .*

"There have been some who said that the Morris Canal was a blue scar across the northern waist of New Jersey. I think, however, that the Morris Canal was a beauty mark, . . . a place where a Sunday walk on the towpath was sheer contentment; a place where there were more fish than fishermen; and an engineering wonder that brought visitors from all over the world. . . .

"The Morris Canal is gone forever. Never again will the sound of the boatmen's conch shell horn echo and re-echo in the valleys and throughout the mountains of New Jersey."

Source: James Lee, *The Morris Canal: A Photographic History* (Easton, Pa.: Delaware Press, 1979).

incline on the Morris Canal; rising 100 feet (30 m), the plane was 1,510 feet (460 m) long to its summit and 1,788 feet (545 m) long overall.

The inclined planes of the Morris Canal originally were powered by overshot wheels. During the winter of 1851–52, Plane No. 9 West was repowered with a Scotch (reaction) turbine as part of a modernization program. The Scotch turbine in principle resembles the common lawn sprinkler. It consists of a horizontal pipe with tangential outlets at the ends, in which the reaction of the escaping water causes the pipe to rotate about its central axis. The Morris Canal wheel itself, located at the bottom of the supply shaft, is a horizontal, circular iron casting, about 7 feet (2130 mm) in diameter, fitted with four "nozzle wings" each measuring 2 feet, 9 inches long (838 mm), giving an overall diameter of 12 feet, 6 inches (317 mm).

The iron penstock made a 90-degree turn down into the supply shaft, then a U-turn up to feed the wheel from below through a 5-foot (1,520-mm) opening in the wheel's annular thrust bearing. After escaping from the nozzles, the water ran off through a 160-foot (49-m) tailrace, or discharge tunnel.

The 1860s was the only prosperous period for the canal. Tonnage—mostly coal, but also grain, wood, cider, vinegar, beer, whiskey, bricks, hay, hides, iron ore, sugar, lumber, and many other commodities—reached a high of almost 900,000 tons (816,462 t) in 1866. Traffic then declined rapidly owing to competition from the railroads, and the canal was abandoned in the 1920s. The canal bed was filled in, and the plane houses, from which the tenders controlled the operation of the planes from their three-story-high perches, were razed. As a safety measure, the turbine supply shafts were filled in; tons of rock and soil were heaped on the Scotch turbines, sealing them in their graves. The story of the discovery and restoration of the Plane No. 9 West turbine after half a century of entombment (see sidebar) is a testament to the difference one person can make in the preservation of history.

Location/Access

Off Highway 519, between State Route 57 and U.S. Route 22 (follow the red-and-white sign for 477 James Lee). Except during the winter months, the turbine can be viewed through a grate across the top of the stone turbine chamber or by walking through the underground tailrace.

FURTHER READING

James Lee, *The Morris Canal: A Photographic History* (Easton, Pa.: Delaware Press, 1979).

———. *Tales the Boatmen Told* (Easton, Pa.: Canal Press, Inc., 1977).

Boyden Hydraulic Turbines, Harmony Mill No. 3

Cohoes, New York

The two Boyden hydraulic turbines at the historic Harmony Mill No. 3 in Cohoes, New York, are among the largest and most powerful ever built to supply mechanical power to a manufacturing plant. Built by the Holyoke Machine Company and installed by the Harmony Manufacturing Company about 1873, they are also among the oldest surviving mill turbines.

The Boyden turbines represent a typical nineteenth-century application of water power, which, by the late 1820s, had begun to advance from the bulky, slow-moving waterwheel to the much more efficient water turbine. By 1827, French engineer Benoît Fourneyron (1802–67) had developed an outward-flow

Uriah Atherton Boyden

Uriah Atherton Boyden (1804–79). *Courtesy National Museum of American History.*

Uriah Atherton Boyden, born in Foxborough, Massachusetts, on February 17, 1804, was an inventor and engineer whose professional focus shifted from railroads—he helped survey the Boston & Providence and Nashua & Lowell early in his career—to hydraulics. Working from an office he established in Boston in 1833, Boyden served as engineer for the Amoskeag Manufacturing Company of Manchester, New Hampshire, designing the power-canal system for that firm's extensive textile mills.

In 1844, at the age of forty, Boyden designed the improved Fourneyron-type water turbine that would carry his name. The prototype unit, a 75-horsepower (56-kW) machine installed at the Appleton Company mill at Lowell, Massachusetts, delivered 78 percent of the power available. Boyden's principal improvement to the earlier Fourneyron turbine was the spiral guide blades, which admitted water to the turbine's runner at a uniform velocity. The Boyden turbine was soon adopted in mills and power plants nationwide.

In his later years, Boyden, who had little formal education, devoted himself to pure science, conducting experiments in light, electricity, magnetism, chemistry, and metallurgy. In 1874 he deposited $1,000 with the Franklin Institute, to be awarded to any resident of North America who could determine whether light and other physical rays were transmitted with the same velocity. The prize was never awarded. Boyden, who never married, lived frugally at a Boston hotel. Upon his death on October 17, 1879, most of his fortune was earmarked for the establishment of an observatory in the Andes at Arequipa, Peru, operated as a department of Harvard University.

Sources: *The Dictionary of American Biography* (New York: Charles Scribner's Sons, 1964); *The National Cyclopaedia of American Biography* (Clifton, N.J.: J. T. White and Co., n.d.).

turbine designed to direct the water, by fixed guide vanes, onto the inner circumference of the rotating wheel. In 1844 Uriah Boyden (see sidebar), an American, improved the Fourneyron turbine by adding a conical approach passage, inclined vanes, and a submerged diffuser that discharged the exiting water more efficiently and ensured that as much of the water's kinetic energy as possible was converted into power at the turbine shaft. (Later, the outward-flow Boyden turbine would be superseded by the even more efficient inward-flow Francis turbine.)

Boyden hydraulic turbines, Harmony Mill No. 3.

Upon completion of the 5-story, 510-foot-long (155-m) south section in 1872, Harmony Mill No. 3—also known as the "Mastodon Mill" for a skeleton unearthed during excavation—was one of the largest textile mills in the United States. It was the pride of the Harmony Manufacturing Company, which had erected its first plant for spinning cotton in 1837. The sprawling Harmony Mills complex drew its power from the Cohoes Company's canals, a sophisticated network of hydraulic canals built between 1834 and 1880 to exploit the water power potential of the Cohoes Falls of the Mohawk River.

With 102-inch (2,600-mm) runners and a horsepower of 800 (600 kW), the turbines were the largest of thirty-two Boyden turbines built by Holyoke in the mid-1870s. Located in the basement at the south end of Harmony Mill No. 3, the two vertical-shaft turbines were connected to a common overhead horizontal shaft by bevel gearing. On this shaft were pulleys that, through leather belts, transmitted the power to each of the mill's five floors, driving the 130,000 spindles and 2,700 looms that produced 700,000 yards (640,000 m) of muslin each week.

The Harmony Mills were liquidated in the 1930s. The two Boyden turbines are intact but no longer in use. Harmony Mill No. 3 today is home to a variety of small industries.

Location/Access

The Boyden turbines are located in the south section of Harmony Mill No. 3. Permission to view them must be obtained from Cohoes Industrial Terminal Inc., 100 North Mohawk Street, Cohoes, NY 12047; phone (518) 237-5000.

FURTHER READING

James B. Francis, *Lowell Hydraulic Experiments* (New York: Little, Brown, 1855).

Norman Smith, "The Origins of the Water Turbine," *Scientific American*, January 1980, 138–48.

Robert M. Vogel, ed., *A Report of the Mohawk-Hudson Area Survey*, Smithsonian Studies in History and Technology, no. 26 (Washington, D.C.: Smithsonian Institution Press, 1973).

Steam

INTRODUCTION by Robert M. Vogel

It has long been a subject of debate, whether the Industrial Revolution made possible the steam engine or vice versa. Surely the practical steam engine could not have been developed without the efficient mining of coal and the smelting and working of metals, mainly the ferrous, and the development of effective prime movers, principally in the form of water-powered machinery. Conversely, it can be equally well argued that none of these developments would have been possible without the steam engine. While the early history of the steam engine is inextricably bound up with the raising of water—almost solely the dewatering of mines and in itself a vital chapter of the Industrial Revolution—it was not until the steam engine became capable of producing continuous rotary power, and thus was able to drive the machinery of factories and mills, that manufacturing developed on a truly industrial scale.

This ability of a prime mover to turn a shaft—independent of the vagaries of flowing water or blowing wind and, most significantly, free of the geographical restraint of a source of falling water—had implications that ultimately reached far beyond the propulsion of factory machinery. As metal-working techniques were refined and, consequently, it became possible to increase the rotational speed of the steam engine, its size could be proportionally reduced resulting in portability. This, in due course, led logically to the steamboat, the steam locomotive, and ultimately a vast array of other mobile steam-powered machinery and vehicles.

But until about 1910, the preponderance of steam power was directed to the driving of stationary machinery in mines, mills, factories, and processing plants in a wide variety of industries. The basic configuration of Watt's rotative beam engine—exemplified by the landmark engine in Sydney—remained essentially unchanged and commercially viable for well over a century. Even with improvements in metallurgy, thermal efficiency, lubrication, and machine design, and even as the direct-connected horizontal steam engine gained in popularity for the mechanical driving of machinery and later generators, the beam engine as conceived by Watt continued to be built by manufacturers principally in Europe throughout the nineteenth century.

The West Point engine at La Esperanza sugar plantation, one of a small handful of American beam engines to survive, is in all respects typical of the breed, differing from a driving engine that might have left the Boulton & Watt shops three-quarters of a century earlier only in mechanical detail and the use of higher steam pressure. It was in every sense a machine of standard design that would have been as much at home driving the line shaft of a textile mill, a large flour mill, or a machine works.

By the middle of the nineteenth century, the horizontal engine had largely displaced the vertical beam type, at least in the United States, as simpler, lighter, and cheaper for the same power. The beam configuration held on as long as it did because, until the widespread use of the metal planer, it was fairly difficult to produce the flat crosshead guides normal to the horizontal engine, whereas the beam engine using the Watt parallel motion to guide the piston rod required only simple pin joints, which were easily produced by turning and boring. Additionally, it was long held by both engine designers and users that the weight of a large horizontal piston would caused undue wear to the bottom of the engine cylinder. While this was true to a certain degree, eventually it was shown that this effect was too slight to offset the many advantages of the horizontal engine. Even in those cases where restricted floor space dictated a vertical engine, toward the end of the nineteenth century the beam engine had given way entirely to the crosshead type with cylinder(s) placed directly above the crankshaft—sometimes referred to as the "marine" or "steam hammer" type. The last American beam engine of any consequence was the renowned Centennial Engine, built in 1876 by George Corliss to power all the exhibits in Machinery Hall at Philadelphia's Centennial Exposition. Although at that time the style was regarded as completely obsolete, it is clear that Corliss employed it there for its monumental visual effect.

Until the turn of the twentieth century, the relatively slow speed of the steam engine was a reasonable match with the speed of most of the driven machinery, so that simple mechanical transmission systems—shafting, gearing, leather flat-belts, and ropes—provided a satisfactory and relatively efficient means for conveying the power from engine to load.

The landmark Harris Corliss engine is representative of the stationary steam engine in the service of directly driving machinery, during the transitional period when mechanical power transmission gradually was giving way to electrical, and electricity was becoming more commonly used in general and for lighting and urban transportation in particular. In the driving of generators—either through belt or rope transmission or by direct connection—the steam engine was called on to operate at increasingly higher speeds, ultimately giving way entirely to the central-station steam turbine as turbine capacity and efficiency increased.

With all prime movers, whatever the operating medium, the story of their development is almost solely one of seeking better efficiency than that currently achievable, a continuing evolution of detail—and usually capacity—in an effort to wring more and more of the potential energy out of the fuel, or falling water. Although the steam engine's single most dramatic leap forward was nothing less than James Watt's improvement of the Newcomen engine, which was the solitary commercial form of steam power in the middle of the eighteenth century, the mechanical engineer's enhancement of efficiency never ceased until the actual demise of the reciprocating engine in the 1940s.

One of the most fertile areas for improving efficiency after Watt's invention of the separate condenser and the expansive use of the steam was the means for controlling the admission of live steam to the engine cylinder. By the 1840s it was recognized that the slide valve, directly moved by an eccentric or cam on the crankshaft, combined with speed regulation by a governor controlling a simple throttle in the steam line, led to considerable thermal loss and, thus, inefficiency. The general awareness of this fact led many inventors to experiment with "releasing" valve gears, which permitted the steam admission valves to be detached from the eccentric at some point after they opened, allowing them to close sharply, and with this point determined either manually by the operator or to some degree automatically under control of the governor. The result was an appreciable advance in efficiency and, thus, reduction in fuel consumption. By further refinement—principally the use of separate semirotary stem and exhaust valves—George Corliss produced an engine that stood well above its contemporaries in economy of operation. The Corliss-type engine as built by Corliss's own works and licensees (and with the expiration of his patents in about 1870 by many of the world's major builders) became the prototype stationary steam engine for efficiency and general excellence of design, reigning supreme until about 1920, at which time practically all engines were designed for electric generating service.

Boulton & Watt Rotative Steam Engine

Sydney, Australia

The Newcomen engine (see "Newcomen Memorial Engine," p. 3), with a chain connecting the walking beam and the piston rod, was capable of performing mechanical work only during the downstroke of the piston. Its construction precluded the possibility of any upward thrust being exerted on the beam. Thus, its usefulness was limited to mine drainage and water supply; even then, its miserably low thermal efficiency (1 percent or less) gave it a voracious appetite for coal.

The Boulton & Watt rotative steam engine introduced the second generation of steam power, whereby the chain connection between the piston and beam of the Newcomen pump was replaced by a set of links forming a "parallel motion." The piston then could push upward on the beam as well as pull down, enabling the beam—with a rigid connecting rod at its opposite end—to drive a crank, resulting in an almost continuous flow of power. It was this rotative, or double-acting, engine that eventually would turn the shafts of industry.

Power House Museum staff inspect the 14-foot flywheel of the Boulton & Watt rotative steam engine. *Courtesy Museum of Applied Arts and Sciences.*

Horsepower

Apart from his improvements of the steam engine, Watt's most enduring achievement was to establish a common unit of power measurement. With the introduction of rotative steam engines came the need to describe the rating of an engine's power to prospective customers. Watt coined the term "Horse-Power," i.e., the number of horses, working at the same time, that would do the same work. Inexplicably, Watt made one horsepower equal to 33,000 pounds (14,966 kg) raised 1 foot (304 mm) high per minute; there is nothing, according to Dickinson and Jenkins in James Watt and the Steam Engine, *to suggest that he determined this value by experiment.*

In Watt's "Blotting and Calculation Book 1782 & 1783" is the notation: "Blackfryars corn mill engine. It is required to work 18 pairs of stones and each pair to grind 6 bushels p. hour and reckoning each bushel p. hour = to one horse and each horse = 33,000 lb I foot high p. minute." While the notation is undated, it is known that Watt finished the drawings of the engine for the Albion corn mill (on the banks of the Thames near Blackfriars Bridge, London) in October 1783.

Within a few years, it was common practice at Boulton & Watt's Soho engine works to refer to engines by "horses"—a "14-horse engine," a "20-horse engine," and so on. In 1814 Watt himself explained his origin of the term "horsepower": "Horses being the power then generally employed to move the machinery in the great breweries and distilleries of the metropolis, where [rotative] engines first came into demand, the power of a mill-horse was considered by them to afford an obvious and concise standard of comparison, and one sufficiently definite for the purpose in view."

The steam engine's transition from pumping duty to its more versatile form came slowly. In 1763 James Watt (1736–1819), an instrument maker, was asked to repair a model of an atmospheric engine used in a natural philosophy class at Glasgow University. The model, Watt observed, consumed such an excessive amount of steam that it could not be kept running for more than a few minutes at a time. Watt recognized that much steam was wasted by heating and cooling the cylinder at every stroke and that the cylinder should be kept as hot as possible the whole time. Watt's discovery led to his patent, in 1769, for a separate condenser. His idea was to condense the steam not in the cylinder below the piston but in a separate vessel connected to it by a pipe with a valve in between.

In 1775 Watt entered into a partnership with businessman Matthew Boulton (1728–1809) of Birmingham, England, whose Soho Manufactory had won wide fame for the quality of its metal work. Boulton & Watt manufactured pumping engines as Watt continued his experiments. Demand was growing for engines that would provide a continuous rotary motion; while the reciprocating motion of existing engines was useful only for pumping, a rotative engine could be used to drive mill machinery. "The people in London, Manchester and Birmingham are *steam mill mad*," Boulton wrote to Watt in June 1781. "I don't mean to hurry you but I think . . . we should determine to take out a patent for certain methods of producing rotative motion from . . . the fire engine."

The basic problem, of course, was to convert the beam engine's oscillation into rotation. This could be readily accomplished with a connecting rod working a crank or, as Watt initially chose in order to avoid a patent dispute, a sun-and-planet (or epicyclic) gear. A serious problem, however, was to smooth out the uneven power cycle of the single-acting engine so that it could be used to drive speed-sensitive machinery, such as that in textile mills. In his 1782 patent for mechanical contrivances to equalize the power stroke, Watt suggested using the power of steam to push the piston both up and down, i.e., a double-acting engine. In conjunction with the flyball governor to regulate the engine's speed, this would result in a smoother output of power, but it required a rigid connection between the beam and the piston rod so that power could be transmitted from the piston on its upward stroke.

In 1784 Watt obtained a patent for his simple but elegant "parallel motion." His invention, which connected piston rod to the end of the beam by an arrangement of links roughly in the form of a parallelogram, allowed the piston to push as well as pull the end of the beam, making the engine double-acting and, therefore, twice as powerful. With justice, Watt could later say, "I am more proud of the parallel motion than of any other mechanical invention I have ever made."

The Boulton & Watt rotative engine helped launch the Industrial Revolution. By 1800, Boulton & Watt had built 496 engines, 308 of which were of the rotative type for use in driving machinery, principally that of textile mills. In 1784 London brewer Samuel Whitbread ordered from Boulton & Watt a rotative engine with a 24-inch (609-mm) cylinder and 6-foot (1,829-mm) stroke. The single-acting engine, 30 feet (9,144 mm) tall and about the same in length, was the first to incorporate parallel motion. It was made double-acting in 1795, doubling its indicated horsepower from about 17 (13 kW) to about 35 (26 kW).

The Whitbread engine, high technology for its day, remained in service until 1887, when it was dismantled. Whitbread donated it to the Sydney Technological Museum in Australia two years later. To celebrate the engine's bicentennial, the museum restored it to steaming condition in 1985.

Location/Access

The Boulton & Watt engine is on display at the Power House Museum, Castle Hill, Sydney, Australia. (The next-oldest survivor, of 1788, is at the Science Museum in London.) The atmospheric engine model, on which Watt first experimented, is preserved in the Hunterian Museum of Glasgow University in Scotland.

FURTHER READING

H. W. Dickinson and Rhys Jenkins, *James Watt and the Steam Engine* (Oxford: The Clarendon Press, 1927).

Richard S. Hartenburg, "Parallel Motions: Certain Combinations of Levers Moving upon Centers" ASME paper no. 83-WA/HH-2 (New York: ASME International, 1983).

Richard L. Hills, *Power From Steam: A History of the Stationary Steam Engine* (Cambridge: Cambridge University Press, 1989).

Samuel Smiles, *Lives of Boulton and Watt* (Philadelphia: J. B. Lippincott and Company, 1865).

Robert H. Thurston, *A History of the Growth of the Steam-Engine*, Centennial Edition, with a supplementary chapter by William N. Barnard (Port Washington, N.Y.: Kennikat Press, 1972).

Hacienda La Esperanza Sugar Mill Steam Engine

Manatí, Puerto Rico

In the nineteenth century, sugar production became Puerto Rico's economic mainstay. Until the rise of the modern sugar plant, or *central,* in the 1890s, sugar cane was processed on individual estates. One of the island's largest producers was Hacienda La Esperanza, a 2,265-acre (917-ha) estate located in the fertile valley of the Río Grande de Manatí, west of San Juan. There, a beam engine once used to power the plantation's crushing mill has been preserved. Countless such engines powered small mills and waterworks in the nineteenth century, but only a few have survived.

Hacienda La Esperanza was established by Fernando Fernández, a Spaniard who arrived in Puerto Rico in the late eighteenth century. Archaeological evidence (there are no written records) suggests the presence of an animal-powered mill early on, superseded by a small steam-powered mill sometime in the 1850s. José Ramón Fernández y Martínez, Fernando's eldest son, inherited the estate and expanded production with the addition of a steam-powered sugar mill in the early 1860s. By 1862, the Hacienda was producing 135,000 pounds (61,224 kg) of moscavado (unrefined) sugar and 500 hogsheads (63 gallons, or 238 l) of molasses annually.

Sugar production begins with the extraction of juice from sugar cane. Next, the juice is clarified (all matter, except sugar and water, is removed); reduced (the water is removed from the sugar); and, finally, purged (cleansed by washing or draining). Hacienda La Esperanza was semimechanized, i.e., it had a steam-powered crushing mill, but evaporation, purging, and packing operations were performed manually (by slaves until emancipation in 1873).

At La Esperanza, an endless chain conveyed sugar cane into the crushing mill, consisting of three cast-iron rollers set horizontally in a cast-iron frame, the

Stationary beam engine on the Hacienda La Esperanza sugar plantation. *Photograph by Fred Gjessing, Library of Congress Collections.*

roller axes forming the vertices of an isosceles triangle. During crushing, the juice drained into large square pans, or collectors. Cast into the base of the crushing mill is the legend "West Point Foundry 1861," the only maker's mark on any of the machinery. The six-column beam engine that powered the mill carries no identifying marks but is presumed to have come, like the crushing mill, from the West Point Foundry in Cold Spring, New York. It is the only West Point beam engine, and one of only eight stationary beam engines produced by any American manufacturer, known to survive.

The engine, distinguished by elaborate Gothic styling, is a noncondensing, drop-valve, side-crank engine with 16-inch (406-mm) bore and 40-inch (1,016-mm) stroke. When turning 20 rpm on steam of 60 psig (413 kPa), the engine developed approximately 25 horsepower (19 kW). The cast-iron beam, 13 feet, 4 inches (4,063 mm) overall, served as a rocking lever connecting the piston rod and crank. Two eccentrics on the crankshaft controlled the engine's drop valves. The Watt-type "parallel motion" guiding the piston rod was easier to maintain in a damp, hostile environment than a simpler crosshead and guides, needing lubrication only at readily protected plain bearings.

To deliver maximum power, the engine had to run at approximately 20 rpm, but to extract cane juice effectively, the mill had to turn much more slowly. Double-reduction gears accomplished this change in speed, permitting the mill rollers to turn at just under 2 rpm. A "Lancashire" boiler fired with wood or *bagasse* (residue cane, following crushing) provided the steam.

The West Point Foundry, established in 1817, turned out steam engines, hydraulic presses, and blowing engines in quantity during the nineteenth century and built the first two locomotives manufactured in America for actual service on a railroad: the *Best Friend of Charleston* (1830) and the *West Point* (1831). Its real fame, however, was as a manufacturer of military supplies for the United States Army and Navy. No records of the foundry, which was closed in 1911, survive; as a defense contractor, West Point regularly disposed of sensitive material by lighting huge bonfires, and virtually the only evidence of the firm's work is in manufacturers' catalogs and technical books of the period.

Beginning in the late nineteenth century, plantation factories making moscavado sugar on individual estates gave way to modern sugar *centrals*. After 1891, sugar cane harvested at Hacienda La Esperanza was sent to the *central*, and the West Point mill and engine were never used again.

Location/Access

Hacienda La Esperanza is located in Manatí, 35 miles (56 km) west of San Juan. It is operated as a living historical farm by the Conservation Trust of Puerto Rico and may be visited with their permission: P.O. Box 4747, San Juan, Puerto Rico 00902-4747; phone (809) 722-5834. (Note: At press time, the engine and mill had been dismantled and were awaiting restoration.)

FURTHER READING

Louis C. Hunter, *A History of Industrial Power in the United States*, vol. 2, *Steam Power* (Charlottesville, Va.: University Press of Virginia, 1985).

Noel Deerr, *The History of Sugar*, 2 vols. (London: Chapman and Hall, Ltd., 1949–50).

Harris-Corliss Steam Engine

Atlanta, Georgia

In 1977 the 350-horsepower (260-kW) steam engine that had powered the woodworking shop of Randall Brothers, Inc., for some eighty years came to a standstill. It was retired not because of age or infirmity but because of U.S. Environmental Protection Agency concern about smoke issuing from the boiler smokestacks.

The Corliss-type engine was built around 1894 by the William A. Harris Company of Providence, Rhode Island. It was exhibited at the Cotton States and International Exposition of 1895 in Atlanta before being sold to Exposition Cotton Mills of Atlanta in 1898. Randall Brothers purchased the engine from Expo-

Harris-Corliss Steam Engine.

sition Mills sometime between 1898 and 1910, using it to drive an electric generator and to power the firm's woodworking machinery, including lumber saws.

William A. Harris (1835–79) joined the Corliss Steam Engine Company in 1856 as a draftsman. He later became chief assistant to George H. Corliss (see sidebar), making the drawings for the inventor's numerous patent applications before leaving to establish his own firm in 1864. Under a licensing agreement with the Corliss Company, Harris manufactured Harris-Corliss engines from 25 to 2,000 horsepower (19 to 1,491 kW), in simple and compound, and condensing and noncondensing styles.

Although no longer in use, the Randall Brothers engine remains in its original location. The company periodically fires up the boilers to show students at the Georgia Institute of Technology a classic stationary steam engine at work.

George H. Corliss and the Corliss Engine

George H. Corliss (1817–88). Courtesy National Museum of American History.

George H. Corliss ranks with James Watt (see "Boulton & Watt Rotative Steam Engine," p. 43) in his contributions to the improvement and refinement of the steam engine. Born in Easton, New York, on June 2, 1817, George Henry Corliss was the only son of Hiram and Susan (Sheldon) Corliss. He began his working life as a store clerk and inspector for William Mowray & Son, textile manufacturers. He attended the Castleton (Vermont) Academy before opening a general store in Greenwich in 1838. Customer complaints about inferior stitching in the boots he sold reputedly led him to design and patent a sewing machine for stitching leather, in turn leading him to experiment with building various types of machinery.

In 1844 Corliss moved to Providence, Rhode Island, where he joined Fairbanks, Bancroft & Company as a draftsman. He left that firm four years later to organize, with John Barstow and E. J. Nightingale of Providence, a new company under the name Corliss, Nightingale & Company. There, in 1848, Corliss invented an improved means for controlling the amount of steam admitted into the cylinder of a steam engine under varying load (Patent No. 6,162). The Corliss engine, with its improved valve gear regulated by the centrifugal governor, revolutionized steam engine design by providing for uniform motion regardless of load and by reducing the waste of steam, which dramatically reduced fuel costs.

Corliss engines are characterized by four cylindrical valves that alternately open and cover ports (openings) at opposite ends of the cylinder: two ports admit steam from the boiler into the cylinder, while two exhaust the spent steam from the cylinder into the condenser (or the atmosphere). This arrangement was more efficient than the conventional piston or slide valve; the steam valves were not cooled by the exhaust, and each valve could be regulated separately.

In 1856 Corliss incorporated the Corliss Steam Engine Company in Providence. The company became the largest and finest maker of stationary steam engines in the world, employing one thousand workers by 1880. With the expiration of his basic patents in 1870, dozens of engine builders worldwide profited from his ideas. The Corliss engine, in myriad forms, became the standard factory prime mover of the late nineteenth century. In 1875 Corliss proposed and built a 700-ton (635-t), 1,400-horsepower (1,044-kW) double engine, the largest in the world, to furnish all power for the U.S. Centennial Exhibition's Machinery Hall in Philadelphia. The engine was acclaimed as the ultimate manifestation of American technological prowess.

During the last three years of his life, Corliss invented special machinery to make interchangeable parts. Among his other inventions were a machine for cutting the teeth of bevel gears, an improved boiler with condensing apparatus, and a pumping engine for water works. In all, he received sixty-seven patents.

Corliss died in Providence on February 21, 1888.

Sources: *The Dictionary of American Biography* (New York: Charles Scribner's Sons, 1964); Louis C. Hunter, *A History of Industrial Power in the United States*, vol. 2, *Steam Power* (Charlottesville, Va.: University Press of Virginia, 1985); *Scientific American*, June 2, 1888.

Location/Access

Open upon application to the owner: Randall Brothers, Inc., 665 Marietta Street N.W., P.O. Box 1678, Atlanta, GA 30371; phone (404) 892-6666.

FURTHER READING

Louis C. Hunter, *A History of Industrial Power in the United States*, vol. 2, *Steam Power* (Charlottesville, Va.: University Press of Virginia, 1985).

"The Manufacture of the Harris-Corliss Engine," *Scientific American*, September 20, 1879, 1–2.

SPECIFICATIONS

Harris-Corliss Engine, Randall Brothers

Cylinder: 16 inches (406 mm) in diameter,
42-inch (1,067 mm) stroke
Steam pressure: 125 psig (862 kPa)
Speed: 90 rpm
Flywheel: 13 feet (3,960 mm) in diameter
Horsepower: 350 (260 kW)

Internal Combustion

INTRODUCTION by Robert M. Vogel

The idea of a prime mover based on the power produced by the explosion of fuel inside a closed cylinder actually is older than the steam engine. In the seventeenth century, the Dutch scientist Christiaan Huygens experimented with a cylinder in which he ignited a small charge of gunpowder, the explosive gases of which blew freely out of the cylinder through a check valve. As the residual gases cooled, a vacuum was formed in the cylinder, causing a piston to be forced into it by atmospheric pressure, raising a weight, and so performing work. The apparatus was merely a laboratory curiosity; it was not a practical device, let alone an "engine," but the concept attracted other experimenters. Not until the middle of the nineteenth century was a commercially practical internal-combustion engine developed by the Frenchman J. J. E. Lenoir. Like many others working at the time, he employed coal gas as the fuel, and it was, in fact, gas that eventually launched the internal-combustion engine into the realm of viable competition with the steam engine as a practical prime mover.

Lenoir's engine was built in commercial quantities even though its efficiency was low and its fuel costs were high compared to a steam engine of like power. This was due to the fact that the combustible mixture of fuel and gas was not compressed before ignition. However, the engine's moderate success as a power plant free of the need for boiler, water supply, coal and ash handling, and the other accoutrements of a steam plant inspired inventors to continue the quest for an efficient internal-combustion engine.

The German Nicholas A. Otto launched the first serious attack on the steam engine's supremacy with the invention and marketing in 1867 of his "free-piston atmospheric" gas engine. Ironically, this first internal-combustion engine with serious commercial potential harked back strongly to Huygens's gunpowder device in that the exploding gas-air mixture drove a piston to the top of a long cylinder but performed no work in its flight. The piston's inertia plus the cooling of the combustion gases at the end of the long stroke produced a strong vacuum in the cylinder. Into this the atmospheric pressure drove the piston, which on its downstroke engaged the engine's shaft through a clutch, thus producing power.

The Otto & Langen engine (Eugen Langen was Otto's business partner) was built in considerable numbers and in sizes up to about 10 horsepower (7.5 kW). Manufactured in Germany and under license in England, it was used to power small factories and shops. The commercial success of the machine went far in spreading the gospel of internal combustion: a prime mover that could be started instantly in the morning (no fire to light; no boiler full of cold water to heat up); run all day with no attention other than occasional oiling (no fire to constantly stoke and tend; no boiler-water level to maintain); and be shut down at day's end with no further attention (no fire to bank; no ashes to haul away; no coal pile to replenish). If the efficiency was somewhat less than that of an equivalent steam engine, a potential gas-engine buyer usually could be persuaded that the extra fuel cost would be more than offset by the floor space no longer devoted to a boiler, its manifold auxiliaries, and coal and ash storage, to say nothing of the cleaner atmosphere resulting from the absence of coal and ash dust.

The position of the internal-combustion engine in the world of commerce and industry was sealed with Otto's invention in 1876 of his four-stroke cycle gas engine. Here was internal combustion that could compete on a nearly equal basis with steam in terms of fuel efficiency and mechanical simplicity. By dividing the engine's working cycle into four separate "events," or strokes, the gas-air charge could be compressed with certainty and perfect control, greatly increasing the engine's thermal, and thus operating, efficiency. At the same time, with the piston directly connected to a crankshaft and the intake and exhaust valves also driven directly by a camshaft, most of the mechanical excesses of the free-piston engine were eliminated, with a simpler, cheaper, and quieter engine the result.

Otto's four-stroke "Silent" engine became the model for the world. Dozens of inventors and manufacturers took up the cause, in direct competition with the steam engine, obtaining internal-combustion patents by the hundreds and producing lines of engines by the score, which burned both liquid and gaseous fuels. By the turn of the century, units of up to 1,000 horsepower (750 kW) were in production. Burning liquid fuel, the internal-combustion engine became completely portable, a simple and logical power plant for small self-propelled vehicles.

Despite the technical and commercial success of this new prime mover, inventors remained aware of two major shortcomings: primarily, that these engines remained inherently somewhat higher in fuel consumption than steam engines; and that the fuel required some form of external ignition to burn and thus expand. The ignition system of all early internal-combustion engines invariably was a weak link, whether by electric spark, open flame, incandescent tube, or by some other means.

In attempting to solve the one problem, the German Rudolf Diesel inadvertently solved the other. His aim was to increase the temperature range between that in the cylinder at combustion and that of the exhaust at the end of the power stroke, in effect extracting more of the energy in the fuel. The diesel engine, which is better called the compression-ignition engine, operates (as finally evolved) on a cycle in which the combustion occurs initially at constant volume and then at constant pressure, sometimes called a dual- or limited-pressure cycle. Although Diesel's engine operated on the four-stroke cycle, modern engines also operate on the two-stroke cycle.

During the compression stroke, only a charge of air is compressed but to such a high pressure and, therefore, temperature that, when a metered quantity of fuel is injected into the cylinder at the end of the stroke, it ignites spontaneously, producing the power stroke. This inherent self-ignition means that the diesel is entirely free of an ignition system.

Diesel's engine was found to be approximately 11 percent more efficient than any other contemporary form of internal-combustion engine, which had the effect of propelling it rapidly into a powerfully competitive position with respect to the steam plant by the turn of the twentieth century. Had the steam turbine and later the uniflow steam engine not appeared on the scene, it is likely that steam power would have passed entirely out of the picture at that time.

Certain applications ultimately emerged for which the diesel engine was ideally suited: small- and medium-sized generating plants, marine and railroad propulsion, and the powering of most commercial vehicles. Although the basic principle of the engine has remained unchanged since the work of Rudolf Diesel, there has been continuous improvement in the engine's details and auxiliary organs, mainly in the means for injecting the fuel into the cylinders. The injection system must convert the fuel from a liquid into a fine mist, it must "meter" the fuel in a precisely measured quantity proportional to the load on the engine, and it must inject it at precisely the right time in the engine's cycle. These tasks are accomplished by a great variety of pumps, injectors, distributors, and combinations of these, with their use determined by engine size, type of service, and manufacturer's design philosophy. The landmark Roosa Master is a leading example of the type that combines the metering and timing functions in a single unit. Its effectiveness and simplicity were instrumental in popularizing the diesel engine in light and medium service.

Roosa Master Diesel Fuel-injection Pump

Windsor, Connecticut

In 1947 high-speed diesel power in the United States was still very limited. Less than 5 percent of all engines being built, even for nonautomotive applications, were diesels. Although diesel power had proven to have real advantages, its price was prohibitive for many applications. A simpler, more compact, and less costly fuel-injection pump was needed before diesels could compete effectively in the small-engine field.

Vernon D. Roosa (1911–), a versatile and prolific inventor, solved the problem in 1939 by designing a simple unit having very few moving parts. With only three critical fits, it was designed for inexpensive production. "Its simplicity is deceptive," *Diesel Power and Diesel Transportation* reported shortly after the pump

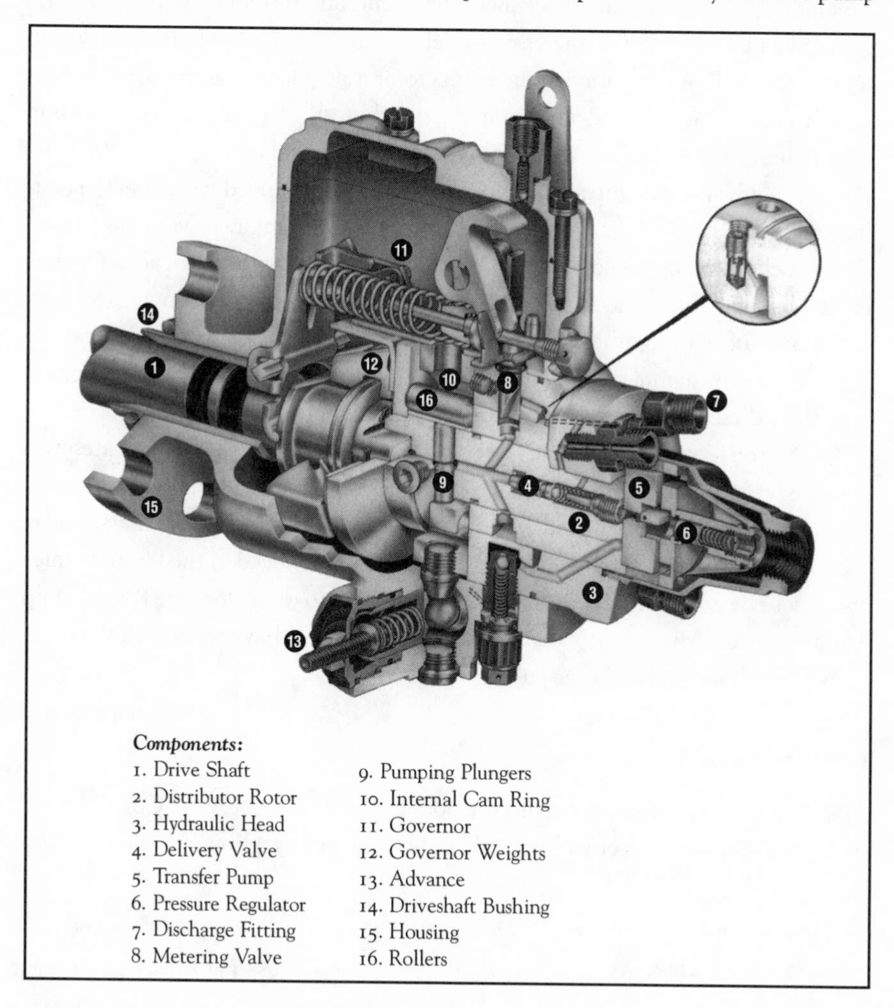

Sectional view of the Roosa Master Diesel Fuel-injection Pump.

was introduced commercially in 1952. "A lot of hard work and good engineering went into its development."

Roosa brought his invention to the Hartford Division of Stanadyne, Inc. Throwing aside the traditional in-line injection pump with its pumping element for each engine cylinder, Roosa instead used a single pumping unit, which distributed the pressurized fuel to each cylinder in turn, combined with inlet metering. The result was a simple, lightweight, and flexible fuel-injection pump that opened new design possibilities in the high-speed, small-engine diesel field.

Roosa's idea was borne out in extensive field testing. Then in March 1952 came the first production order: five hundred "Roosa Master" Model A pumps for Hercules Motor Corporation's Oliver Cletrac tractors. By 1956, Continental Motors, Budd Engine, and Waukesha Motor were using the rotary-distributor pump. Stanadyne engineers, meanwhile, continued working to make the pump even simpler, more versatile, and less expensive, eventually introducing Model B with its sand-cast housing, the die-cast Model D, and in 1958, Model DB, which incorporated all of the basic features of its forerunners into one standard housing together with automatic advance and electric shut-off. A single delivery valve in the center of the rotor provided improved part-load regularity.

The advanced features of the Model DB pump extended the diesel's operating range and made it competitive with spark-ignition engines. Most important, the Model DB pump could be mounted either horizontally or vertically, allowing engine builders to use the same basic engine block for both diesel engines and the conversion of spark-ignition designs to diesel operation. This paved the way for farm-equipment manufacturers, already making their own spark-ignition engines, to turn to diesel-engine production with a minimum of tooling costs.

By 1961, virtually every diesel tractor built in the United States was equipped with a Roosa Master pump. Allis Chalmers, Ford, International Harvester, John Deere, J. I. Case, and Minneapolis Moline were all pump users. Today more than 90 percent of farm and industrial tractors produced in the United States are diesel-powered; the opposite was true in the mid-1950s, before the Roosa pump was introduced. Since 1952, Stanadyne and its licensees have manufactured more than 23 million Roosa Master pumps worldwide.

Location/Access

The Roosa pump is displayed in the lobby of Stanadyne Auto Corp., Diesel Systems Division, 92 Deerfield Road, Windsor, CT 06095; phone (203) 525-0821.

FURTHER READING

"Roosa Master: A Fuel Injection Pump with Ideas," *Diesel Power and Diesel Transportation* 30 (November 1932): 36–39.

Electrical Power Production

Water

INTRODUCTION by R. Michael Hunt

Water has been used as a power source from the earliest times, and the simple waterwheel dipping into the river flow to extract power to drive the grinding stones in a mill is a familiar example. In the early nineteenth century, water was the only source of power available on a large scale, and the first factories—often for manufacturing textiles—flourished in areas such as the northeastern United States, where water power was abundant. By the time Edison produced his first dynamos in the 1870s and 1880s, efficient hydraulic machinery was available to generate the new power: electricity.

One of the first applications using water power to generate electricity was in Appleton, Wisconsin, in 1882, which took advantage of a 33-foot (10-m) drop in the Fox River. Although relatively crude at first, by 1888 the system was serving many area businesses and residential customers with metered, twenty-four-hour service.

At first, only customers in the immediate vicinity of electricity generating plants could be connected because electricity was sent out of the generating station at the same voltage that it was to be used by the customer. This meant that currents were large and that losses became large as transmission distances increased. The solution was to increase the transmission voltage, decrease the current, and reduce the losses. But the early Edison stations generated direct current, or DC, which was difficult to raise or lower in voltage. However, alternating current, or AC, is easily "transformed" from one voltage to another, and eventually became the preferred system. In 1895, Folsom Power House No. 1 demonstrated that AC electricity could be transmitted to Sacramento, California, 24 miles (38.6 km) away.

The first hydroelectric plant in North Carolina started near Clemmons in 1898, and transmitted power to Winston and Salem, 10 miles (16.1 km) away. With the new power source, industrialization of the state proceeded apace. As in the Folsom station, waterwheels are gone, replaced by efficient reaction turbines. In a reaction turbine, the water flow is turned by vanes, or blades, on the outside of a wheel, or "runner." Because the water is forced to change direction, a reaction force is generated against the vane, and this turns the wheel to drive the generator. The whole turbine runs full of water, so there is no splashing to waste energy.

At the turn of the century, Sault Sainte Marie, Michigan, also seemed poised for industrial expansion. The locks constructed in the 1850s had bypassed the rapids on the St. Mary's River, and ships loaded with copper from the Keweenaw Peninsula and iron from the Menominee Range now passed through from Lake Superior on their way south. There was still plenty of water flow to exploit through the 20-foot (6.1-m) drop of the river, so the Michigan–Lake Superior Power Plant was built, with its seventy-nine turbines totaling 40,000 horsepower (29.8 MW). But the markets for manufactured goods were too distant, and the hoped-for industry did not come.

Another type of hydraulic turbine, the Pelton wheel, was developed in the western United States, where large water-pressure "heads" are available from mountain streams. The Pelton is extremely compact for its power output and is an impulse machine. It is a modern-day waterwheel in which the water squirts from a nozzle to hit shaped cups or buckets mounted on the circumference of a wheel, and the wheel moves from the force of the water striking it. At Child, Arizona, a three-Pelton-wheel station began operation in 1909, running under the then enormous head of 1,075 feet (327.7 m) of water.

In the eastern United States, the output of many hydroelectric plants was dictated by the flow in the river. To enable the Rocky River plant in Connecticut to generate when electricity was needed instead of when water was available, pumps were provided to pump water back uphill into a reservoir for use later—a water storage—battery of sorts. In 1928 its reservoir was the largest for pumped storage in the world.

Until World War I, reaction turbines were designed with fixed blades. These operate with maximum efficiency at one specific flow and, thus, one specific power. It had long been known that a turbine with variable blade angles would operate at high efficiency over a wide range of powers, but suitable designs were not available. In 1929 the first variable-blade turbine of Kaplan design in the United States was installed at York Haven, Pennsylvania.

At another pumped-storage plant, the Hiwassee Dam, Unit 2, in Murphy, North Carolina, the technology had progressed beyond that of Rocky River, so that the turbine could be driven in reverse as a pump and separate pumps were not required. In 1956 this was a premier machine in an era of superlatively large machines—it had the largest Francis-type reaction turbine runner ever built, driven by the world's largest electrical machine. Its builder was Allis-Chalmers, of West Allis, Wisconsin.

A point to ponder: in reading about the steam engines, turbines, pumps, and generators of the last one hundred years, three company names keep recurring: General Electric, Westinghouse, and Allis-Chalmers. Forty years ago Allis-Chalmers was a giant; today, it is no more. Twenty years ago George Westinghouse's East Pittsburgh Works was humming with the manufacture of motors and generators; today, it is shuttered and silent, and the company he founded struggles for survival. Only General Electric thrives, with great changes in the products it offered fifteen years ago. The successes of the past are no guarantee of the future.

Vulcan Street Power Plant

Appleton, Wisconsin

The present Vulcan Street Power Plant, a replica of the original plant of 1882, was built in 1932 for the Golden Jubilee celebration of what was then billed as the "World's First Hydro-Electric Central Station." Subsequent research suggests that this claim must be qualified as "the first Edison hydroelectric central station to serve private and commercial customers in North America." Still, in addition to the American Society of Mechanical Engineers, two other organizations—the Institute of Electrical and Electronics Engineers and the American Society of Civil Engineers—have recognized the Vulcan Street Power Plant as a landmark in the history of technology.

Some 150 Edison-installed electric light plants already were at work in residences, mills, stores, offices, and on ships by 1882. Edison's Pearl Street Station in New York, with the capacity to operate 7,200 lamps at 110 volts, went into service on September 4 that year, while England had put a 3,000-lamp station on line a bit earlier, in January.

In July 1882, Appleton financier H. J. Rogers, president of the Appleton Paper & Pulp Company, purchased the Edison patent-licensee rights for Wisconsin's Fox River Valley. The Western Edison Light Company sent an engineer, P. D. Johnson, to Appleton to explain the lighting system to a group of Appleton businessmen. Convinced that the investment was a good one, Rogers and a handful of other investors ordered two Edison Type "K" dynamos, each with a generating capacity of 250 lamps, or about 12.5 kilowatts. By mid-August, Edward T. Ames, Western Edison's erector and electrician, arrived in Appleton to install the equipment.

Making use of the Fox River's 33-foot (10-m) drop, Ames connected one generator to the waterwheel of a pulp mill that belonged to Rogers. This mill, another of Rogers's mills, and Rogers's nearby residence all were wired for electric light. The first test of the system, on September 27, failed, but three days later the lights went on. The *Appleton Post* declared them "as bright as day." But the generator was driven by the same waterwheel that drove the pulp mill, the speed of which fluctuated; sometimes the voltage was so high the lamps burned out—an expensive fault, since lamps cost $1.60 each. Accordingly, in November, both generators were belted to a dedicated turbine in a separate powerhouse, a modest structure on Vulcan (now South Lawe) Street, resulting in a steady voltage.

The speed of the Appleton installation was accomplished by sacrificing safety and reliability features now taken for granted. The equipment was crude. There were no voltage regulators, voltmeters, or ammeters; operators used their eyesight to gauge the proper brightness of the light. There was no lightning protection and no fuses; when storms caused short circuits, the plant had to be

Interior of the Vulcan Street Power Plant showing Edison generator and drive.

shut down until the problem was found. There were no customer meters—customers were charged so much per lamp per month—and service was available only from dusk to dawn. Still, electric light was a great popular success, and by November 1882, several additional homes were lighted by the Edison system. Early the following year, Appleton's Waverly House became the first hotel in the Midwest to boast electric light.

By the end of 1886, the Appleton Edison system served almost one hundred residential, commercial, and industrial customers. That year, the Appleton Edison Light Company built a new 190-kilowatt plant with all the advanced features of the Edison system, including regulating devices, fuses, and the three-wire distribution system, one of the world's earliest. Customer meters were introduced in 1888; and twenty-four-hour service, in 1890. Electric service in Appleton was now as modern as anywhere in the world.

Location/Access

A replica of the Vulcan Street Plant is located at 807 South Oneida Street, Appleton, WI 54915, next to the general offices of Wisconsin Michigan Power Company.

FURTHER READING

Louise P. Kellogg, "The Electric Light System in Appleton," *Wisconsin Magazine of History* 6 (December 1922): 3–8.

Forrest McDonald, *Let There Be Light: The Electric Utility Industry in Wisconsin, 1881–1955* (Madison, Wisc.: American History Research Center, 1957).

Folsom Powerhouse No. 1

Folsom, California

The first electric power plant in central California was constructed on the American River in the Sacramento Basin. The old Folsom Powerhouse still shelters the machinery that in 1895 generated 3,000 kilowatts of electricity for the city of Sacramento, 24 miles (38.6 km) to the southwest. Although Folsom was not the first hydroelectric plant in the country, its transmission line was three times as long as the better-known plant at Niagara Falls (1895), and it demonstrated the commercial feasibility of electrical transmission over long distances.

The story of Folsom begins with the Gold Rush. Not all the adventurers of that period sought gold; some, envisioning industrial complexes like those in New England, sought to exploit the water power. One such émigré was Horatio Gates Livermore of Maine, who, with his two sons, acquired control of the Natoma Water and Mining Company in 1862. The company bought 9,000 acres (3,642 ha) of land, including water rights, on the American River and in 1866 set out to build a dam to provide a holding area for logs and to furnish water for power and for the irrigation of orchards and vineyards. Following protracted delays, work on the dam—much of it using convict labor from the new Folsom prison in exchange for water power privileges and certain grants of land—began in 1888 and was completed in 1893.

By the late 1880s, Horatio P. Livermore (the elder Livermore had died in 1879) understood that water power might be used more efficiently if it was converted to electricity; Folsom power could even be used to operate the Sacramento street railways. In 1892 Livermore incorporated the Sacramento Electric Power & Light Company to build a powerhouse and construct a long-distance power line to the capital city. Water from the Folsom Dam was diverted by canal to the site of a new powerhouse, a distance of almost 2 miles (3.2 km). General Electric supplied the electrical equipment, and S. Morgan Smith supplied the hydraulic machinery.

The power originated with four 30-inch (762-mm) McCormick reaction turbines, each having a capacity of 1,100 horsepower (820 kW) operating at 300 rpm under a head of 55 feet (16.8 m). To each shaft was coupled a 750-kW General Electric three-phase generator, the largest of their type yet constructed. From the generators, the current passed to the generator switchboard, then to nine step-up transformers—each of 250-kW capacity—in the second story of the powerhouse. There, the voltage was raised from 800 to 11,000; the current then passed through marble switchboards to the bare copper wires of the double high-tension transmission line, which followed the highway from Folsom to Sacramento, a distance of 24 (38.6 km) miles.

At 4 A.M. on July 14, 1895, a 100-gun salute marked the first transmission

One of the four dual McCormick turbines at Folsom Powerhouse. *Courtesy Folsom Lake State Recreational Area.*

of power from Folsom to Sacramento. On September 9 of that year, an "electrical carnival" celebrated what, to date, was the longest commercial power transmission ever effected. By 1896, in addition to the city's street railways, electricity from Folsom was being used for manufacturing and for lighting the city.

The generators, although still intact, were removed from service in 1952 after fifty-seven years of continuous duty. In 1958, following construction of the new Folsom Dam as part of the massive Central Valley power project, the Pacific Gas & Electric Company donated the Folsom Powerhouse to the state of California.

Location/Access

The Folsom Powerhouse is open during the summer daily from 9 A.M. to 5 P.M.; and during the spring and fall, on Saturday and Sunday from 9 A.M. to 5 P.M. Tour reservations and information are available from the Folsom Lake State Recreation Area, 7806 Folsom-Auburn Road, Folsom, CA 95630; phone (916) 988-0205.

FURTHER READING

Charles M. Coleman, *PG&E of California* (New York: McGraw-Hill, 1952).

"The Folsom-Sacramento Electric Power Transmission Plant," *Engineering News* 35 (7 May 1896): 302.

"The Sacramento-Folsom Power Transmission Line," *Electrical World* 30 (6 April 1895): 433–34.

Idols Station, Fries Manufacturing & Power Company

near Clemmons, North Carolina

Idols Station, privately developed and put into service in 1898, was the first hydroelectric plant in North Carolina. Transmitting alternating current at 10,000 volts, the station supplied power to factories and public utilities in Winston and Salem, some 13 miles (20.9 km) away. (The two towns did not become one municipality until 1913.) From this small-scale, low-head station, North Carolina's

Interior of the powerhouse of Idols Station. *Photograph by Bill Yoder.*

production of hydroelectric power grew rapidly, contributing to its rapid industrialization in the first quarter of the twentieth century.

The Yadkin River flows southeast through North Carolina and South Carolina before entering the Atlantic Ocean. In 1891 industrialist Henry Elias Fries chartered the Fries Manufacturing & Power Company to harness Douthit's Shoals in Forsyth County for the generation of hydroelectric power. Fries issued capital stock in 1897 (inventor Thomas A. Edison and street-railway innovator Frank J. Sprague were among the original investors) and engaged two Providence, Rhode Island, firms—C. R. Makepeace & Company, mill engineers and architects, and Lewis & Claflin, electrical engineers—to design the plant.

Idols Station takes its name from its location on the site of a former ferry crossing. It was designed as a "run-of-the-river" plant to avoid flooding the low-lying land adjacent to the river. A low, curved gravity dam of rubble stone, 482 feet (147 m) long and 10 feet (3 m) high, impounded a small reservoir 35 acres (14 ha) in area. With its limited storage capacity and low head, the station was designed to provide a modest 2,000 horsepower (1,491 kW), with its initial generating equipment providing just half that amount.

The station was equipped with eight 54-inch- (1,370-mm-) diameter McCormick vertical turbines manufactured by S. Morgan Smith of York, Pennsylvania. These were designed to deliver 165 horsepower (123 kW) each when running under 9 feet (2,743 mm) of head. Bevel gearing connected the turbines to a horizontal drive shaft consisting of two sections; by means of a coupling, either four or eight turbines could be employed.

The station's electrical equipment, manufactured by the Stanley Electric Manufacturing Company of Pittsfield, Massachusetts, consisted of a single three-phase generator with an output of 750 kW at 166 rpm, wound to deliver 12,000 volts to the transmission line without the use of step-up transformers. The transmission line carried the power to a substation near the Arista Cotton Mill in Salem.

About 1903, the plant's capacity was expanded with the addition of a second line of eight turbines and a second generator. In 1913 Fries Manufacturing & Power was absorbed by the Southern Public Utilities Company, a forerunner of Duke Power Company. In 1914 Duke Power replaced the plant's original machinery with six 300-horsepower (224-kW) Allis-Chalmers vertical, 54-inch (1,370-mm) Francis-type turbines. Mounted directly above each turbine was an Allis-Chalmers 2,300-volt, 74-amp, 90-rpm, three-phase generator. The former generator room was converted to a transformer room to step up the voltage for transmission to the Duke system. This machinery, with minor modifications, remains in service today.

Location/Access

Idols Station is located on the Yadkin River, one-quarter mile (0.4 km) west of State Road 3000. Duke Power Company-Winston-Salem, 1405 South Broad Street, Winston-Salem, NC 27127; phone (704) 875-4332.

FURTHER READING

"The Transmission Plant of the Fries Manufacturing & Power Company," *American Electrician* 10 (October 1898): 447–50.

Michigan Lake Superior Power Company Hydroelectric Plant

Sault Sainte Marie, Michigan

Cheap power to lure industry. That was the premise on which an investors group led by entrepreneur Francis H. Clergue planned and built the largest low-head hydroelectric plant in the world at Sault Sainte Marie, Michigan. When it opened in 1902, the Michigan Lake Superior Power Company hydroelectric plant was the longest in the world and, in design capacity (40,000 horsepower, or 29,828 kW), second in size only to Niagara's Adams Station in the United States. However, it exceeded even Niagara in the volume of water for which it was designed: 30,000 cubic feet (850 m^3) of water each second—a significant proportion of the outflow of Lake Superior—could pass through its eighty penstocks.

Sault Sainte Marie, commonly known as the "Soo," is located on the south bank of the St. Marys River, across from its Canadian counterpart, Sault Sainte

Marie, Ontario. Over the course of about a mile (1.6 km), the St. Marys, which connects Lake Superior with Lake Huron, drops some 20 feet (6 m). Earlier attempts to exploit the rapids of the St. Marys for power had ended in failure. When Clergue, who had developed a hydropower plant and pulp mills on the Canadian side of the river, offered to buy the rights to a partially completed power canal at Sault Sainte Marie, he revived the economically depressed community's long-held hopes of becoming a great manufacturing city.

Clergue appointed Hans von Schon (1851–1931) as chief engineer. A German-born engineer of wide experience, von Schon previously had directed a topographical survey of the St. Marys River. The hydropower development von Schon planned was influenced by three factors: the need to limit the canal right-of-way through the city to 400 feet (122 m), dictating a narrow, deep channel; the decision to build a single 40,000-horsepower (29,828-kW) powerhouse rather than a half dozen smaller ones as first contemplated; and the decision to lease about half of the projected power output, as well as a portion of the powerhouse itself, to the newly organized Union Carbide Company for the manufacture of calcium carbide. (Calcium carbide is a hard, brittle, crystalline compound of calcium and carbon. It is made by heating calcium oxide and coke, charcoal, or anthracite coal in an electric furnace. When water is added to calcium carbide, acetylene—a gas widely used in the welding and cutting of metals—is produced.) The final design incorporated a number of unusual features, including a stone-and-steel powerhouse 1,368 feet (417 m) long—the longest in the world—and a timber-lined power canal of unprecedented scale. The canal, more than 2 miles (3.2 km) long, was 200 feet (61 m) wide and 22 feet (7 m) deep.

The Michigan Lake Superior Power Company hydroelectric plant on opening day, October 25, 1902. *Edison Sault Electric Company photograph, Library of Congress Collections.*

The decision to design the plant for carbide manufacture fixed the unit output per penstock at 500 horsepower (373 kW), the rating of Union Carbide's early Horry furnaces. Thus, to develop an output of 40,000 horsepower (29,828 kW) required a minimum of eighty penstocks, an unusually large number considering that the typical powerhouse at the turn of the century had five to ten penstocks. Von Schon chose 33-inch (838-mm) horizontal-shaft double turbines, arranged in tandem, placing four runners in each penstock to secure the desired output. Forty-one horizontal Jolly-McCormick turbines, each developing 564 horsepower (420 kW) at 180 rpm, were mounted in steel draft cases designed and built by the Webster, Camp & Lane Company of Akron, Ohio. (Thirty-seven others were added from 1915 to 1916, for a total of seventy-nine turbine units.) The original electrical equipment consisted primarily of 375-kW alternating-current generators coupled to each turbine shaft.

Sault Sainte Marie's bid to become a major industrial center unfortunately was never realized; cheap power was not enough to attract new industry to a location remote from both markets and raw materials. From the outset, the Michigan Lake Superior Power Company was plagued with a succession of financial, legal, and technical problems. Union Carbide, through its subsidiary Michigan Northern (later Carbide) Power Company, assumed control of the plant in 1913 and operated it for the next half century, selling power to the adjacent Union Carbide plant. When Union Carbide decided to close its Soo factory in 1963, the Edison Sault Electric Company, a local utility, purchased the hydropower plant, which is still operating, a testament to the durability of von Schon's design.

Location/Access

The Edison Sault powerhouse is located on the St. Marys River four blocks east of Ashum Street, the city's main thoroughfare. Although readily visible from the street, it is not open to the public. The American Soo Locks, one of Michigan's major tourist attractions, are located nearby.

FURTHER READING

"The Jolly-McCormick Turbines at the 'Soo,'" *Iron Age* 70 (20 November 1902): 1–4.

Terry S. Reynolds, "The 'Soo' Hydro: A Case Study of the Influence of Managerial and Topographical Constraints on Engineering Design," *IA: The Journal of the Society for Industrial Archeology* 8 (1982): 37–56.

"Water Power Development by the Lake Superior Power Co., at St. Mary's Falls, Mich.," *Engineering News* 40 (4 August 1898): 68–71.

U.S. Department of the Interior, National Park Service, Historic American Buildings Survey/Historic American Engineering Record, *Sault Ste. Marie: A Project Report,* by Terry S. Reynolds (Washington, D.C.: U.S. Government Printing Office, 1982).

Childs-Irving Hydroelectric Project

Childs and Irving, Arizona

In the late nineteenth century, a cattleman in Arizona Territory scouting water for his herd stumbled onto a gushing spring in the desolate but beautiful Verde Valley, 70 miles (113 km) north of Phoenix. The heavy mineral content of the water gave everything it touched a fossilized appearance; hence, the water source was dubbed Fossil Spring, and its runoff, Fossil Creek. The creek's drop of some 1,600 feet (487 m) during the course of its 10-mile (16-km) journey to the Verde River suggested the potential of developing it for hydropower to serve the copper mines of Jerome and Prescott.

The source of Fossil Spring is believed to be a large area to the south of the Grand Canyon. Rainfall soaks into the ground, passes through sedimentary formations (hence, the mineral content) capped by an impervious layer of lava, then comes up through a volcanic fissure, forming the spring. Winter or summer, in wet or dry years, the flow does not vary appreciably, averaging 28 million gallons (106 million liters) a day.

With a contract in hand for power sales to the United Verde Copper Company, the Electric Operating Construction Company began construction of a generating plant at Childs in 1907. The following year, the Arizona Power Company (now Arizona Public Service Company) was organized and assumed the assets of the earlier company. The plant location was chosen because of a small flat known as Dry Lake (now Stehr Lake), which could be used as a reservoir for regulating the flow of water into the penstock.

Mule teams carried all materials to the remote site. The largest piece of apparatus, the generator stator, required a twenty-six-mule team. Interestingly, the steel for the lower end of the penstock came from the Krupp Works in Germany because no U.S. company could produce steel pipe strong enough to withstand the water pressure. The watercourse from spring to lake consisted of approximately 7 miles (11 km) of concrete flumes and tunnels, providing a static head of 1,075 feet (327 m) for three Pelton-wheel-powered generators turning at 400 rpm to produce 9,000 horsepower (6,700 kW) at 44,000 volts. From Childs, a double-circuit 44,000-volt transmission line went to Prescott via Mayer and Poland Junction, with intermediate taps to a number of mines.

Towers for the transmission line presented a problem. It was impossible to transport wooden poles by mule, and steel towers had not yet been developed. Steel windmill towers furnished by the U.S. Wind Engine & Pump Company of Batavia, Illinois, were adapted for the line. All three generators went on line in 1909. A new copper smelter at Clarkdale required additional power and led to the construction, between 1914 and 1916, of another power plant at Irving that had a single Allis-Chalmers reaction-type Francis turbine of 2,100 horsepower (1,566 kW).

Childs-Irving hydroelectric powerhouse (center building), 1976.

In 1919 a 75-mile (121-km) transmission line was built from Sycamore to Phoenix. In the 1920s the state capital, then a city of forty-four thousand people, received 70 percent of its electric power from the Childs-Irving hydroelectric stations. Both stations are still active. Annual output from the Childs plant is 23.4 million kWh, that from the Irving plant 10.8 million kWh. Despite their age and low output—the region they serve today is also served by newer installations having many times their generating capacity—the plants are economical to operate and maintain.

Location/Access

The Childs-Irving plants are open upon application to the Arizona Public Service Company, P.O. Box 53999, M/S 8510, Phoenix, AZ 85072-5399; phone (602) 250-2888.

Rocky River Pumped-storage Hydroelectric Plant

New Milford, Connecticut

The 148-mile- (238-km-) long Housatonic River drops some 650 feet (198 m) during its journey through southwestern Connecticut to Long Island Sound. From earliest colonial days, it was a source of power. In the early twentieth century, the Connecticut Light & Power Company owned and operated two hydroelectric stations on the Housatonic: Bulls Bridge (1913), with an installed capacity of 7,200 kilowatts, and Stevenson (1914), with an installed capacity of 18,750 kilowatts. Together, they generated a combined output of almost 26,000

General view of Rocky River Pumped-storage Hydroelectric Plant showing powerhouse and penstock.

kilowatts. But only a small portion of that output—some 10,000 kilowatts—could be counted on as firm capacity.

The flow of the river varies from season to season and from year to year. To regulate the flow, and thereby the firm hydropower capacity of the river, it would be necessary to store water in times of high flow and return it to the river when the flow was low. The obvious method for accomplishing this was to dam the river and release water as needed. However, there were no suitable dam sites on the Housatonic. In 1926 CL&P proposed a plan whereby water would be pumped from the Housatonic, stored in a lake, then returned to the river during periods of low water flow. This would be the first pumped-storage hydroelectric plant in the nation. (Pumped-storage hydroelectric plants were common in Europe—the world's first was built at Zurich, Switzerland, in 1882—but thus far were untried in the United States.)

CL&P selected a site for the storage reservoir on the Rocky River, a tributary that joins the Housatonic just above New Milford. The runoff from natural drainage would furnish part of the water required to fill the reservoir; the rest would be pumped from the Housatonic in times of high water. Some 6,000 acres (2,428 ha) of the Rocky River Valley were flooded, requiring the relocation of more than 100 homes, 31 miles (50 km) of highways, and six cemeteries. Construction of the main dam and power plant, under the direction of the United Gas Improvement Company of Philadelphia, began in 1926.

The main dam, about 100 feet (30 m) high and 1,000 feet (305 m) long, was located on the Rocky River about a mile (1.6 km) above its confluence with the Housatonic. An earth-filled structure with a concrete-and-timber core wall, the dam created a storage reservoir 10 miles (16 km) long that was named Lake Candlewood, after a nearby mountain. With more than 60 miles (97 km) of shoreline and a capacity of almost 6 billion cubic feet (170 million m³), it was the largest pumped-storage reservoir in the world. Five smaller dams were built elsewhere at low points in the rim of the basin.

A 3,300-foot (1,006-m) canal delivered water from the reservoir to a 15-foot- (4,570-mm-) diameter wood-stave pipe about 1,000 feet (305 m) long. Passing through a surge tank at the end of the pipe, the water entered the penstock, the inside diameter of which tapered from 13 to 11 feet (3,960 to 3,350 mm) as it dropped down the hillside another 670 feet (204 m) to the powerhouse. Just outside the powerhouse was a Y-connection: one branch for the generating unit, the other for the two pumping units.

Inside the powerhouse, electricity was generated by a single 33,300-horsepower (24,832-kW), vertical-shaft Francis turbine direct-connected to a 30,000-kilowatt generator. The station was equipped with two 54-inch (1,370-mm), 8,100-horsepower (6,040-kW), vertical-shaft centrifugal pumps with a capacity to deliver 112,500 gallons (425,800 liters) per minute to the Lake Candlewood reservoir against a maximum head of 240 feet (73 m). The pumps could be used to discharge water into the reservoir whenever the generating unit was not in use. When installed, the centrifugal pumps were the largest in the United States.

With the completion of the Rocky River Pumped-storage Hydroelectric Plant in 1928, the river's firm hydropower capacity (Stevenson, Bulls Bridge, and Rocky River plants combined) was boosted from 10,000 to 50,000 kilowatts—an increase of 500 percent. Now more than sixty years old, the Rocky River plant continues to provide customers with electricity more economically than an oil-fueled plant. Seasonal peak loads occur during the winter months; the Rocky River hydroelectric plant operates for extended periods, requiring drawdown from the reservoir. In spring, during a period of low system loads, pumping resumes (using the output of steam plants that would otherwise be idle, thus minimizing operating costs) in order to have the reservoir full again by June 1. In 1951 the pumps were modified to allow them to operate in reverse when water was being returned to the Housatonic, thereby boosting the station's output to 31 megawatts.

Location/Access

The plant occupies both sides of U.S. Route 7, 1.2 miles (1.9 km) north of U.S. Route 202. Connecticut Light & Power, Rocky River Station, 41 Park Lane Road, New Milford, MA 06776; phone (203) 355-6554.

FURTHER READING

E. J. Amberg, "Power from Pumped Water," *Electrical World* 91 (12 May 1928): 959–65.

William W. K. Freeman, "Pumped-Storage Hydro-electric Plants," *American Society of Civil Engineers Proceedings* 54 (November 1928): 2,457–75.

Joel D. Justin, "Rocky River Hydro-electric Development," *American Society of Civil Engineers Proceedings* 55 (March 1929): 690–98.

Kaplan Turbine

York Haven, Pennsylvania

When the Metropolitan Edison Company decided to expand its hydroelectric plant at York Haven, Pennsylvania, in 1928, it purchased a Kaplan turbine. Built by the S. Morgan Smith Company of York, Pennsylvania, and put into service on April 5, 1929, this was the first adjustable-blade propeller turbine in the United States, and the first of four Kaplan turbines installed at the York Haven plant. The Kaplan turbine was quickly recognized as an important advance in hydraulic-turbine design, providing maximum economy for low and variable heads. By 1930, S. Morgan Smith had built sixteen Kaplan turbines for nine low-head hydroelectric projects in the United States and Canada.

Patents for an adjustable-blade turbine had been issued as early as 1867. But Dr. Viktor Kaplan (1876–1934) of Brünn, Austria-Hungary (now Brno, Czech Republic), was the first to realize the advantages of adjusting both the runner vanes and the wicket gates simultaneously in order to maintain high efficiency at all loads. He filed his first patent application in Europe about 1913 and in the United States in 1914.

The Kaplan turbine resembles a ship's propeller. It differs from other wicket-gate turbines in the shape of the top plate and in the construction of the runner and shaft. The top plate is shaped to form a vane-free transition space

The Kaplan turbine, which resembles a ship's propeller, is shown temporarily removed from its housing for maintenance in 1973.

between the wicket gates and the runner; water leaving the gates in a radial direction is deflected to flow axially through the runner. The angle of the movable blades is adjusted by a hub-mounted servomotor controlled by hydraulic pressure lines in the bore of the shaft. The angle of the blades changes simultaneously with each change of gate opening, so that the most efficient gate and vane angle coincide no matter what the load, resulting in exceptionally high part-load efficiencies.

In Europe, Kaplan turbines were widely adopted following the First World War; by 1928, some 150 were in use in projects having maximum heads ranging from 7 feet (2 m) to 49 feet (15 m). In the United States, S. Morgan Smith secured exclusive rights to the Kaplan patent in 1927 and began commercial development, selling its first unit to Metropolitan Edison in 1928.

The four Kaplan turbines installed at York Haven each developed 2,970 horsepower (2,215 kW) at 200 rpm with a 26-foot (8-m) head. The first unit, however, required manual adjustment of the blades, a decided disadvantage because the turbine had to be shut down to change the position of the blades, resulting in a loss of output. Metropolitan Edison subsequently ordered three automatically adjustable turbines, identical in size and output to the first unit. These joined the plant's earlier fixed-blade turbines, using the Susquehanna River to power twenty generators that produced a total output of approximately 20,000 kilowatts. (The second Kaplan turbine built by S. Morgan Smith, which was also the first in the United States to have automatically adjustable blades, was put in service by the Central Power & Light Company at its Lake Walk plant near Del Rio, Texas, in May 1929.)

Kaplan turbines quickly proved their worth for developments with low and variable flows of water. Adjustable blades deliver higher efficiencies under widely varying load and flow conditions than are possible with fixed-blade designs. In addition to offering increased total output, they also made it possible to reduce the number of units in a powerhouse, reducing the size and thereby the cost of the powerhouse itself.

Location/Access

The York Haven Power Company is owned by Metropolitan Edison Company, P.O. Box 16001, Reading, PA 19640-0001; phone (717) 848-7278.

FURTHER READING

C. L. Dowell, "At Last America Accepts the Kaplan Turbine," *Power Plant Engineering* 33 (1 July 1929): 757–60.

George A. Jessop and C. A. Powell, "Greater Efficiency for Low-Head Hydro," *Electrical Engineering* 50 (February 1931): 118–21.

B. E. Smith, "The Kaplan Adjustable-Blade Turbine," Transactions of the American Society of Mechanical Engineers (Hydraulics) 52 (1930): 137–41.

Hiwassee Dam Unit 2 Reversible Pump-Turbine

Murphy, North Carolina

The Hiwassee Dam pump-turbine, installed in 1956 as part of an expansion of the Tennessee Valley Authority's power network, was the world's largest reversible pump-turbine and the first pump-turbine installed in the United States for the purpose of storing electrical energy in a pumped-storage hydroelectric plant.*

Located on the Hiwassee River in southwestern North Carolina, the Hiwassee Dam and power plant was built by the TVA between 1936 and 1940 as a flood-control and electrical-generating facility. The initial power installation, placed in service in May 1940, consisted of a single 80,000-horsepower (59,656-kW) Francis turbine driving a generator with a rated output of 57,600 kW at 190-foot (58-m) head. Space was provided in the powerhouse for later installation of a second, identical unit.

All pump-turbine installations operate on the same principle: they use low-cost, off-peak power to pump water into a reservoir, from which it can then be drawn to furnish high-cost, peak power. The need to increase system capacity during peak periods made it economically attractive for the TVA to install a reversible pump-turbine. A single hydraulic machine would operate in one direction as a turbine and in the reverse direction as a pump. A direct-connected electrical machine would serve as a motor for pump operation and as a generator for turbine operation. During periods of peak power demand (December through March), the pump-turbine would function as a conventional turbine-generator, adding 59,500 kW of rated capacity to the system. During off-peak periods, espe-

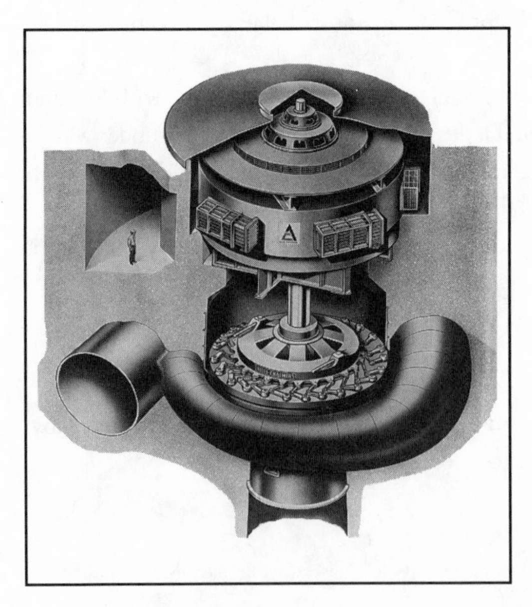

Diagram of Hiwassee Dam Unit 2 reversible pump-turbine. *Courtesy Allis-Chalmers.*

* An earlier pump-turbine, installed in 1954 at the Flatiron Power and Pumping Plant in Colorado, was used primarily for irrigation rather than electrical energy storage. It was much smaller than the Hiwassee pump-turbine and, lacking wicket gates, provided no control of turbine power output.

SPECIFICATIONS

Hiwassee Dam Unit 2 Reversible Pump-Turbine
Manufacturer: Allis-Chalmers Manufacturing Co.

	As Turbine	*As Pump*
Type:	Vertical Francis	Centrifugal
Diameter of runner, intake:		266 in. (676 cm)
Direction of rotation:	Clockwise	Counterclockwise
Rated horsepower:	80,000 (60,000 kW)	102,000 (76,000 kW)
Rated head:	190 feet (58 m)	205 feet (62 m)
Rated discharge:	4,180 cfs (118 m^3)	3,900 cfs (110 m^3)
Rated speed:	105.9 rpm	105.9 rpm
Efficiency at rated head and discharge:	90%	90%

cially periods of minimum rainfall, the unit would operate as a pump to lift water from Appalachia Lake into Hiwassee reservoir against an average operating head of 205 feet (62 m). In this way, surplus electric power would be stored as additional water for reuse during the next peak-load period.

The reversible pump-turbine, built by Allis-Chalmers, was placed in operation in May 1956. It incorporated the largest Francis-type runner ever built; with a diameter of 266 inches (6,756 mm), it had to be fabricated and shipped in three sections and bolted together on site. The motor-generator, also furnished by Allis-Chalmers, was equally impressive. With a rated horsepower of 102,000 (76,061 kW) at 106 rpm, it was the world's largest electrical machine, some 50 percent larger than the generators at Grand Coulee.

Prior to the Hiwassee installation, pumped-storage plants used separate pumps and conventional turbines for storage and generation (see "Rocky River Pumped-storage Hydroelectric Plant," p. 69). The Hiwassee pump-turbine demonstrated to electric-power companies worldwide that reversible pump-turbines could be used to efficiently store electrical energy during periods of low power demand to meet later peak-load demands. The Hiwassee installation served as a prototype for the construction of subsequent pumped-storage facilities. Today, reversible pump-turbines have almost completely supplanted the use of separate pumps and turbines.

Location/Access

At the time of publication, only the lobby is open to the public (Route 4, Box 170, Murphy, NC 28906).

FURTHER READING

L. R. Sellers and J. E. Kirkland, Jr., "Pump-Turbine Addition at TVA Hiwassee Hydro Plant," *Electrical Engineering* 75 (March 1956): 263–69.

Steam

INTRODUCTION by R. Michael Hunt

Say "electric light," and the standard reply will be "Thomas Edison." But Edison had to do more than perfect and commercialize the light bulb in 1879. No use having light bulbs if no one is generating electricity. No use generating electricity if there is no wiring or switches to connect it. No, Edison had to invent the whole system of electric power generation.

By the late nineteenth century, many of the technological building blocks were already in place. Dynamos—machines to generate electricity—were available, but they were small and inefficient. The reciprocating steam engine had matured through the century as a power source for industry, which had rapidly used up the available water power to turn its mills and machines. These engines were reliable and available to turn the dynamos.

Edison quickly sold the enthusiastic public on electric power. By 1882 he had improved and enlarged his dynamos so that each "Jumbo" dynamo could be directly coupled to its single-cylinder steam engine, then he installed six of these 240-horsepower (179-kW) machines in the first central electric power station on Pearl Street in New York City. The Pratt Institute Power Plant of 1900, an example of this type of plant, still operates.

In 1891 Edison introduced another increase in generating unit size, with 640-horsepower (477-kW), triple-expansion, marine-type engines driving direct-coupled dynamos of 200 kilowatts each at either end. Central station electric power generation was on its way.

As the twentieth century dawned, the limitations of the steam engine—large size and slow speed—were obvious. In 1884 in England, Charles Parsons had demonstrated a practical steam turbine driving a dynamo, and this became the new "engine." In the turbine, rows and rows of windmill-like blades were mounted on a rotating shaft, enclosed in a casing with rows of stationary blades interspersed between the moving blades. Steam introduced into the casing expanded through the movable and stationary blades, thus turning the shaft. In 1903 General Electric introduced a 5,000-kilowatt turbine of its own design, the largest in the world, based on Curtis patents. The smooth-turning, high-powered steam turbine had arrived, and the giant reciprocating engine was on its way into history.

The GE Curtis steam turbine had a vertical shaft, with the generator above the turbine. This allowed for very compact arrangements, such as that of the Georgetown plant in Seattle, but became impractical as turbines and generators grew in size. Meanwhile, Westinghouse had taken out a license from Parsons and was building turbine-generators with horizontal shafts. This arrangement became the standard for power generation.

In the early years of the twentieth century, coal-fired boilers still were fired by men shoveling coal into the fireboxes. Engineers soon realized that if coal was pulverized, it could be blown into the fire mechanically. After four years of experimentation, pulverized-coal firing was demonstrated at the Oneida Street plant in Milwaukee in 1918, with a 5 percent improvement in efficiency over the hand-fired units. The pulverized-coal-fired boiler not only allowed much larger quantities of coal to be burned in a boiler of given size (thus increasing its steam-power output) but also proved to be easily converted to natural gas or oil firing when times demanded.

By the 1920s, steam pressures had inched up from the Curtis turbine's 175 psig (1,207 kPa) to around 300 psig (2,068 kPa). The advantage of pressure is that each pound of steam contains more heat at higher pressures, and the turbine can be made more efficient. In a 1925 breakthrough, the Edison Electric Illuminating Company of Boston opened its Edgar Station operating at 1,200 psig (8,274 kPa), a world first. The station was vastly more efficient than its contemporaries and became the model for plants worldwide.

The next jump in turbine generators was in size. The Edgar station generated 85,000 kW; four years later, Commonwealth Edison Company opened its State Line Unit No. 1 in Hammond, Indiana, generating 208,000 kW with a single machine. For fifty years, this was the largest turbine generator in the world. (The importance of prime-mover size is that useful work tends to increase faster than losses with size, all else being equal, so efficiency improves.)

In 1935, the Port Washington, Wisconsin, plant was built, incorporating pulverized-coal firing, high pressures, and a new design of superheater (a device that puts more heat energy into the steam). Smashing efficiency records, it became "America's premier station."

It appears now that electricity will continue to be generated in large central power stations. Combined cycle plants—in which a combination of gas turbines, and boiler and steam turbines produce electricity very efficiently—will likely increase. There will also be more use of wind and solar power, but these are capital-intensive and the institutional cost of money has a major effect on how quickly they will spread. Fuel cells will also develop, but probably only for specialized applications where their low-pollution advantages offset their low-power density. Looking back, we see how the Pearl Street Station set the stage for more than one hundred years of technology.

Edison "Jumbo" Engine-driven Dynamo

Dearborn, Michigan

Thomas Edison's incandescent lamp of 1879 would have been of scant consequence without the development of a practical large-scale electricity generation and distribution system. An essential part of such a system was an efficient and reliable prime-mover-driven generator. The Edison "Jumbo" dynamo, now located in Greenfield Village at the Henry Ford Museum, generated power in the first large-scale central electric station in the United States.

With financial backing from Western Union, the Edison Electric Light Company was formed in October 1878. Edison set to work in his Menlo Park, New Jersey, "invention-factory" to develop a practical incandescent lamp and an efficient dynamo. After months of agonizing trials, on October 21, 1879, Edison sealed a carbonized-cotton filament in an evacuated bulb. The lamp glowed for more than forty hours, casting a feeble, reddish glow. On the closing nights of the year, crowds of visitors arrived at Edison's laboratory to witness improvised demonstrations put on with a single dynamo and a few dozen lights. They came away astounded.

Simultaneously, working with chief assistant Francis Upton, Edison set to work on a constant-voltage dynamo. Numerous trials resulted in a bipolar dynamo—nicknamed the "long-waisted Mary Ann"—in 1879. Edison attached a power station housing eleven dynamos driven by a central steam engine to his

Edison "Jumbo" engine-driven dynamo. *Photograph from the Collections of Henry Ford Museum & Greenfield Village.*

Menlo Park machine shop and wired laboratory buildings and half a dozen neighboring houses. The awkwardness and inefficiency of so many small dynamos with their belting led Edison to design a larger dynamo that was to be directly coupled to a 120-horsepower (89-kW) Porter-Allen engine ordered from the Southwark Foundry & Machine Company of Philadelphia. The Porter-Allen engine arrived in January 1881, and engine and dynamo were mounted together to form a self-contained generating unit. Tests of the coupled engine-dynamo proved it to be less than perfect; for one thing, the armature developed high internal temperatures, threatening to destroy the insulation of the armature conductors. Nevertheless, Edison was able to demonstrate how an abundant supply of current could be produced at a reasonable cost. He named the massive unit and its successors "Jumbo," after P. T. Barnum's famous circus elephant, for good reason: they weighed 27 tons (24 t) in all, there were twelve massive field coils, and the Siemens-type armature and its shaft were more than 10 feet (3 m) long!

In the meantime, Edison searched for a suitable location for his first central power station, finally finding an old commercial building at 257 Pearl Street, New York City, in a squalid section near the financial district. He gutted the interior of the building and erected an iron superstructure independent of the building to support the generating machinery. Steam was supplied by eight Babcock & Wilcox boilers occupying the basement. Above these were six improved "Jumbo" dynamos directly connected to six Porter-Allen steam engines, each of 240 horsepower (179 kW). Finally, on September 4, 1882, without fanfare, Edison's Pearl Street Station began commercial operation, transmitting power through some 14 miles (22.5 km) of underground conduit.

Earlier trials had shown that the governors of the Porter-Allen engines failed to regulate properly—due to vibration of the building frame—causing the engines to seesaw fitfully. Edison appealed to engine designer Gardiner Sims for a steam engine with a mechanical governor that would function unaffected by vibration. The Pearl Street Station limped along on one dynamo until the Armington-Sims engines arrived in November; these were substituted for the Porter-Allen engines, correcting the earlier trouble. By the end of the year, 193 buildings with more than 4,000 lamps had been connected to the first large-scale central power station in the United States.

Edison's Pearl Street Station operated successfully until January 2, 1890, when fire partially destroyed it. Jumbo No. 9 was the only one of the six original dynamos to survive. It was put back into operation and worked in conjunction with belt-driven generators and engines installed as temporary equipment until 1893, when it was sent to the World's Columbian Exposition in Chicago. In 1904 it was exhibited at the Louisiana Purchase Exposition in St. Louis, and in 1924 it was displayed at the Grand Central Palace in New York to mark the fortieth anniversary of the American Institute of Electrical Engineers.

In 1930 Jumbo No. 9 was presented to Henry Ford for his new museum of industry and technology. The machine was completely rebuilt for the fiftieth anniversary celebration of the original Pearl Street Station. Still fully operational, it is a monument to the inventor whose successful carbon-filament lamp, together with his system of electrical distribution, moved America and the world into the Electrical Age.

Location/Access

Jumbo No. 9 is exhibited in a partial replica of the original Detroit Edison "A" Station in Greenfield Village at the Henry Ford Museum, 20900 Oakwood Boulevard, Dearborn, MI 48124; phone (313) 271-1620. Hours: daily, 9 A.M. to 5 P.M. Admission fee.

FURTHER READING

"Description of the Edison Steam Dynamo," *Transactions of the American Society of Mechanical Engineers* 3 (1882): 218–25.

"The Edison Electric Lighting Station," *Scientific American*, August 26, 1882, 1–2.

Matthew Josephson, *Edison: A Biography* (New York: McGraw-Hill Book Company, Inc., 1959).

Marine-type, Triple-expansion, Engine-driven Dynamo

Dearborn, Michigan

The success of Edison's Pearl Street Station and the host of similar installations that followed it stimulated the electrical lighting industry in the United States. In the 1880s, manufacturers sprang up to produce everything from lamps to generators. It was a period of great engineering advances—and great legal battles over patent rights.

In the early years, the typical central power station used many small, high-speed generators of the bipolar type. The arrangement worked reasonably well, with efficient, slow-speed engines like the Corliss driving the generators through speed-increasing belt transmission systems. Such stations usually were arranged with the engines on the ground floor and the dynamos on the floor above. But as the capacity of the generators (and their size) increased, this arrangement became limiting. Between 1889 and 1892, the Edison General Electric Company built some 200-kW monsters that weighed nearly 20 tons (18 t). Belt drives, meanwhile, were plagued by all the defects Edison had noted in 1880: they were dangerous, wasteful of space, and inefficient. The solution to these problems was to build multipolar generators of large diameter that would operate efficiently at speeds between 100 and 150 rpm. These could be directly coupled to the engines, as had been done at Pearl Street, but the power output of these new machines could be far greater.

Marine-type, triple-expansion, engine-driven dynamo as it appeared in 1989, before its restoration and relocation inside the Henry Ford Museum. Today, the engine and dynamo operate with the assistance of an electric motor. *Photograph from the Collections of Henry Ford Museum & Greenfield Village.*

On December 15, 1891, the Edison Electric Illuminating Company of New York put the first of its new marine-type, triple-expansion, engine-driven generators into operation at its Duane Street Station. (Almost simultaneously, similar units began operation at the company's Twenty-sixth Street Station; these were described in detail in the March 1892 issue of *Power.*) With an output of 400 kilowatts, this revolutionary unit represented the true beginning of large-scale electric power generation in the United States. Built by the Dickson Manufacturing Company of Scranton, Pennsylvania, the engine was designed by John Van Vleck, Edison chief engineer, and J. W. Sargent, of the Dickson Company, with assistance from English engine builders David Joy and S. F. Prest. The generators were supplied by the Edison General Electric Company of Schenectady, New York.

The choice of a marine-type (vertical) engine made a great deal of sense. The requirements of an urban power station were not unlike those of a high-speed ocean liner, i.e., reliable, continuous power produced in a compact space. The 625-horsepower (466-kW) engine, with cylinders 18, 27, and 40 inches (46, 69, and 102 cm) in diameter and a stroke of 30 inches (762 mm), was designed to be compact and reliable. The steam chests were placed at the sides of the cylinders (the usual practice was to place them between the cylinders), reducing the overall length of the engine by about 40 percent, while the valve gear was of the Joy type with motion derived directly from the connecting rods rather than eccentrics, decreasing the number of working parts and increasing reliability. The ends of the main shaft carried the armatures of two 200-kilowatt Edison multipolar dynamos. Each dynamo fed one side of a standard Edison three-wire DC system. The armature was ring-wound on the surface, and copper brushes collected current from the outside ends of the rings. The machines each had fourteen field coils and poles, allowing efficient power output at the rated speed of 130 rpm.

The vertical, triple-expansion, engine-driven generator operating at slow speed represented the first phase of large-scale central power generation. Within fifteen years, both the reciprocating steam engine and direct current were obsolete. In 1884 Charles Parsons tested his first steam turbine; prophetically, it was used to drive a high-speed electric generator. Turbines—compact, simple, and capable of operating at optimal generator speeds—soon proved to be the ideal power source for generators.

Location/Access

The marine-type, triple-expansion, engine-driven dynamo is on display at the Henry Ford Museum. (See "Edison 'Jumbo' Dynamo," above, for location and hours.)

FURTHER READING

"Vertical Triple Expansion Engine at Twenty-Sixth Street Edison Station," *Power* 12 (March 1892): 1–2.

Pratt Institute Power Plant

Brooklyn, New York

After making his fortune as a member of the Standard Oil organization, oil merchant Charles Pratt retired in 1874 to devote himself to philanthropic enterprises. After much study, he founded the Pratt Institute in Brooklyn for the technical education and manual training of young men and women. It opened on October 17, 1887, with a class of twelve students. Enrollment had grown to three thousand by the time of Pratt's death in 1891.

The institute's physical plant was to be thoroughly modern, with steam heat, incandescent and arc lamps, and elevators. In the late 1890s, the addition of two new buildings—including the Renaissance-style Pratt Institute Free Library, the first public library in Brooklyn—increased the electrical load and forced an overhaul of the original power plant equipment. Three new steam engines and generators were installed in 1900. These are still in service, qualifying the Pratt power

The Pratt Institute Power Plant features three General Electric direct-current generators. The control panel is at the left. *Photograph by David Sharpe, Library of Congress Collections.*

plant as the oldest generating plant in the northeastern United States powered by single-cylinder steam engines, a rare survivor of a once-common technology.

The horizontal single-cylinder steam engine, often called a mill engine, was the workhorse of the late nineteenth century. Sturdy, uncomplicated, and occupying scant space compared to the traditional beam engine, it found unlimited application. Factories, office buildings, schools and other institutions, and streetcar lines commonly generated their own power, and in the late nineteenth and early twentieth centuries, the simple single-cylinder engine driving a dynamo at a leisurely 250 to 300 rpm reigned supreme.

The Pratt engines were furnished by the Ames Iron Works of Oswego, New York, during the summer of 1900. As built, they exactly matched a description of "a new automatic engine" designed by E. J. Armstrong and published in the *American Machinist* in October 1893. With 14-inch (356-mm) bore and 12-inch (305-mm) stroke, the engines were designed for operation with steam at 100 psig (689 kPa). They were direct-connected to General Electric 75-kW generators. Initially, the engines were equipped with balanced slide valves. Sometime in the 1920s, they were converted to outside-admission piston valves. Speed is controlled by inertia governors mounted in the flywheel, which vary the valve travel (and, therefore, the steam admission opening) to maintain a constant rpm of about 270.

Today, the engines operate on 120 psig (827 kPa) steam. The electrical output—120 volts DC—supplies a small amount of the light and power loads on campus, belying an Edison Company report of 1929 that the engine-generator units were "nearing the end of their useful life."

Location/Access

Grand Avenue, between Willoughby and DeKalb avenues. The power plant, which normally operates only during the heating season, is open to the public. Visitors are advised to write or call ahead: Pratt Institute Power Plant, 200 Willoughby Avenue, Brooklyn, NY 11205; phone (718) 636-3694.

FURTHER READING

"A New Automatic Engine," *American Machinist* 16 (12 October 1893): 1–2.

5,000-kilowatt Curtis Steam Turbine-Generator

Schenectady, New York

When it was built in 1903, the 5,000-kilowatt Curtis steam turbine-generator was the most powerful in the world. It stood just 25 feet (7.6 m) high—compared to 60 feet (18.3 m) for a reciprocating engine-generator of like capacity—and required but a fraction of the reciprocating engine's floor area. The compact, high-speed turbine—"radical in economy, simplicity and efficiency" in the words of designer William Le Roy Emmet—conclusively demonstrated the steam turbine's value as a practical source for large amounts of power and stimulated the growth of modern electrical generation in large central stations nationwide.

As president of Commonwealth Electric (now Commonwealth Edison) Company of Chicago, Samuel Insull was responsible for building one of the earliest steam turbine generating stations, the Fisk Street Station. He equipped it with three General Electric 5,000-kilowatt vertical Curtis steam turbine-generators, then the most powerful in the world.

The Curtis steam turbine represents the ideas of two men: patent lawyer and inventor Charles G. Curtis (1860–1953) and engineer William Le Roy Emmet (1859–1941). Curtis approached General Electric early in 1897 with a proposal to build a turbine with a new kind of wheel that had a succession of concave buckets, which revolved by the force of steam striking them. General Electric agreed to offer Curtis the facilities of its Schenectady works for further experimentation provided GE could purchase the patent rights if the turbine proved a commercial success. (Curtis eventually received $1.5 million for his patent rights, retaining the rights to nonelectric marine applications.)

Two years later, when the turbine experiments had yet to bear fruit, GE called

The Curtis steam turbine-generator on display at General Electric's Schenectady plant, ca. 1910.

in lighting engineer William Emmet to study the problem. Taking the bucket and nozzle arrangement from Curtis (i.e., two stages with three rows of buckets in each), Emmet designed two small units of 500 kW and 1,500 kW. He next built a 5,000-kW turbine—about twice as big as the largest Parsons turbine in England—which was purchased by Insull. The turbine-generators Insull ordered from General Electric in 1902 required one-tenth the space and weighed one-eighth as much as reciprocating engines of comparable output. Central-station executives nationwide clamored for the new turbines, whose biggest selling point, next to efficiency, was that generating capacity could be expanded within existing buildings.

SPECIFICATIONS

5,000-kilowatt Curtis Steam Turbine-Generator

Electrical output: 5,000 kW, 3-phase, 25 Hz at 9,000 volts
Speed: 500 rpm
Turbine inlet conditions: 175 psig (1,207 kPa), 150F (66C) superheat
Turbine outlet conditions: 28 inches (711 mm) mercury vacuum

The Curtis turbine is pressure- and velocity-compounded. Each of the turbine's two pressure-compounded stages consists of a nozzle, three rows of stationary turning vanes, and four rows of moving buckets attached to one wheel.

The Curtis steam turbine-generator is of special significance in the history of electrical power generation, for it spelled the end of the cumbersome but magnificent reciprocating engine-generators and spurred the development of ever-larger turbines of increased efficiency. In 1909 the original Curtis turbines at Fisk Street Station were replaced by improved Curtis units of 12,000 kilowatts. The pioneer Curtis unit was returned to the place of its birth, General Electric's Schenectady plant, and displayed as a monument to technological achievement.

The original configuration of the early Curtis machines was vertical, with the generator above the turbine. As turbines grew larger and turbine speeds became higher, though, horizontal-shaft machines that could draw upon the building's foundation for greater lateral support became the standard.

Location/Access

The Curtis turbine-generator is located outside Building 263 at the General Electric Company plant in Schenectady, New York. Direct questions to: GE Power Generation, 1 River Road, Schenectady, NY 12345; phone (518) 385-3072.

FURTHER READING

William Le Roy Emmet, *The Autobiography of an Engineer* (New York: The American Society of Mechanical Engineers, 1940).

[William Le Roy Emmet], "The Curtis Steam Turbine," *Electrical World and Engineer* 41 (11 April 1903): 609–12.

John Winthrop Hammond, *Men and Volts: The Story of General Electric* (Philadelphia: J. B. Lippincott Company, 1941).

J. C. Thorpe, "A 100,000-Kilowatt Steam-Turbine Station," *Power* 26 (December 1906): 715–28.

Georgetown Steam Plant

Seattle, Washington

Seattle's former Georgetown Steam Plant contains the best preserved examples of the world's first large-scale steam turbines. By 1907, when the plant went on line, the vertical Curtis steam turbine-generator had established itself as a practical and compact prime mover capable of producing large amounts of power (see "5,000-kilowatt Curtis Steam Turbine-Generator," p. 84). Manufactured by the General Electric Company between 1902 and 1913, Curtis turbines made possible the widespread marketing of electricity for domestic and industrial use, and marked the beginning of the end of reciprocating steam engines for that purpose.

Located on 18 acres (7 ha) of land on the Duwamish River, the Georgetown Steam Plant was designed and built for the Seattle Electric Company by Stone & Webster of Boston, with Frank B. Gilbreth (1868–1924) serving as consultant.

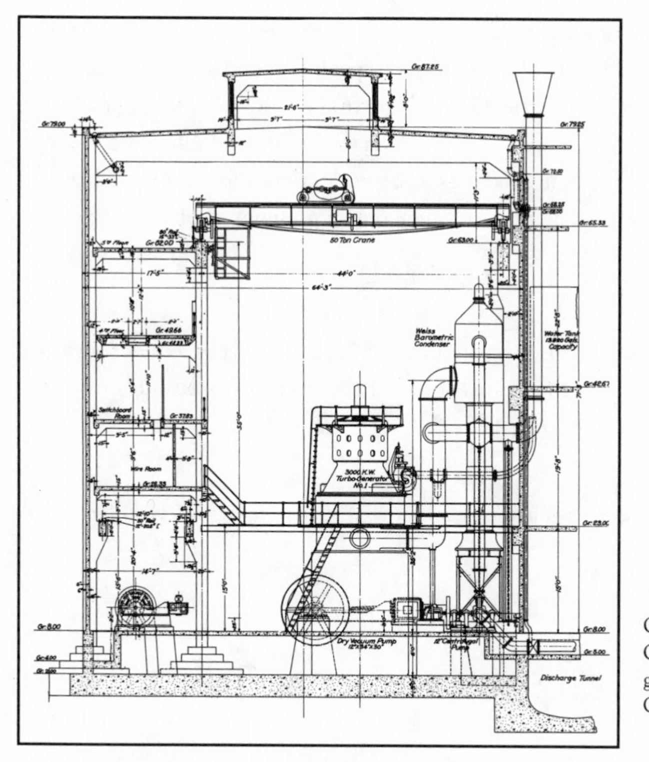

Cross-section view of the Curtis steam turbine-generators at the Georgetown Steam Plant.

The reinforced-concrete station, envisioned as the first unit of a much larger plant, was designed ready for expansion in the future. It was initially equipped with two Curtis vertical-shaft turbine-generators, one of 3,000-kW and the other of 8,000-kW capacity, manufactured by GE. According to *Engineering Record*, one of the plant's most significant features was that generating equipment with a capacity of 11,000 kW occupied a floor space only 64 by 78 feet (20 by 24 m) in size, thus underscoring just one aspect of the steam turbine's economy compared with much larger reciprocating engines.

The new plant was intended to provide Seattle Electric with additional peak-load capacity. Two 500-kW motor-generator sets supplied 600-volt direct current to the city's street railway system, while two 500-kW, 13,800- to 2,300-volt water-cooled transformers furnished current for the Georgetown neighborhood. The plant's fourteen Stirling water-tube boilers were oil-fired, though the boiler plant was designed for either oil- or coal-fired operation.

In 1912 the Puget Sound Power & Light Company purchased Seattle Electric, consolidating all of the electric companies in the Seattle area except for the municipal utility. Five years later, the Georgetown plant was expanded with the addition of a third Curtis turbine-generator, this time a horizontal type of 10,000-kW capacity, which was simpler and more compact than its predecessors. But by the late 1920s, the Georgetown plant was outdated, and in 1930 Puget Sound Power built a new steam plant at Renton, Washington. The Georgetown Steam

Plant was relegated to standby duty, supplying power when there was not enough water to allow the hydroelectric plants to meet peak demand.

In 1951 the City of Seattle Light Department (now Seattle City Light) purchased the properties of Puget Sound Power & Light, including the Georgetown Steam Plant. The plant made its last production run during the winter drought of 1964. Today it is being redeveloped as a museum of electric power.

Location/Access

The Georgetown Steam Plant is located at 6066 Thirteenth Avenue South (off South Hardy Street), 4 miles (6.4 km) south of the downtown business district.

FURTHER READING

"An 11,000-kw. Turbo-Generator Station in Seattle, Wash.," *Engineering Record* 57 (6 June 1908): 721–24.

East Wells (Oneida) Street Power Plant

Milwaukee, Wisconsin

The early years of central-station electric power production were plagued by the growing pains of a new technology. Efficiencies were low—about one-third those of modern stations; outages were frequent and expected; coal was declining in quality even as it was rising in cost. During this period, John Anderson and Fred Dornbrook were getting their educations at sea—Anderson as a marine engineer in the British Navy, Dornbrook as a marine engineer on a lake steamer. Later, as the two top mechanical engineers of The Milwaukee Electric Railway & Light Company (TMER&L), they agreed that the hardest work either had ever done aboard ship was hand-stoking the boiler.

Out of their discussions (Dornbrook later gave the credit to Anderson) came the idea of grinding the coal to a fine powder and feeding it into the furnace with large blowers. Anderson foresaw two advantages: pulverized coal could be delivered into a furnace more efficiently and economically than lump coal, while crushing the coal would increase its burning surface many hundred times, insuring complete and thereby more efficient combustion.

Burning pulverized coal was hardly a new idea. Attempts had already been made to burn pulverized coal in locomotives, without success. In 1914 Anderson received permission to conduct experiments on the use of pulverized fuel. Anderson and his team (Dornbrook, W. E. Schubert, and Ray Mistele) concurrently studied the pulverizing process and the efficient burning of coal. Early in 1918, TMER&L management approved a trial installation at the Oneida (later renamed

Early twentieth-century view of the Oneida Street Power Plant, site of the first successful use of pulverized coal in utility boilers.

East Wells) Street Power Plant. Equipment for drying and pulverizing the coal was installed, and an experimental boiler was placed in service in May.

The boiler on which the first tests were performed was an Edge Moor three-pass, water-tube boiler, equipped with a Foster superheater. The coal feeders and burners were of the "Lopulco" type, manufactured by the Locomotive Pulverized Fuel Company of New York. Preliminary operation was not without problems. An insufficient air supply, for example, caused high furnace temperatures, which in turn caused ash particles to fuse into slag and accumulate between the tubes, on furnace walls, and in the ash pit. The inventive engineers designed a new furnace with a larger combustion chamber and auxiliary air openings equipped with dampers. To prevent the accumulation of slag, they raised the point of fuel admission to the furnace, thereby raising the flame path above the base of the pit so that ash particles dropping from the flame were not fused; ash, in the form of powder and small slugs of slag, could be easily raked from the pit.

On August 12 and 13, 1918, the engineers ran a final efficiency test, with encouraging results. The pulverized-coal boiler showed a gross efficiency of 85.22 percent, compared to a maximum stoker-fed boiler efficiency of 80.54 percent. "The ease of controlling the fuel, feed, and drafts," Fred Dornbrook reported in *National Engineer*, "the ability to take on heavy overloads in a brief time, [the] thorough combustion of the coal, and the uniform high efficiency obtainable under normal operation makes pulverized coal a most satisfactory form of fuel for central station uses." In addition, the pulverized-coal boiler, fed automatically by screw conveyors and blowers, required very little attendance. "The day of the roughneck fireman is gone," John Anderson observed following a test of five 468-horsepower (349-kW) boilers at the Oneida Street plant.

The Oneida Street experiments were widely publicized and closely watched by combustion engineers and central-station executives. The tests conclusively proved the efficiency of pulverized coal and resulted in changes that eventually became standard in steam-electric plants worldwide. The Oneida Street innovations were incorporated into TMER&L's Lakeside Power Plant (1920) in Milwaukee, the first in the nation designed to burn pulverized coal.

According to historian Forrest McDonald, "the development of pulverized fuel and its attendant developments constituted a monumental achievement, ranking with Edison's lamp and multiple distribution system, Stanley's transformer, and Parsons' steam turbine as one of the four fundamental technological developments that made low-cost central station service possible."

Location/Access

The East Wells Power Plant, 108 East Wells, Milwaukee, WI 53202, was retired in 1982 and is now occupied by the Milwaukee Repertory Theater, phone (414) 224-1761. One of the historic pulverized-coal boilers has been preserved in situ and sectioned longitudinally as a permanent public exhibit.

FURTHER READING

John Anderson, "Pulverized Coal Under Central-Station Boilers," *Power* 51 (2 March 1920): 336–39,

Fred Dornbrook, "Pulverized Fuel in the Oneida Street Plant of the T. M. E. R. & L. Co.," *National Engineer* 22 (October 1918): 535–39.

Forrest McDonald, *Let There Be Light: The Electric Utility Industry in Wisconsin, 1881–1955* (Madison, Wisc.: The American History Research Center, 1957).

"The New Lakeside Pulverized-Coal Plant, Milwaukee," *Power* 52 (7 September 1920): 358–60.

Edgar Station, Edison Electric Illuminating Company

Weymouth, Massachusetts

In the early 1920s, steam pressures on the order of 300 psig (2,068 kPa) were common to the electric utility industry. But new materials like chrome-molybdenum steel, offering superior heat resistance, promised substantial gains in efficiency. When Boston needed a new electric station, Irving E. Moultrop (1865–1957), assistant superintendent of construction (later chief engineer) of the Edison Electric Illuminating Company of Boston, took a bold step by planning the

The 1,200-psig turbine-generator units of the Charles L. Edgar Station served as a model for high-pressure power plants worldwide.

first 1,200-psig (8,300-kPa) steam plant in the world. The 1,200-psig boiler and turbine would work with a more conventional 350-psig (2,400-kPa) steam-turbine system; the latter, in effect, would operate on the exhaust steam of the high-pressure unit.

Named after Charles L. Edgar, president of Boston Edison for thirty-two years, the station was designed and built by Stone & Webster of Boston under Moultrop's direction. The 1,200-psig boiler was of the cross-drum type with water tubes 15 feet (4,570 mm) long and 2 inches (51 mm) in diameter, spaced 4 inches (101 mm) on centers. The drum was a 32-foot-(9,753-mm) long steel forging of 4-foot (1,219-mm) diameter and 4-inch (101-mm) wall thickness made in the gun works of the Midvale Steel Company, Philadelphia. To check for flaws, the castings used for valve parts and pipe fittings were X-rayed—a pioneering application of this now-standard procedure.

A new record for economy was established when the first phase of construction went into commercial service at the end of 1925: performance records showed that high-pressure steam resulted in a 12 percent increase in efficiency and substantial savings in the cost of fuel. Two years after it went on line, the Edgar Station was extended. The completed installation comprised four high-pressure boilers supplying steam to two 10,000-kW, high-pressure turbines, which exhausted through reheaters in the boilers to the throttle of a 65,000-kW main generating unit, giving a gross output of 85,000 kW. The Edgar Station became a model for high-pressure power plants all over the world.

The once-pioneering equipment of the Edgar Station has since been overtaken by technology. A 1947 plant addition featured an even more efficient turbo-generator, while the boiler firing was changed from coal to pulverized coal and oil; later, it was again modified to allow the use of natural gas as an alternative fuel. The earliest equipment was removed following the plant's retirement in 1971.

Location/Access

Bridge Street at the Fore River, Weymouth, Massachusetts. No public access.

FURTHER READING

I. E. Moultrop, "Story of First 1,200-lb. Steam Plant," *Power* 67 (24 April 1928): 713–18.

I. E. Moultrop and E. W. Norris, "High-Pressure Steam at Edgar Station," *Transactions of the American Society of Mechanical Engineers* 50 (1928): 32–40.

State Line Generating Unit No. 1

Hammond, Indiana

For a quarter of a century—from 1929 to 1954—Unit No. 1 of the State Line Station, with its rating of 208 megawatts, was the largest turbine-generator in the world. Located on the Illinois-Indiana border, it represented "reasonable preparation" for the future power needs of the burly industrial district stretching from Chicago to Gary, Indiana, according to Samuel Insull (1859–1938), the man responsible for this marvel.

Interior view of the State Line power plant. *Courtesy Commonwealth Edison Company.*

As president of Commonwealth Edison Company, Insull set out to create a monopoly of service in the Chicago region, advocating the supply of electricity from large central stations. By 1919, his stations covered most of Illinois and extended into neighboring states. In 1929 Insull announced plans for the million-kilowatt State Line Station. (In 1903 Insull had installed a 5,000-kilowatt Curtis steam turbine-generator, then the most powerful in the world, in Chicago's Fisk Street Station; see p. 84.)

The State Line power plant was located on 90 acres (36 ha) of fill on the Lake Michigan shore, a site providing ample water and readily supplied with coal by rail or water. Sargent & Lundy served as consulting engineers; and Graham, Anderson, Probst & White, as architects. The General Electric Company furnished the 208-MW turbine-generator, designated Unit No. 1. It consisted of one high-pressure turbine of 76 MW and two low-pressure turbines of 62 MW each; the latter also drove two 4 MW auxiliary generators, giving the triple cross-compound unit a total capacity of 208 MW when turning 1,800 rpm. Six Babcock & Wilcox boilers furnished 450,000 pounds (205,000 kg) of steam per hour at 650 psig (4,500 kPa) and 730F (387C), burning 1.5 tons (1.2 t) of coal per minute.

In December 1953 State Line Unit No. 1 was retired as champion by a 213-MW turbine-generator in Ohio. It has since been dismantled.

FURTHER READING

Samuel Insull, *Public Utilities in Modern Life* (Chicago: Privately printed, 1924).

Forrest McDonald, *Insull* (Chicago: The University of Chicago Press, 1962).

"State Line Station Officially Opened," *Power* 70 (29 October 1929): 670–74.

Port Washington Power Plant

Port Washington, Wisconsin

When The Milwaukee Electric Railway & Light Company's Port Washington Station was dedicated during the first week of September 1935, more than 36,000 visitors came to see it. Hailed as "America's premier station," Port Washington consisted of one boiler, one turbine, one set of transformers, one 132-kilovolt transmission line, and one set of auxiliaries, combining, in the words of *Power Plant Engineering*, "the utmost in heat economy with low unit cost." The plant quickly smashed generating efficiency records, producing a kilowatt-hour with less than 10,700 Btu's (11,286 kJ), or about four-fifths of a pound of coal—this at a time when the U.S. average was about 16,000 Btu's (16,876 kJ).

Located on the west shore of Lake Michigan, 28 miles (45 km) north of Milwaukee, the new plant was at the north end of a 132,000-volt transmission loop around the Milwaukee metropolitan district. Its design was the result of The

Interior view of the Port Washington Power Plant showing pulverized-coal feeders.

Milwaukee Electric Railway & Light Company's two decades of pioneering work with pulverized coal (see "East Wells Street Power Plant," p. 88), high-pressure steam, and radiant superheaters, and followed the company's celebrated Lakeside plant (1920), the first power station in the world designed to burn pulverized coal.

The Port Washington Station consisted of one 80,000-kilowatt turbo-generator and one boiler designed to operate at 1,320 psig (9,100 kPa) with a maximum steam temperature of 850°F (454°C). Fired by pulverized coal, the three-drum, bent-tube boiler had a total heating surface of 44,087 square feet (4,096 m^2) and a capacity of 690,000 pounds of steam per hour (313,000 kg/h). The tandem-compound turbine, built by the Allis-Chalmers Company of West Allis, Wisconsin, drove an air-cooled generator operating at 1,800 rpm and 22,000 volts. Laid out on the unit system (i.e., one boiler supplies one turbine), the Port Washington plant was designed so that its initial capacity of 80,000 kW could readily be expanded with five additional units—to 480,000 kW—in the future.

The plant produced electricity for the first time on October 14, 1935, coinciding with the celebration of Port Washington's centennial. Following an initial shakedown period, performance records showed that the power plant was turning out kilowatt-hours at higher efficiency than any other in the world—a record it held until 1948, when newer plants finally surpassed it. Additional 80,000-kilowatt units were added in 1943, 1948, 1949, and 1950, bringing Port Washington's total generating capacity to the present 400,000 kilowatts.

Location/Access

Contact: Wisconsin Electric Power Company, 231 West Michigan, Milwaukee, WI 53203; phone (414) 221-2345.

FURTHER READING

"Port Washington Station," *Power Plant Engineering* 39 (November 1935): 636–37.

Thomas Wilson, "Port Washington Ties In," *Power* 79 (November 1935): 585–89.

Internal Combustion

INTRODUCTION by Euan F. C. Somerscales

The gas turbine has always fascinated mechanical engineers because it appears to be the ideal prime mover. It is a rotating machine, like the steam turbine, which means that it can be perfectly balanced. Unlike the reciprocating engine, it takes in air continuously, not intermittently. In other words it is a continuous-flow machine; consequently, its power output can be much larger than that of the reciprocating engine of the same size. Finally, it is an internal-combustion engine, which allows it to use the energy released by the combustion of the fuel more efficiently than does the steam engine or steam turbine.

Surprisingly, the gas turbine has a much longer history than is generally realized. The first patent was issued in 1791, but apparently a working machine was never built. The patent described all the elements of a gas turbine: a compressor to raise the pressure of the intake air, a combustion chamber in which the fuel was burned to heat the compressed air, and an expansion turbine in which the energy in the heated air was converted to work before it was discharged to the surroundings.

Early attempts (1900–1910) to build gas turbines were unsuccessful for two reasons. First, the efficiency of the available rotary compressors was so low that it took all the work produced in the expansion turbine to drive them, with the result that no surplus energy was available to drive an electrical generator or other load. Second, the metals that were available for constructing the blades of the expansion turbine were not able to withstand the temperature of the hot gases leaving the combustion chamber. As a consequence, the power that was produced and supplied to the compressor was insufficient to raise the air pressure to an adequate level. That did not stop engineers from trying to construct a gas turbine, and some notable attempts were made in France.

To circumvent the difficulties arising from the compressor, Hans Holzwarth built a number of successful gas turbines in Germany between 1908 and 1933. In his machines, the compression of the intake air was a result of burning the air and fuel in a closed chamber before releasing it into the expansion turbine. These machines had intermittent action, like the reciprocating engine, which meant that they did not have the advantages produced by the continuous flow of the air.

The last of the gas turbines designed by Holzwarth was built by the Brown Boveri Company of Switzerland. Their experience with that machine, together with their substantial experience with steam turbines, convinced them that engineering knowledge, particularly concerning compressor design, had advanced sufficiently by 1933 to justify a reconsideration of the continuous-flow gas turbine. They were able to interest the Sun Oil Company in purchasing a gas turbine to provide large volumes of hot gas for use in the Houdry catalytic converters that were installed at the company's Marcus Hook, Pennsylvania, refinery in 1936. This turbine was not required to produce the work necessary to drive an electrical generator, so it represented a very appropriate first step in the development of a power-producing machine. The opportunity to do this came in 1939, when the first commercial gas-turbine-driven electric generator was placed in service at Neuchâtel, Switzerland. The significance of this machine was recognized in 1988, when it was designated a Historic Mechanical Engineering Landmark.

While Brown Boveri was making its very public entry into the manufacture of gas turbines, other engineers in Germany, England, and the United States were working secretly in the late 1930s on gas turbines to replace the aircraft piston engine. One of these groups was located at the Schenectady plant of the General Electric Company. They had originally been interested in a gas turbine to power a locomotive but had switched to aircraft gas turbines at the request of the government when it appeared likely that the United States would be drawn into World War II. They produced a turbojet engine that flew in 1946 and a turboprop engine, which was plagued with difficulties and did not fly until 1949.

Although the wartime years were a hiatus in General Electric's efforts to produce a locomotive gas turbine, the experience obtained in the development of the two aircraft gas turbines proved to be invaluable. Following the end of the war, the Schenectady engineers returned to their work on the locomotive power plant, and by 1949 they had an operating machine. As things turned out, however, it was first used—in that year—for electric power production at the Belle Isle station of the Oklahoma Gas & Electric Company. Later in the same year, Alco-GE gas turbine locomotive No. 50, using an essentially identical turbine, was tested on a number of American railroads.

The Neuchâtel and the Belle Isle turbines were the forerunners of a long series of gas turbines of increasing power and efficiency, and the trend appears to be continuing. The gas turbine came into the world with difficulty, but its present thriving state is a monument to the skills of mechanical engineers and their contributions to the welfare of society at large.

Neuchâtel Gas Turbine

Neuchâtel, Switzerland

Although the gas turbine was described in a patent granted to John Barber of England in 1791, its use as a prime mover for the production of electricity remained little more than an inventor's dream until the advent in the 1930s of high-efficiency compressors and modern alloys able to withstand high temperatures.

Gas turbines are internal-combustion engines, such as conventional spark-ignition or diesel engines, except that they use rotating compressors and expansion turbines instead of pistons reciprocating in cylinders. In an internal-combustion engine, air is drawn into the engine and compressed, fuel is added to the air before or after compression, and the mixture is burned, raising the temperature of the gas. The hot gases are then expanded—i.e., their temperature and pressure are lowered by withdrawing the energy supplied by the burning fuel to drive the load.

In a gas turbine, these same processes are carried out in separate components, namely, a compressor, a combustion chamber, and an expansion turbine. Air is compressed from atmospheric pressure (14.5 psia, or 100 kPa) to about 60 psia (or 413 kPa), mixed with the fuel, and burned in a continuous-flow combustion chamber. The hot gases enter the turbine at a temperature of about 1,000°F (538°C) and exhaust to the flue or stack. About 75 percent of the turbine output is used to drive the compressor, while the remaining power is available for the generation of electrical energy. The gas turbine appeared to be the ideal prime mover because of its internal combustion, which eliminated the need for a steam plant with its many complex and expensive auxiliaries; because of the universal presence of its working medium, air; and because it is a rotating machine, like the steam turbine.

The pioneer Neuchâtel gas turbine, still in service today.

In 1939 Brown Boveri & Company of Baden, Switzerland, pioneered the construction of gas turbines to generate electrical power by installing the first commercial unit in an underground emergency standby power station at Neuchâtel, Switzerland. The unit's simplicity and its independence of water facilities made it ideal for this class of service.

The single-shaft, simple-cycle gas turbine at Neuchâtel consists of a compressor, turbine, and generator arranged in line and directly coupled, similar to large steam turbines. Rotating at 3,000 rpm, it has a power output of 15,400 kW, of which 11,400 kW is absorbed by the compressor. The remaining 4,000 kW drives the generator. Official tests made prior to installation indicated a thermal efficiency of 18.04 percent when operating with a turbine-inlet temperature of 1,067°F (574°C).

Following the installation at Neuchâtel, Brown Boveri installed gas turbine units at generating plants in Iran, Peru, Venezuela, Egypt, and Luxembourg. But because of its low efficiency, the gas turbine could not compete with the steam turbine for base-load power generation. Until recent years, the gas turbine was used primarily for peak-load electrical generation, as power plants for offshore oil-drilling platforms, and for aircraft propulsion. Today, however, the gas turbine increasingly is used for base-load electrical generation in so-called STAG cycles (combined steam and gas). Hot gases leaving the gas turbine are used to generate steam, which is then supplied to a steam power plant; both the gas turbines and the steam turbines drive generators to produce electrical power. The thermal efficiency of the latest such plants is about 51 percent, compared to about 43 percent for contemporary steam turbine plants.

The pioneer unit at Neuchâtel, installed to meet the power requirements of vital industries in time of war, remains in service today.

FURTHER READING

Adolph Meyer, "The Combustion Gas Turbine: Its History, Development, and Prospects," *Institution of Mechanical Engineers Journal & Proceedings* 141, no. 3: 197–212.

R. Tom Sawyer, *The Modern Gas Turbine* (New York: Prentice-Hall, Inc., 1945).

S. A. Tucker, "Now Gas Turbines That Work," *Power* 83 (June 1939): 58–61.

Belle Isle Gas Turbine

Schenectady, New York

Installed in 1949 at the Arthur S. Huey (later renamed Belle Isle) Station of the Oklahoma Gas & Electric Company near Oklahoma City, this was the first gas turbine used to produce commercial power in the United States. It represented the transformation of the aircraft gas turbine engine, which seldom ran for more

The Belle Isle gas turbine following its removal from service in 1980.

than ten hours at a stretch, into a reliable and long-life prime mover. The low-cost, simple-cycle gas turbine provided quick additional capacity, arousing considerable interest in the U.S. electric-utility industry and leading to widespread adoption of similar units. Between 1966 and 1976, American utilities installed more than fourteen hundred gas turbines with outputs over 3,500 kilowatts, accounting for some 9 percent of total electric output nationwide.

Gas turbines using constant-pressure combustion were built independently by Sanford Moss in the United States and by René Armengaud and Charles Lemale in France and demonstrated in 1903. Neither was successful because material properties placed limits on the temperature of the gas entering the expansion turbine, resulting in reduced thermal efficiency in comparison with other prime movers.

In 1936 the Sun Oil Company installed a gas turbine at its Marcus Hook refinery near Philadelphia to supply air to the Houdry catalytic cracking process. The turbine, constructed by Brown Boveri & Company of Baden, Switzerland, did not perform useful work, but the knowledge Brown Boveri gained led to the construction of the first commercial gas turbine-driven electric generator at Neuchâtel, Switzerland (see Neuchâtel Gas Turbine, p. 97) in 1939. Both machines were characterized by comparatively low temperatures at the inlet to the expansion turbine (875–950°F, or 468–510°C, at Marcus Hook; 1,020°F, or 548°C, at Neuchâtel); and modest compression ratios (3:1 at Marcus Hook, 4:1 at Neuchâtel). Significant increases in compression ratios, expansion-turbine inlet temperatures, and power came as a result of the intense effort during World War II to produce a practical gas turbine for military aircraft propulsion.

In the late 1930s, engineers at General Electric's Schenectady plant began to study the application of gas turbines to locomotives. With the outbreak of war in Europe, the team, under the direction of Alan Howard (1905–66), turned its attention to aircraft engines, producing both a turboprop engine (a gas turbine that drives an aircraft propeller) and a turbojet engine (which produces a high-speed gas jet). Drawing on their experience with aircraft engines, the engineers returned to their original gas-turbine locomotive project following the end of hostilities. One of these locomotive gas turbines was slightly modified for stationary use—to drive an electric generator—and sold to Oklahoma Gas & Electric. It was installed in the utility's Belle Isle Station in 1949.

The gas turbine was housed in a separate building added to one side of the existing 51-MW steam plant and coupled to a conventional 4,000-kW, 3,600-rpm generator and exciter. An abundance of low-cost, high-Btu natural gas made the installation attractive. On July 29, 1949, the 3,500-kW unit started delivering power to the company's distribution system. By September 1953, when it was removed from service for an overhaul, it had operated for 30,000 hours—more than any other gas turbine in the world—at less maintenance cost than steam turbines and boilers.

Together with a second unit installed in 1952, the Belle Isle turbine served Oklahoma Gas & Electric for thirty-one years. It was withdrawn from service in 1980 when the station was closed, and returned to GE's Schenectady plant for display.

Location/Access

The gas turbine is located outside Building 262 at the General Electric Company plant in Schenectady, New York. Direct questions to: GE Power Generation, 1 River Road, Schenectady, NY 12345; phone (518) 385-3072.

FURTHER READING

J. W. Blake and R. W. Tumy, "3,500-kW Gas Turbine Raises Station Capability by 6,000-kW," *Power* 92 (September 1948): 64–71.

Joel W. Blake, "Belle Isle Gas Turbine: After 30,000 Hours," *Power* 98 (August 1954): 75–79.

Alan Howard, "Design Features of a 4,800-HP Locomotive Gas-Turbine Power Plant," *Mechanical Engineering* 70 (April 1948): 301–6.

SPECIFICATIONS

Belle Isle Gas Turbine

Output: 3,500 kW
Speed: 6,700 rpm (step-down gearing drove the alternator at 3,600 rpm)
Compressor: 15-stage, axial-flow with 6:1 pressure ratio
Combustors: 6
Fuel: natural gas
Expansion turbine: 2-stage
  Inlet temperature: 1,400°F (760°C)
  Exit temperature: 780°F (415°C)
Weight: 64,000 lbs. (29,000 kg)
Length: 18 feet (5,486 mm)
Width: 9 feet (2,743 mm)

Nuclear

INTRODUCTION by R. Michael Hunt

Following the announcement by two German physicists in 1938 that they had observed the splitting of the atomic nucleus, many scientists realized that a bomb of enormous power could be constructed utilizing this "fission." By 1943 the Allies were at war with Germany, Italy, and Japan and were in an all-out effort to produce the bomb before it could be developed by the enemy. That same year Enrico Fermi, expatriate Italian physicist, demonstrated a self-sustaining nuclear fission reaction with an "atomic pile" of natural uranium and graphite under the Stagg Field Grandstand at the University of Chicago.

Even in the infancy of the nuclear age, it was recognized that the power released in this reaction could be harnessed for peaceful purposes, such as the generation of electricity. But the first news that the public had of this new technology was the explosion of the atomic bomb over Hiroshima, Japan, at 8:15 A.M. on August 6, 1945.

After the war, there were amazing prophesies about the new "atomic power." Atomic-powered airplanes would fly around the globe nonstop, and atomic-powered cars would fuel up just once a year. Electricity would become too cheap to meter. But again the next big step was not civilian but military.

Admiral Hyman G. Rickover became convinced that the submarine could have almost unlimited range if powered by a nuclear reactor. Cleverly playing off the Navy Department and the Atomic Energy Commission, he asked industry for proposals for nuclear propulsion systems. Westinghouse Electric of Pittsburgh, Pennsylvania, won with its pressurized water reactor concept, installed in the USS *Nautilus* in 1955. In this, the nuclear reactor is contained in a very strong steel vessel and is cooled by, and therefore heats up, water. By keeping the water in this primary circuit under very high pressure, it is prevented from boiling. Pipes carry this hot water into heat exchangers—also cooled by water—in which steam is created. The steam drives turbines that turn the propellers and a turbo-generator, which makes electricity for the boat. The cooled primary water returns to the reactor, and the cycle continues.

In 1954, President Dwight D. Eisenhower announced the Atoms for Peace program to accelerate the peaceful uses of atomic energy, and construction of the nation's first commercial power reactor was begun. Based on the Westinghouse submarine technology, the power plant was constructed at Shippingport, Pa., on the Ohio River near Pittsburgh and operated by Duquesne Light Company. The pressurized water reactor has since become the preferred reactor type for electric power generation throughout the world.

As a technology, nuclear power seems two-edged. Properly controlled, maintained, and running smoothly, the nuclear reactor is one of the most environmentally benign ways to make electricity. Nuclear waste disposal is an issue, but it seems to be one requiring a decision on methodology rather than practicality. But its birth in the bomb, the complexity of its technology, and the invisibility and potential long-term latent effects of nuclear radiation have made us fearful. Intrinsically safe power-reactor designs, which do not rely so heavily on motors, pumps, valves, etc., for safety, are on the drawing boards. Only time will tell if they will be built in the United States.

Shippingport Atomic Power Station

Shippingport, Pennsylvania

The development of nuclear power plants for the generation of electricity was a cornerstone of President Dwight D. Eisenhower's Atoms for Peace plan, a proposal to give the world access to nonmilitary benefits of nuclear fission. In the 1950s there was no obvious or single direction leading toward the production of economical power reactors. Consequently, the U.S. Government joined private industry in developing a variety of prototypes, including light- and heavy-water-cooled reactors, gas-cooled reactors, and other systems. When plans for a proposed aircraft carrier powered by a large-scale, light-water-cooled reactor were canceled, the Naval Reactors Branch of the Atomic Energy Commission (AEC), led by Captain Hyman G. Rickover, redirected its efforts toward a civilian reactor for the production of electric power.

The AEC's decision to build the first full-scale power reactor was a first step toward establishing a new industry. In 1953 the AEC invited proposals for investment in the project. Of the nine offers received, that from the Duquesne Light Company of Pittsburgh was by far the best. The company offered to build the plant on a site it owned in Shippingport, Pennsylvania, a sleepy village on the Ohio River 25 miles (40 km) northwest of Pittsburgh. Duquesne offered to build the

Shippingport Atomic Power Station prior to its closure in 1982.

turbogenerator plant and operate and maintain the entire facility; Westinghouse would serve as general contractor for the power plant, which the AEC would own.

On Labor Day (September 6) 1954, ground was broken for the nation's first commercial power reactor. Speaking from Denver via radio and television, President Eisenhower announced his administration's Atoms for Peace plan, then, waving an "atomic wand," set a bulldozer in motion at the Shippingport site. He said of the plant, expected to produce enough power for one hundred thousand people: "In thus advancing toward the economic production of electricity by atomic power, mankind comes closer to fulfillment of the ancient dream of a new and better earth. . . . I am confident that the atom will not be devoted exclusively to the destruction of man but will be his mighty servant and tireless benefactor."

The Shippingport Atomic Power Station was designed and built under the guidance of Captain Rickover, who paid frequent visits to the site to check progress and confer with project managers John W. Simpson of Westinghouse and John E. Gray of Duquesne Light. The project team had the seemingly Herculean task of building the plant in twenty-four months' time, by March 1957, though labor strikes and steel shortages pushed completion back to December of that year.

The Shippingport project was directed toward advancing the basic technology of light-water- (i.e., ordinary-water-) cooled reactors through its design, development, building, testing, and operation as part of a public utility system. The reactor was housed in four interconnected containment vessels of reinforced-concrete and steel buried below ground. It consisted of a primary system containing the nuclear reactor and the water that circulated through the reactor core to cool it, and a completely isolated (and thereby uncontaminated) secondary system containing light (demineralized) water. As it flowed, the water in the primary system absorbed heat from the fissioning nuclear fuel. The primary system was kept under high pressure—2,000 psig (13,780 kPa)—to prevent the water from boiling. (Rickover and others feared the corrosive effects of boiling water around the reactor core.) The heated water flowed to four heat exchangers through which the water of the secondary system circulated. Here, the secondary-system water was converted to steam, providing the energy to drive the single turbine and its generator.

The seed-and-blanket reactor core was chosen. The "seed" consisted of enriched uranium, which leaked neutrons into a "blanket" of natural uranium comprising 95,000 fuel elements, the heart of the reactor. It was housed in a pressure vessel—25 feet (7,620 mm) high and 10 feet (3,048 mm) in diameter made of 8-inch- (203-mm-) thick carbon steel walls—that approached the very limits of steel fabrication at that time.

The Shippingport reactor achieved criticality on the morning of December 2, 1957, fifteen years to the day after Enrico Fermi had achieved the world's first nuclear chain reaction in Chicago. Two weeks later, the turbine-generator was synchronized with Duquesne Light's distribution system. On Wednesday, December 18, 1957, just after midnight, the first electricity was fed into the power grid, which carried it throughout the greater Pittsburgh area. On December 23, the reactor attained its full capacity of 60 megawatts.

Though construction delays, escalating costs, design redundancies, and expensive test equipment had combined to push the plant's generation costs to as much as ten times those of existing fossil-fuel stations, its engineering achievements were notable. Shippingport, a nuclear power "laboratory," represented a fundamentally new conception of reactor design specifically for the production of electric power. The plant performed almost flawlessly from the first day of operation, establishing itself as a resource of information on reactor technology for a fledgling industry. Over the next six years, hundreds of engineers and technicians learned the rudiments of reactor technology at Shippingport, while hundreds of articles about its design and operation appeared in technical journals. The plant's excellent performance and the information it provided contributed to the adoption of light-water-reactor technology by nations worldwide. Today, about 80 percent of the world's reactors are light water. Their heritage, historian William Beaver has written, "can be traced to Shippingport."

The Shippingport station operated with its first core until 1964. A second core increased the plant's generating capacity to 100 megawatts. From 1976 to 1977 the plant was modified with installation of a light-water breeder reactor core, designed to produce more uranium-235 than was used to produce energy. The pioneering plant was shut down in October 1982. Since then, it has achieved another historic first, becoming the first nuclear power plant to be completely dismantled. The reactor vessel and other contaminated components are now stored at the Hanford, Washington, reservation.

Location/Access

The Shippingport Atomic Power Station has been dismantled.

FURTHER READING

William Beaver, *Nuclear Power Goes On-Line: A History of Shippingport*, Contributions in Economics and Economic History (New York: Greenwood Press, 1990).

Richard G. Hewlett and Jack M. Holl, *Atoms for Peace and War, 1953–1961: Eisenhower and the Atomic Energy Commission* (Berkeley, Calif.: University of California Press, 1989).

Geothermal

INTRODUCTION by Euan F. C. Somerscales

Energy stored within the interior of the earth is available for exploitation. Generally the heat flow is too small for practical use, but there are locations, typically associated with volcanoes, hot springs, fumaroles (steam jets), and other phenomena, where the heat flow is large enough to be useful. This geothermal energy was used by the Romans two thousand years ago to heat their baths, and certain towns in France have used geothermal water for domestic heating since the Middle Ages. Warm-water spas are reputed to have important therapeutic properties.

Geothermal energy was first used as a source of power in 1904 in Italy. Prince Piero Ginori Conti used steam issuing from the ground at Larderello in Tuscany to drive a steam engine connected to an electrical generator. The steam at the Geysers in California was similarly harnessed in the mid-1920s.

The steam available at Larderello and the Geysers contains relatively little moisture and is consequently easy to use in steam engines and steam turbines. Most geothermal sources yield a mixture of steam and hot water, and such liquid-dominated sources are more difficult to use as a steam source for engines and turbines. Nevertheless, small power plants were located at such sources in Japan in 1925 and in 1951 but have since been abandoned. Large-scale exploitation of sources of this type first took place in 1958 at the Wairakei plant in New Zealand.

As of 1980, 1,072 MW of electrical power were being produced at the Geysers, 391 MW at Larderello and other Italian sites, and 281 MW at Wairakei and other New Zealand locations. Various countries, including Japan, Iceland, and the former Soviet Union are making substantial use of geothermal power. With a view to encouraging wider use of this energy source, the United Nations has convened a number of international conferences on the topic, and as a result, many other countries have begun to assess the possibilities of exploiting their geothermal resources.

The fluids produced by geothermal sources are complex, including, besides steam and substantial amounts of water, dissolved minerals and gases. These constituents have to be removed to a greater or lesser extent before the fluid can be used in a steam turbine. Much effort has been applied to the development of suitable separation methods. In some cases the geothermal fluid has been used to heat a secondary fluid, which is then supplied to the turbines. This was, in fact, the procedure originally adopted at the Larderello plant in Italy, but it was subsequently abandoned. Such indirect methods incur an efficiency penalty because the secondary fluid cannot reach the temperature of the geothermal fluid.

The mineral and gaseous constituents of the geothermal fluid can present problems of disposal after the available energy has been extracted because discharge of wastewater into rivers can degrade the water quality for downstream users or release hazardous materials into the environment. Where this is a problem, reinjection of the water into the ground has been tried, but its effect on the fluid issuing from the source is uncertain. Noise and fumes are also associated with the use of geothermal energy. So it is not as benign as might at first appear. In spite of this, the anticipated increase in fuel prices as a consequence of the decrease in available oil will make geothermal energy more attractive, particularly to developing countries that have such a source available. The mechanical engineer can therefore expect to have many interesting problems to solve before this type of energy is exploited to its full potential.

The Geysers Unit 1, Pacific Gas & Electric Company

near Healdsburg, California

Hunting grizzlies in the mountains between Cloverdale and Calistoga in 1847, explorer-surveyor William Bell Elliott came upon a startling sight: puffs of steam rising from the canyon of Big Sulphur Creek. The awestruck hunter thought he had come upon the gate of hell. What Elliott had seen, in fact, were puffs of geothermal steam, called fumaroles. "The Geysers," as the area came to be called, quickly acquired fame for its hot springs, fumaroles, and steam vents, and in 1851 a hotel was built there. (The name "Geysers" is a misnomer, as no geysers occur here.) More than a century after Elliott's discovery, the natural steam would be put to work, powering the first commercial plant in the United States to generate electricity from geothermal steam.

Geothermal steam originates in the magma (molten rock) of the Earth's interior and from slow radioactive decay in solid rock formations. Although the thickness of the Earth's crust averages 20 miles (32 km), in some places it is thinner or there are weak spots. Such regions are marked by volcanic activity and, in areas where trapped bodies of subterranean water exist, by the presence of hot springs, geysers, or fumaroles.

In the 1920s, J. D. Grant of Healdsburg began drilling wells with the hope of harnessing the steam for the generation of power. Further development was restrained by the relatively cheap fossil fuel available for steam-electric power generation and the need for improved materials, especially stainless steel, that

The Geysers Unit 1. *Courtesy Pacific Gas & Electric Company.*

could stand up to the corrosive effects of the hydrogen sulfide in the steam. In 1955 Magma Power Company leased 3,200 acres (1,295 ha) of land from the Geysers Development Company and drilled its first well. A year later, Magma contracted with the newly formed Thermal Power Company to drill additional wells and aid in marketing the steam.

In 1958 Pacific Gas & Electric Company signed a contract with Magma-Thermal to build a steam-electric power plant and agreed to construct additional generating facilities as Magma-Thermal developed the necessary steam supply. PG&E's geothermal complex began modestly. On September 25, 1960, the Geysers Unit 1, with a net capacity of 11 megawatts, came on line.

The principal differences between a geothermal power plant and other power plants are threefold: (1) there is no boiler; (2) steam pressures are lower (100 psig, or 689 kPa, at the Geysers Unit 1); and (3) the steam contains much larger quantities of noncondensible gases, which must be removed from the condensers to maintain vacuum.

At Unit 1, superheated steam is obtained from a network of wells 7,000 to 10,000 feet (2,134 to 3,048 m) deep. The steam, as it comes from the well, contains particulates (rock dust, for example), which must be removed by whirling them off in centrifugal separators. The steam is then piped to the turbine-generator—a 12,500-kW General Electric unit installed in 1924 at PG&E's Sacramento Power Plant that was modified to permit greater steam flow—where it is expanded through six stages to drive an electric generator of standard design, then exhausted into a barometric condenser. Condensing steam and cooling water mix, and the mixture drops down a barometric leg into a hot well, from which it is pumped to a cooling tower where it is cooled by evaporation. A 10-mile (16-km), 60-kW transmission line connects the plant to the PG&E system.

Noncondensible gases—these average 0.75 percent by weight and include methane, hydrogen, hydrogen sulphide, and ammonia—are cleansed from the steam by steam-jet ejectors. Originally, the gases were exhausted into the atmosphere at high velocity. In the 1970s, Unit 1 was retrofitted with an incinerator and a chemical water-treatment system that uses a vanadium solution to cleanse the waste gases of malodorous hydrogen sulfide. Condensate, formerly discharged into adjacent Big Sulphur Creek, today is reinjected into the steam field through injection wells, helping to replenish the steam reservoir and allaying environmental concerns.

From its small beginning in 1960, the Geysers of PG&E has grown to a complex of 19 geothermal units in Sonoma and Lake counties. In 1984 natural steam from below the Earth's surface was harnessed to produce a record 7.1 billion kilowatt-hours of electricity, enough to meet the needs of more than a million customers. Unit 1, the grandfather of geothermal energy in the United States, continues to produce power today.

Location/Access

The Geysers Unit 1 is located on Geysers Road near Healdsburg, 95 miles (153 km) north of San Francisco.

FURTHER READING

E. T. Allen and Arthur L. Day, *Steam Wells and Other Thermal Activity at "The Geysers" California* (Carnegie Institution of Washington, 1927).

A. W. Bruce, "Natural Steam Source Harnessed," *Electrical World* 153 (27 June 1960): 46–50.

Albert W. Bruce and Ben C. Albritton, "Power From Geothermal Steam at the Geysers Steam Plant," *Journal of the Power Division, Proceedings of the American Society of Civil Engineers* 85 (December 1959): 23–45.

Power Transmission

INTRODUCTION by Euan F. C. Somerscales

The utilization of power requires its generation, its transmission to some point of use, and its application. As an example, the automobile engine generates power from burning gasoline, transmits the power by a driveshaft, and applies the power, through the wheels, to move the vehicle. Rotary motion is common to each one of these sequences of processes. The engine crankshaft rotates, the driveshaft rotates, and the wheels rotate. However, all these components must be guided and restrained in some way: a rotating shaft has no natural discipline; it must be imposed. At the point of restraint—the "bearing," as the mechanical engineer calls it—one part will rotate inside another. Our own experience tells us that at the bearing there is friction and generation of heat. This heat represents a loss of power. However, the loss can be minimized by lubricating the bearing with a semisolid material: grease or a liquid oil.

Although bearings are a concept dating from the invention of the wheel in prehistoric times, it is only recently—somewhat more than one hundred years ago—that mechanical engineers have understood how to design bearings that minimize the loss of energy due to friction. This may, at first sight, seem like a limited contribution to the welfare of the human race, but it takes little reflection to appreciate that friction is present whenever one surface moves over another. Because of the universality of this loss and the recognition that the energy has come, in most cases, from a finite and renewable source, a reduction in friction can be seen as a substantial benefit.

Robert H. Thurston (1839–1903) of Cornell University, and one of the nineteenth century's leading engineering educators, probably was the first American engineer to recognize the possibilities of designing bearings rationally. In 1885 he published *A Treatise on Friction and Lost Work*. As a consequence he was consulted by other engineers faced with problems in lubrication, and he drew his students into this work by assigning them projects connected with questions arising from his consulting activities. One of these students was a bright and impecunious fellow by the name of Albert Kingsbury (1862–1943). Kingsbury's life seems to be typical of that of many American engineers working at the turn of the century. For financial reasons he alternated between college and work as a machinist, but this was no bad thing. It gave Kingsbury and others like him a practical sense about mechanical devices that was critical to his invention of an entirely new type of bearing, the Kingsbury thrust bearing. This story is too long to tell here, but there is an intriguing related tale that must be mentioned. As so often is the case in technical work, the story of the thrust bearing is one of simultaneous and independent invention. A bearing essentially identical to Kingsbury's was invented, quite independently, by an Australian engineer, A. G. M. Michell (1870–1959). Even more interesting is the common origin of these inventions. Both Kingsbury and Michell were led to the design of their bearings by the theoretical work of Osborne Reynolds (1842–1912), professor of engineering at the University of Manchester in England, and one of the most distinguished of engineering scientists. In 1886, Reynolds had shown by pure mathematical analysis how to design the type of bearing that Kingsbury and Michell had invented, but by some oversight of fate he did not take the extra step and invent the practical device.

The pivoted-pad bearing, as the Kingsbury and Michell bearings should properly be called, is used extensively in steam turbines and water turbines. In machines of both these types, the shafts carry substantial thrusts along their axis, arising, in the case of the steam turbine, from unbalanced steam forces. In the water turbine, which is arranged with the shaft vertical, the thrust bearing carries the weight of the generator rotor and turbine wheel; in a modern machine, this would amount to several hundred tons. Prior to the introduction of the pivoted-pad bearing, the axial thrust in these applications usually was carried by a so-called multicollar thrust bearing. With this type of bearing, collars that are integral with the shaft are arranged uniformly along its length, and a horseshoe-shaped bearing pad is located between each pair of collars. This device can take up a substantial length of shaft (several meters, in the case of marine steam turbines), but the comparable pivoted-pad thrust bearing is typically limited to a length of less than a meter.

The transmission of power is not the stuff of drama; nevertheless, its role in modern society is critical, and the story, not without human interest, has lessons for today about engineering research and practice. Three people working independently were able to revolutionize the transmission of power. Like so many of the contributions of the mechanical engineer to modern life, they are unknown to a wider public.

Kingsbury Thrust Bearing

Holtwood, Pennsylvania

The first Kingsbury thrust bearing was put into service on June 22, 1912, under the 10,000-kilowatt Unit 5 at the Holtwood hydroelectric station of the Pennsylvania Water & Power Company. The 48-inch- (1,219-mm-) diameter bearing has been at work ever since, effortlessly carrying 410,000 pounds (186,000 kg) at a speed of 94 rpm.

All rotating machinery must use bearings to maintain the correct location between stationary and revolving parts, and to maintain the correct relative position of the shaft and its supporting structure. Specifically, a thrust bearing maintains the relative axial location of a shaft and its supporting structure. Helicopter rotors, for example, and boat and airplane propellers need thrust bearings on their shafts. So do water and steam turbines, which must operate continuously for long periods of time—usually several years—with no maintenance.

The Kingsbury thrust bearing was the brainchild of Pittsburgh mechanical engineer Albert Kingsbury (1863–1943). Kingsbury's idea was deceptively simple: instead of roller bearings, a series of adjustable bearing surfaces would carry the weight, gliding, as they did so, over a continuous film of oil.

Patented in 1910 (No. 947,242), the Kingsbury thrust bearing consists of a stationary cast-iron ring (called a "runner"), a cup-shaped frame or collar (to

Albert Kingsbury (right) and Frederick A. Allner, who later became a vice president of the Pennsylvania Water and Power Company, inspect the Unit 5 thrust bearing in 1937.

contain the lubricant), the shaft, and a segmental ring-bearing member comprised of several wedge-shaped bearing shoes (usually six, as in the case of Holtwood Unit 5) that are identical in size. Each shoe is loosely bolted through a tapped hole at its midpoint so that it can rock a bit. As the shaft rotates, a film of oil is forced between the stationary ring and the shoes, where the pressure is highest. The oil actually supports the weight—there is no physical contact between the runner and the shoes—resulting in extremely low friction and almost no mechanical wear.

Until the advent of the Kingsbury thrust bearing, units like Holtwood represented the upper limit of hydroelectric turbine size; even then, roller thrust bearings commonly used in such installations wore out quickly and had to be repaired or replaced with annoying (and expensive) frequency. Kingsbury bearings could support one hundred times the load of roller bearings with negligible wear and were rapidly adopted for hydraulic and steam turbine use. Eventually, Pennsylvania Water & Power put them on all ten Holtwood units.

When Holtwood Unit 5 was rebuilt for sixty-cycle service in 1950, the original Kingsbury bearing was found still to be in perfect condition. The bearing was inspected again in 1969 with the same result. "Not a single part has ever been replaced," reads a plaque attached to the unit in recognition of Albert Kingsbury's singular mechanical achievement, which made possible the design of much larger hydroelectric units, including those of the Tennessee Valley and Bonneville power authorities. Kingsbury thrust bearings have also found wide application on the propeller shafts of ocean liners.

Location/Access

A model of the Kingsbury bearing is mounted on Unit 5, Pennsylvania Power & Light Company, 405 Old Holtwood Road, Holtwood, PA 17532; phone (717) 284-4101.

FURTHER READING

Richard F. Snow, "Bearing Up Nobly," *American Heritage of Invention and Technology* 4, no. 1 (Spring/Summer 1988): 4–5.

"A Thrust Bearing for High Unit Pressures," *American Machinist* 38 (13 March 1913): 444–45.

"Water Wheel Thrust Bearing," *Engineering Record* 67 (11 January 1913): 44–45.

Minerals Extraction and Refining

INTRODUCTION by Robert M. Vogel

One of our earliest organized industrial activities—if not the earliest—was the extraction from the earth of a variety of useful minerals and their separation from the worthless components of their ores. Since the time of the Bronze and Iron ages, vast amounts of human energy and great ingenuity have gone into locating, and then digging, hacking, drilling, fire-setting, crushing, blasting, hauling, hoisting, separating, washing, smelting, and—by an almost endless array of other methods, systems, and treatments—obtaining those metallic and chemical substances needed or simply desired by people in their continual forward march. As in literally every other undertaking involving any degree of mechanical technology, the role of the mechanical engineer in this service gradually evolved from that of the millwright, inventor, and general "artificer," and skilled miner, smelter, and refiner.

It can easily be argued that despite the increasing introduction of aluminum and plastics into the products and engineering works of today's world, we remain in the Iron Age. The great bulk of all machinery, transportation systems, and the works of construction, down to even the finest instruments, are based on cast iron, steel, or some alloy of steel. Historically, this was true even in the "Wooden Age," for with the exception of the simplest wooden implements, all objects, devices, and structures of wood or timber incorporated ferrous elements to join, reinforce, resist wear, or provide a cutting edge. As iron does not occur in its native, metallic state, the smelting of its ores has been a challenge from antiquity. Although historically a variety of primitive means were used to extract the metal from its ore, not until the eighteenth century was there anything like widespread use in Europe of the blast furnace, then as now the most practical and efficient method of iron-ore reduction. Here, fuel in the form (then) of charcoal and a flux of limestone or shells were burned with the ore at high temperature in a vertical, cylindrical furnace under a continuous blast of air. This caused the metallic iron to separate from the other constituents and sink to the furnace bottom to be run off into molds as pig iron.

The principal improvement in blast-furnace technology was the substitution for the charcoal of coke made from soft coal. This occurred first in the early eighteenth century although, for reasons having to do mainly with the chemical composition of different ores, charcoal pig iron was produced in commercial quantities in all industrial nations until late in the nineteenth century. Charcoal smelting persisted far longer in the United States and Sweden than in Great Britain for the simple reason that the British timberlands had been depleted early and heavily; moreover, Great Britain was blessed with a nearly inexhaustible reserve of coal.

The landmark ironworks at Saugus, Cornwall, and Ringwood all were charcoal-fueled throughout their history, although smelting with anthracite was briefly and unsuccessfully attempted at Cornwall.

Second in importance only to the extraction of iron from its ore in the production of the ferrous metals is the conversion of iron to steel. Historically, a relatively small percentage of the pig iron produced was converted into the malleable wrought iron and, to an even lesser extent, into the stronger and harder steel. With the invention in the mid-nineteenth century of the Bessemer and open-hearth processes for producing steel cheaply and in large quantities, the world entered the Age of Steel, and by about 1890, wrought iron was being produced only in limited volume for specialized products.

By about 1950, open-hearth steel had almost totally displaced Bessemer, and yet today hardly an open-hearth furnace operates anywhere in the industrialized world, that process in turn having been almost totally eclipsed by the faster and cheaper basic-oxygen method.

While the ferrous metals can be regarded as the basis of the (ongoing) Industrial Revolution, no other metal has had as powerful an impact on the course of world history as gold. Gold's intrinsic value, the consequence of its peculiar physical and chemical properties, has led not only to the epochal searches and the lasting effects on society but to an extensive, highly specialized technology for the extraction of the meal both from the earth and from its ores when not found in the native state. A vast arsenal of mechanical, hydraulic, and chemical processes has been developed for these purposes. The landmark stamp mill at the Reed gold mine is typical of the machinery once used in the gold fields to reduce gold-bearing rock to a fineness that permitted ultimate separation of the metal from the worthless elements, while the Alaska Gold Dredge typifies the mining on a massive scale of low-yield alluvial deposits. These great floating processing plants combined the technologies of dredging, materials handling, and ore concentrating into an ingenious, highly specialized machine dedicated to extracting a minuscule quantity of a very valuable substance from a huge volume of totally worthless stuff, the value of the product justifying the enormous capital costs of the dredge and of its operation.

At the other end of the scale is the engineer's challenge to mine a material such as coal, where the unit value is relatively low but where there remains the problem of handling great volumes of both the material itself and a useless overburden that must be removed to reach it. While that has always been a problem in the mining of coal by the traditional means of shafts and adits that more or less followed the coal seams, with the introduction of large-scale, open-pit mining in the twentieth century the entire complexion of the process changed. Open-pit coal mining amounts to little more than excavating massive volumes of overburden and (usually) somewhat lesser volumes of coal. With this process, mining technology essentially became one of building larger and larger mechanical shovels, such as "Big Brutus," in an effort to reduce costs through sheer economy of scale. Today, stripping shovels, invariably electrically powered, are among the largest movable objects on land, operating as they do entirely free of dimensional and weight constraints.

The discovery of the Pennsylvania oil fields in 1859 commemorated by the Drake Well landmark, signaled the appearance of a new source of energy that has in many areas—perhaps only temporarily—displaced coal. This has given rise to a mammoth industry that continues to expand in scale and technological refinement to the present day. Whereas the force driving the petroleum industry in the nineteenth century was the need for illuminating oils to replace the waning supply of whale oil, by the end of the century there was an entirely new series of expanding markets for the many fractions found in the crude. An exponentially increasing demand for oil, paralleling the growth of automobile and truck usage, spawned technologies in the extraction of oil and its refining. Both had earlier roots. Techniques for the drilling of deep wells for water and brine extraction were well developed by the time Drake's exploit had inspired exploration in many parts of the world. Under pressure to break the crude oil into its various commercially useful fractions, engineers and chemists quickly developed a variety of refining processes based on earlier chemical technology. Simple stills, such as that at Newhall, with capacities measured in hundreds of gallons, rapidly evolved into massive, full-fledged refineries as demand for illuminants, fuels, solvents, lubricants, and other petroleum-based products exploded in this century.

A radically new source of energy and the requirement for an equally nontraditional means of "refining" its mineral basis emerged with stunning suddenness in the years just prior to and during World War II. With the discoveries that in atomic fission there lay undreamed of energy potential and that there were means for effectively harnessing that energy, the Atomic Age was born. The "effective harnessing" was anything but simple, however. Under wartime pressures to produce an atomic bomb, an entirely new industry was developed. "Atomic piles" that converted the enriched natural uranium into the plutonium fuel needed in the bomb were erected at the landmark Hanford B Reactor.

The vast potential of atomic energy is tempered, of course, by the problems—real and perceived—in its use on a large scale for the production of electrical power, and it seems unlikely that it will entirely replace the fossil fuels in the near future.

Saugus Ironworks

Saugus, Massachusetts

The Saugus Ironworks was the first successful integrated ironworks in North America and a prototype of American industry. It was promoted by John Winthrop the Younger and the Company of Undertakers of the Iron Works in New England, a group of some twenty Englishmen and several Massachusetts residents organized for the purpose of developing an ironworks in the American colonies. Winthrop and his successor, Richard Leader, built two plants between 1644 and 1647 to convert bog iron into cast and wrought iron. At Braintree, south of Boston, was a blast furnace and a forge. On the Saugus River, northeast of Boston near Lynn, was a complete ironworks. Undertaken only a quarter century after the landing of the Pilgrims, it was an impressive technological achievement for an early colony.

The ironworks at Saugus—or "Hammersmith," as the plant was called—copied those in England, from which its builders came. It consisted of a huge furnace; a forge comprising two fineries, a chafery, and a hammer; a rolling and slitting mill; an extensive water-power system; and workers' housing. The furnace was a shell of fieldstone, about 26 feet (7.9 m) square at the base and 21 feet (6.4 m) high, with outer walls sloping inward as they rose and a barrel-shaped hollow core, the inwardly sloping lower section of which—the "bosh"—supported the charge of ore, flux, and fuel. Below the bosh was a square crucible lined with refractory sandstone. The furnace was surmounted by a short stack with a side opening at its base for charging the raw materials from a timber bridge connecting the furnace to the adjacent hillside.

The Saugus Ironworks National Historic Site is a reconstruction of the first successful integrated ironworks in North America.

Leather bellows, operated by a waterwheel, supplied the air blast through a tuyere (or nozzle) in the furnace base. Under the influence of heat, the ore (iron oxide) was reduced by the charcoal fuel (carbon) to metallic iron. The heavy molten mass collected in the hearth at the base of the furnace. Impurities that were coalesced by the flux floated on top of the molten iron and were drawn off as slag. To tap the iron, the side openings in the hearth, temporarily sealed with fire clay, were pierced. The resulting iron was either cast directly into articles—such as pots, skillets, and firebacks—or it was cast into slablike "pigs," which were later reheated and processed into a variety of cast- or wrought-iron articles. About 3 tons (2.7 t) of bog ore and 265 bushels (9,338 l) of charcoal were required to make a single ton (0.9 t) of iron.

In the forge house, equipped with water-powered bellows and a hammer, were the two fineries and a chafery. In the fineries, the iron pig was melted and the residual carbon removed by oxidation to produce wrought iron. It was then worked by hammer into a rough rectangular bar, or bloom. Blooms were reheated in the chafery, then forged by power hammer into long rectangular bars, the principal product of the ironworks. In the nearby rolling and slitting mill, some of the bar iron was further reduced to flats, then cut lengthwise into rods and bundled for sale and eventual reduction to nails.

The water-power system at Hammersmith consisted of a stone-and-earthen dam across the Saugus River, from which water was directed by a 1,600-foot (488-m) canal to a reservoir. From the reservoir, a race channeled the water to wheels that powered the furnace bellows, the forge, and the rolling and slitting mill. Archeological excavation has shown that the furnace bellows were driven by a six-spoked overshot wheel between 16 and 17 feet (4,877 and 5,181 mm) in diameter and about 2 feet (610 mm) wide. The size and type of the other wheels is not definitely known.

The Saugus Ironworks was not financially successful, nor was it as long lived as many of its successors. The ironworks changed owners a number of times and suffered from lack of capital, high production costs, and competition with iron imported from England. Production came to an end about 1670. Even before its demise, however, Hammersmith had begun to provide some of the impetus for other ironmaking ventures, including much of the skilled labor without which they might never have gotten under way. By 1700, ironworks had been established in Massachusetts, Connecticut, Rhode Island, and New Jersey; most were started or staffed by the workers who had once worked at Braintree or Hammersmith.

Following extensive (if imperfect) archeological investigation between 1948 and 1954, the American Iron and Steel Institute and the First Iron Works Association, Inc., undertook the reconstruction of the Saugus Ironworks, which today includes a furnace, forge, blacksmith shop, and rolling and slitting mill. The ironworks house, built about 1646 for Richard Leader, is the only structure from the original complex that survives.

Location/Access

The Saugus Ironworks National Historic Site, administered by the National Park Service, is located at 244 Central Street near Saugus Center, Saugus, MA 01906; phone (617) 233-0050. Admission free.

FURTHER READING

E. N. Hartley, *Ironworks on the Saugus* (Norman, Okla.: University of Oklahoma Press, 1957).

Cornwall Iron Furnace

Cornwall, Pennsylvania

The iron industry in the American colonies began in New England in the seventeenth century (see "Saugus Ironworks," p. 119), but it did not show real growth until the eighteenth century. Then Pennsylvania took the lead, owing to its seemingly inexhaustible deposits of iron ore, endless timberlands, great deposits of limestone, plentiful water power, and great pool of ironmasters and skilled workers. Ironmaking flourished there between 1720, when iron was first produced at the Coalbrookdale Furnace in Berks County, and the Revolutionary War. By the start of war, there were about twenty furnaces in Pennsylvania, more than in any other colony. One of these, the Cornwall Iron Furnace, is the only one of several hundreds of American charcoal-fueled blast furnaces to survive fully intact.

The Cornwall Furnace owed its existence to the renowned Cornwall Ore

The Cornwall Iron Furnace as it appeared ca. 1860.

Banks, located a few miles south of Lebanon. This was an extraordinarily rich deposit of magnetite iron ore that, until development of the Lake Superior ores, was the most important source in the United States. Peter Grubb of Cornwall, England, purchased 300 acres (121 ha) of the iron-rich land in 1732, and by 1742 his Cornwall Furnace was in full operation. Grubb died in 1745 and his sons inherited the estate. Ironmaster Robert Coleman purchased the entire site between 1785 and 1798. Cornwall flourished under the Coleman family's long stewardship, finally going out of blast in 1883.

The Cornwall Furnace had two distinctly different lives. When constructed in the mid-eighteenth century, it was entirely typical of American iron furnaces of that period, consisting of a squat stone stack 20 feet (6,096 mm) square at the base, 11 feet (3,352 mm) square at the top, and 30 feet (9,144 mm) high. A pair of wood-and-leather bellows, driven by an overshot waterwheel powered by a nearby stream, provided the cold-air blast. The resulting iron was cast into pigs or else directly into pots, firebacks, and other domestic articles. During the Revolutionary War, Cornwall cast munitions and salt pans (for making salt from sea water, to make up for the wartime embargo).

From 1856 to 1857, the furnace was entirely rebuilt. The ancillary buildings were rebuilt in stone, and the furnace was strengthened and enlarged to 28 feet (8,534 mm) square at the base and 21 feet (6,400 mm) square at the top, although its capacity was not increased. A vertical blast engine, consisting of two wooden blowing cylinders, or "tubs," and driven by a 20-horsepower (15-kW) steam engine, replaced the bellows. The single-cylinder engine with a 9-inch-by-26-inch (228-mm-by-660-mm) cylinder, together with a pair of plain cylindrical boilers set in the throat of the furnace and heated by the stack gases, were fabricated by the West Point Foundry, Cold Spring, New York.

This modernized furnace is the one we see today. Neither anthracite coal nor coke were ever introduced as the smelting fuel, although both had found limited use in other furnaces of the region. Likewise, the cold blast was retained, although many contemporary ironmasters contended (as was subsequently proved) that the hot blast increased production efficiency. After 1860, charcoal iron was principally used for the production of specialty steels for tool and allied industries, and for those parts of railroad rolling stock most subject to impact and stress reversal, such as locomotive and car wheels and axles.

Eventually, improved methods of steelmaking, especially the open-hearth furnace, resulted in the ready availability of steels equal to those produced from charcoal iron and led to the total demise of the old process. When the Cornwall Furnace went out of blast in 1883, it remained in the Coleman family, which continued to operate other, more modern furnaces in the area. The old site was preserved as a monument to earlier generations of Coleman ironmasters. In 1931 Margaret C. Buckingham, great-granddaughter of Robert Coleman, deeded the furnace and its ancillary structures to the Commonwealth of Pennsylvania.

Location/Access

The Cornwall Iron Furnace, administered by the Pennsylvania Historical & Museum Commission, is located about 28 miles (45 km) east of Harrisburg on Rexmont Road at Boyd Street, Box 241, Cornwall, PA 17016; phone (717) 272-9711. Facilities include the charcoal house, furnace complex, ironmaking exhibits, and a picnic area. Hours: Tuesday–Saturday, 9 A.M. to 5 P.M.; Sunday, noon to 5 P.M. Admission fee. The Cornwall iron mine, once the largest open-pit mine in the East, was abandoned (about 1973) and now is filled with water. Two nineteenth-century miners' villages are within a half mile (0.8 km) of the furnace.

FURTHER READING

Greville Bathe, *An Engineer's Miscellany* (Philadelphia: Patterson & White Co., 1938). (See Chapter 6, "The Old Cornwall Furnace," pp. 61–77.)

Arthur Cecil Bining, *Pennsylvania Iron Manufacture in the Eighteenth Century*, Publications of the Pennsylvania Historical Commission, Vol. 6 (Harrisburg, Pa.: Pennsylvania Historical Commission, 1938).

Paul F. Paskoff, *Industrial Evolution: Organization, Structure and Growth of the Pennsylvania Iron Industry 1750–1860*, Studies in Industry and Society, no. 3 (Baltimore: The Johns Hopkins University Press, 1983).

Ringwood Manor Iron Complex

Ringwood, New Jersey

Among the most colorful figures associated with the colonial iron industry was the German-born ironmaster Peter Hasenclever (1716–93), who established the first large-scale colonial ironworks at Ringwood in the Ramapo Mountains of northern New Jersey. Beginning in the 1740s, and continuing until the 1870s, Ringwood turned out tools for war and peacetime.

Hasenclever, with two partners, purchased the Ringwood Ironworks in 1764. Ringwood had been established in 1742 by the Ogden Family of Newark following the discovery of iron ore but then had been allowed to decay. Hasenclever rebuilt the ironworks, recruiting skilled German laborers to work it. By 1766, Ringwood consisted of one furnace, four forges burning eleven fires, and a stamping mill turning out about 20 to 25 tons (18 to 22 t) per week. Meanwhile, Hasenclever assembled one of the largest business empires in the American colonies, acquiring more than 50,000 acres (20,000 ha) of land in New Jersey, New York, and Nova Scotia, and building the Charlotteburg and Long Pond ironworks (New Jersey) and Cortlandt Ironworks (New York).

In 1767 the American Company (as the now-expanded investors' group was called) appointed Hasenclever manager of all of the company's properties. But it

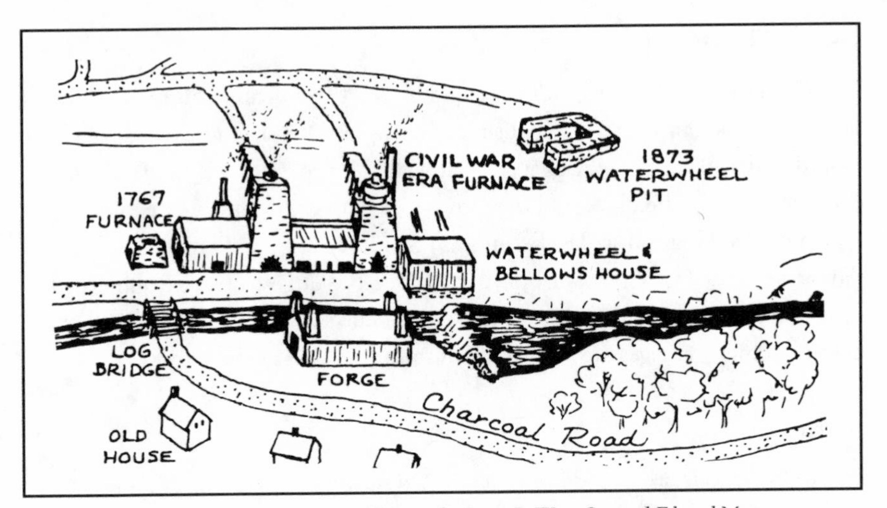

Ringwood Manor Ironworks, ca. 1870. *Drawing by Louis P. West, Sr., and Edward Morgan, March 1970.*

abruptly replaced him, leading to an investigation by the royal governor of New Jersey, William Franklin. An investigating committee inspected the Ringwood works in 1768, finding it in excellent order; but, though the Franklin Committee praised Hasenclever's resourcefulness and the efficiency of the works, the defeated ironmaster departed for London in 1769, never to return.

Robert Erskine arrived from England to manage the properties in 1771. Erskine was sympathetic to the American cause, and despite shortages of workers (and, inexplicably, despite its English ownership), devoted the Ringwood works wholly to making iron for American military use. His nephew, Ebenezer Erskine, succeeded him following his death in 1780, but by war's end, the once-thriving Ringwood works was once again idle.

In 1807 Martin J. Ryerson of Pompton, New Jersey, acquired the properties at Ringwood and Long Pond. Ryerson, an experienced ironmaster, resumed iron production at Ringwood, making round shot during the War of 1812. Ryerson demolished the original ironmaster's house and built a small Federal-style house that forms the west wing of the present Ringwood Manor. Hard times and poor business acumen eventually led Ryerson's sons, who inherited Ringwood, to sell it to New York industrialist and financier Peter Cooper in 1853. The property almost immediately was transferred to the Trenton Iron Company. Abram S. Hewitt (1822–1903), Trenton business manager and Cooper's son-in-law, moved there with his family in 1857.

Hewitt again had Ringwood thriving. During the Civil War, Ringwood turned out mortar carriages and other equipment, but following the panic of 1873, the ironworks languished and was never active again. The Hewitts, meanwhile, transformed Ringwood from a celebrated ironworks to a distinguished family seat. Abram Hewitt, the last ironmaster at Ringwood, died in 1903. His son, Erskine Hewitt, deeded the property to the State of New Jersey in 1936.

Location/Access

Ringwood State Park, on Sloatsburg Road 2½ miles (4 km) north of Ringwood, is administered by the New Jersey Division of Parks and Forestry, Box 1304, Ringwood, NJ 07456. Besides collections of furniture and decorative arts in the manor house, the Ringwood grounds are littered with artifacts of the iron complex, including the hammerhead and anvil of an early water-powered bloomery. The New Jersey Highlands Historical Society, based at Ringwood, maintains archives and a library related to the New Jersey iron industry during the period of 1740–1940. Hours: Wednesday–Sunday, 10 A.M. to 4 P.M. Admission fee. Phone (201) 962-7031.

FURTHER READING

James M. Ransom, *Vanishing Ironworks of the Ramapos: The Story of the Forges, Furnaces, and Mines of the New Jersey-New York Border Area* (New Brunswick, N.J.: Rutgers University Press, 1966).

Drake Oil Well

Titusville, Pennsylvania

When Edwin L. Drake (1819–80) drilled the first oil well in America in 1859, he initiated a new industry and technological changes more revolutionary than anyone could have predicted. A rich source of concentrated energy and abundant chemical compounds, petroleum would support sweeping changes in illumination, lubrication, power generation, transportation, and chemistry. "Few events have so transformed the face of civilization," reads the plaque marking the Drake Well as a Historic Mechanical Engineering Landmark.

Oil Creek in northwestern Pennsylvania was a busy lumber region in 1851 when Dr. Francis Beattie Brewer moved to the village of Titusville to join his father's lumber firm, Brewer, Watson & Company, and found an oil spring on company land. In 1853 he carried a sample of petroleum on a trip to Hanover, New Hampshire, where Dr. Dixi Crosby of the Dartmouth Medical School pronounced it valuable but of limited use since the oil could not be obtained in large quantities. A few weeks later, New York lawyer George H. Bissell saw the oil in Dr. Crosby's office and became interested in its potential for use as an illuminant. Bissell and his business partner, Jonathan G. Eveleth, purchased the land from Brewer, Watson & Company in November 1854 and organized the Pennsylvania Rock Oil Company of New York the following month. With the infusion of money from a group of New Haven capitalists, including banker James M. Townsend, the company was reorganized as the Pennsylvania Rock Oil Company of Connecticut in 1855 but made little progress owing to dissension among stockholders and nationwide financial panic. While matters were at a standstill, Bissell was attracted by an advertise-

How Oil Wells Were Drilled

One of the best descriptions of how oil wells were drilled in the nineteenth century was written by J. H. A. Bone:

> *The exact spot being determined, a huge derrick is erected immediately over it. This is a square frame of timbers, substantially bolted together, making an enclosure about forty feet [12 m] high, and about ten feet [3 m] at the base, tapering somewhat as it ascends. This is generally boarded up a portion of the distance to shelter the workmen. A grooved wheel or pulley hangs at the top, and a windlass and crank are at the base. A short distance from the derrick a small steam engine, either stationary or portable, is fixed, and covered with a rough board shanty; a pitman rod connects the crank of the engine with one end of a large wooden walking-beam, placed midway between the engine and the derrick, the beam being pivoted on its center about twelve feet [3.7 m] from the ground. The walking-beam is a rude imitation of that of a side-wheel steamer. A rope attached to its other end passes over the pulley at the top of the derrick, and terminates immediately over the intended hole. A cast-iron pipe, from 4½ to 5 inches [114 to 127 mm] in diameter, is driven into the surface ground, length following length until the rock is reached. In the older wells the ground was dug out to the rock, and a wooden tube put in it. The earth having been removed from the interior of the pipe the actual process of boring or drilling is commenced. Two huge links of iron, called "jars," are attached to the end of the rope. At the end of the lower link a long and heavy iron pipe is fixed, and in the end of this is screwed the drill, about three inches [76 mm] in diameter, and a yard [914 mm] long. When all is ready the drill and its heavy attachments are lowered into the tube and the engine is set in motion. With every elevation of the derrick end of the walking-beam, the drill strikes the rock, the heavy links of the "jars" sliding into each other and thus preventing a jerking strain on the rope. The rock, as it is pounded, mixes in a pulverized condition with the water constantly dripping into the hole, and assumes a pasty form. After a while the drill is hoisted out and a sand-pump dropped into the hole. The sand pump is a copper tube, about five feet [1.5 m] long, and a little smaller than the drill, having a valve in its bottom opening upwards and inwards. As the tube is dropped into the hole the pasty mass rushes into it through the valve and remains there. When this has been done several times the tube is hoisted out and emptied, the operation being repeated until the hole is clear, when the work of drilling recommences. It is evident that as the drill is not round at the point, but with a chisel-shaped edge, the hole would not be round unless some other means were adopted. This is partially accomplished by the borer, who sits on a seat about six or eight feet [1828 or 2400 mm] above the hole, and holds a handle fixed to the rope, giving the latter a half twist at every blow. By this means a nearer approach to a cylindrical hole is attained. But the hole must be as nearly round as possible, and therefore the tools are taken out, and a "rimmer," or "reamer," sent down, which cuts down the irregularities of the hole. . . .*

When the hole has been sunk to a sufficient depth and "strike oil," the next thing is to extract the oil from the well. If a flowing well has been struck, all the trouble on this head is saved, as the oil and gas rush out in a stream, sometimes with such violence that the men have to make their arrangements with considerable rapidity, or the precious fluid runs to waste. The first business is to tube the well. An iron pipe, with a valve at the bottom like the lower valve of a pump, is run down the entire depth of the well, the necessary length being obtained by screwing the sections firmly together. If the oil does not flow spontaneously, a pump-box, attached to a wooden rod, also made of sections screwed into each other is inserted in the tube, and the upper end of the rod attached to the "walking-beam." The well is now ready for pumping.

Source: Bone, J. H. A., *Petroleum and Petroleum Wells* (Philadelphia: J. B. Lippincott & Company, 1865), 262–64.

---

ment for "Kier's Petroleum, or Rock Oil" depicting a derrick over a salt well from which the oil, sold for medicinal use, was obtained as a by-product; it occurred to him that petroleum might be drilled in the same way.

On March 23, 1858, a group of New Haven capitalists led by Townsend organized a new company called the Seneca Oil Company, assumed the lease to the Titusville property, and tapped Edwin L. "Colonel" Drake, a onetime dry-goods clerk and conductor on the New York & New Haven Railroad, to serve as general agent. They sent Drake to Titusville in 1858 to drill for oil. After numerous delays, he began drilling near Oil Creek, driving sections of pipe 10 feet (3,048 mm) long through the shifting sand and clay. Progress was slow, often less than 3 feet (914 mm) a day. On Saturday afternoon, August 27, as Drake and his crew were about to quit work, the drill dropped into a crevice at a depth of 69 feet (21 m) and slipped down six inches (152 mm). They pulled out the tools and went

Drake Oil Well, ca. 1860.

home. Late the next day, driller William A. "Uncle Billy" Smith peered into the pipe and saw a dark brown liquid. They had struck oil!

While no one knows for certain the exact method Drake used, drillers of the period commonly erected a wooden derrick about 40 feet (12 m) high, tapering at the top, where a pulley was fastened. Over the pulley ran a cable to which was attached a weighted drill with a chisel point. The other end of the cable was attached to a walking beam, a heavy timber beam pivoted at the center. One end of the beam was driven up and down by a small steam engine; the other end of the beam was attached to the drilling rope. With the walking beam set in motion, the drill repeatedly struck the earth, boring a cylindrical hole. As drilling progressed, sections of metal pipe were pushed into the ground, one after the other, to maintain the hole through unstable strata until rock was reached (see sidebar).

By 1865, oil fever had taken hold of the formerly drowsy village of Titusville. With the drilling of other pioneer wells, the petroleum industry began to take shape, spawning allied industries: barrel factories, refineries, engine and boiler works, and oil-well supply companies. Boom towns mushroomed as thousands of wells were sunk. Typical was Pithole City, where between May and September 1865 the population jumped from a single farm family to fifteen thousand, with hotels, theaters, churches, and lecture halls. Not everyone struck it rich—oil country was hit hard by the Civil War and declining prices—but the foundation had been laid for a great industry.

Drake left Titusville in 1863, eventually losing everything by speculating in oil stocks. In 1873 the Pennsylvania Legislature granted Drake a pension of $1,500 annually in recognition of the important contribution he had made to the economic development of the Commonwealth. Pennsylvania's oil production peaked in 1891, with 31.4 million barrels (4.9 billion l).

Location/Access

A working replica of Drake's well and derrick marks the site of the first commercial oil well. The Drake Well Museum (administered by the Pennsylvania Historical & Museum Commission, RD #3, Box 7, Titusville, PA 16354; phone (814) 827-2797) off Route 8, Titusville, includes documents and artifacts related to the discovery. Hours: Monday-Saturday, 9 A.M. to 5 P.M.; Sunday, noon–5 P.M. Admission fee. In Woodlawn Cemetery, Route 8, Union City, is the Drake Memorial, a cut-stone monument with a bronze statue of *The Driller* erected in 1901. Drake's body was brought here from Bethlehem in 1902.

FURTHER READING

Paul H. Giddens, comp. and ed., *Pennsylvania Petroleum, 1750–1872: A Documentary History* (Titusville, Pa.: Pennsylvania Historical and Museum Commission, 1947).

Harold F. Williamson and Arnold R. Daum, *The American Petroleum Industry*, vol. 1, *The Age of Illumination, 1859–1899* (Evanston, Ill.: Northwestern University Press, 1959).

Pioneer Oil Refinery (California Star Oil Works Company)

Newhall, California

The pursuit of petroleum at Titusville, Pennsylvania, following Edwin L. "Colonel" Drake's celebrated strike in 1859 set off a period of frantic competition and laid the foundation for a great industry. The demand for oil—for lubrication and illumination —heightened just as the supply of whale oil decreased, spurring exploration as far away as California. There, four wells were drilled in Pico Canyon in the foothills of the Santa Susana Mountains in the early 1870s. To produce a salable product, the Los Angeles Petroleum Company built a small refinery near Lyons Station in 1873, but the venture, for financial and technical reasons, was unsuccessful and the company went out of business.

In 1876 the California Star Oil Works Company was organized. Meanwhile, better drilling methods imported from the oil fields of Pennsylvania had greatly increased production from the Pico wells, dictating a new refinery. In 1876 the California Star Oil Works Company established the first commercially successful refinery in the West at Andrews Station near the present town of Newhall.

J. A. Scott supervised construction of the refinery, which was located near the route of the newly constructed Southern Pacific Railroad between San Francisco and Los Angeles. Completed in August 1876, the refinery consisted of three stills. Two of them, of 15- and 20-barrel (1,789- and 2,385-l) capacity, had been moved from the earlier Lyons Station refinery; the third still, of 150-barrel

The Pioneer oil refinery.

(17,886 l) capacity, was new. A fourth still of 150-barrel capacity was added a short time later. Initially, a 1.5-mile (2.4-km) pipeline brought crude oil down from the well sites to the canyon's storage tanks; from there, it was hauled by wagon to the refinery. By 1879, a 7-mile- (11.3-km-) long, 2-inch- (50-mm-) diameter pipeline connected Pico Canyon with the refinery. Oil flowed to the stills by gravity from storage tanks set on a hillside.

The California Star refinery at Andrews Station turned out several products, including illuminating oil, lubricating oil, and small quantities of benzene. But kerosene was the main product, and two grades—"Lustre" and "Prime White"—found a profitable niche in the San Francisco market just as Eastern kerosene was rising sharply in price. Kerosene production at Andrews Station averaged 750 gallons (2,838 l) per day.

In 1880 California Star built a much larger refinery at Alameda in the San Francisco Bay area. The older refinery at Andrews Station was phased out by 1888, having produced 90,000 barrels (10,732 kl) during its twelve-year career. In the 1930s, the Standard Oil Company of California carefully restored the two largest stills and opened the historic site to the public. Up in Pico Canyon, meanwhile, Pico No. 4—the oldest working oil well in the West—still operates, producing one barrel (119 l) of crude a day.

Location/Access

The historic refinery is located off Pine Street, less than a quarter mile (0.4 km) south of San Fernando Road in Newhall, California.

FURTHER READING

Harold F. Williamson and Arnold R. Daum, *The American Petroleum Industry*, vol. 1, *The Age of Illumination*, 1859–1899 (Evanston, Ill.: Northwestern University Press, 1959).

Reed Gold Mine Ten-stamp Mill

Stanfield, North Carolina

The Carolina Piedmont, not the American West, was the site of the first U.S. gold rush. Tradition has it that in 1799, twelve-year-old Conrad Reed found a yellow lump about the size of a smoothing iron in Meadow Creek. It served as a doorstop until 1802, when John Reed, the boy's father, unaware of its value, sold it for $3.50. Reed soon found other large nuggets and formed a partnership to exploit his farm's wealth. The success of the Reed Gold Mine—by 1845, the mine's production was estimated at $1 million—started a statewide gold hunt. Between 1799 and 1930, more than $23 million worth of gold was mined in North Carolina.

The Reed Gold Mine ten-stamp mill, built in 1895 by the Mecklenburg Iron Works of Charlotte.

The Reed Gold Mine, designated a state historic site in 1971, exhibits a ten-stamp mill built by the Mecklenburg Iron Works of Charlotte in 1895. It is virtually identical to that which pounded the gold-bearing ore into dust—the first step in separating the gold from the "gange," or useless component of the ore—at the Reed Mine in the 1890s. The mill was moved to the Reed Gold Mine from the Coggins Mine in Montgomery County in 1974 and restored to operating condition.

The ten-stamp mill has its origin in sixteenth-century Germany. Resembling an oversize mortar and pestle, the stamp mill consists of a heavy oak frame containing two sets of five heavy iron stamps, each weighing 750 pounds (340 kg). Operating within vertical timber guides, the stamps are raised by a 5-inch-(127-mm-) diameter camshaft and dropped from a height of 5 to 7 inches (127 to 177 mm) into iron mortar boxes. (The distance varies according to the size of the crushed ore.)

The mortar boxes, each 14 inches wide by 60 inches long (355 mm by 1,524 mm), are constantly supplied with fresh ore and water; as the mill works, the crushed particles float out of the mortar box through a fine brass or tin-plate screen. The camshaft rotates at 35 rpm, dropping a stamp first in the left-hand mortar box, then in the right-hand box, alternating in an irregular but constant pattern from box to box. At 350 strokes per minute, the ten-stamp mill could crush 10 tons (9 t) of ore in twelve hours.

Originally steam-powered, the Reed Gold Mine ten-stamp mill is now powered by an electric motor connected by belt and pulley to the lineshaft. It is believed to be the only such mill to survive east of the Mississippi. The last underground excavation at the Reed Gold Mine was recorded in 1912.

Location/Access

The ten-stamp mill is located off State Route 200 at Reed Gold Mine State Historic Site, 9621 Reed Mine Road, Stanfield, NC 28163, which preserves the site of the first authenticated discovery of gold in the U.S. (1799). Several shafts have been retimbered and are open to visitors. Phone (704) 786-8337. Hours: April–October: Monday–Saturday, 9 A.M. to 5 P.M.; Sunday, 1 P.M. to 5 P.M.; November–March, Tuesday–Saturday, 10 A.M. to 4 P.M.; Sunday, 1 P.M. to 4 P.M. No admission charge; donations accepted.

FURTHER READING

C. G. Warnford Lock, *Practical Gold-Mining* (London and New York: E. & F. N. Spon, 1889).

Fairbanks Exploration Company Gold Dredge No. 8

near Fox, Alaska

Following the discovery of gold on Pedro Creek in 1902, dozens of mining camps were established to exploit the rich placer deposits of the Fairbanks district. Early prospectors commonly used drift mining (also called deep placer mining) methods, sinking a vertical shaft to bedrock; driving "drifts," or underground galleries, from the bottom of the shaft along the top of the bedrock; thawing the frozen ground with steam points; breaking up the material with picks and shovels; and hoisting it by steam engine to the surface for washing in sluices, searching the sand and gravel for "colors."

The supply of readily accessible gold had been exhausted when the Fairbanks Exploration Company, a subsidiary of the United States Smelting, Refining & Mining Company, began to prospect near Fairbanks in 1924. The firm acquired large blocks of already-worked claims on Cleary and Goldstream creeks and built

Fairbanks Exploration Company Gold Dredge No. 8.

the 90-mile- (144-km) long Davidson Ditch to deliver water from the Chatanika River. The water, together with completion of the Alaska Railroad in 1923, made large-scale gold production possible.

Gold-bearing gravel occurs in stream beds buried beneath up to 80 feet (24 m) of frozen muck—decayed moss and vegetable matter. The first step was to drill holes, spaced from 200 to 400 feet (60 to 120 m) apart, to determine what areas could be profitably dredged. Next, using jets of water from hydraulic "giants" (see "Joshua Hendy Iron Works," p. 161), the surface layer of muck gradually was removed, exposing successive layers to thaw. "Hydraulicking" usually was done two or more years in advance of dredging, depending on the depth of the overburden. Next, the gold-bearing gravel was thawed by driving pipes in triangular formation, spaced 16 to 32 feet (5 to 10 m) apart, down to bedrock and forcing cold water through them. After two to four months of thawing, dredges floating on ponds scooped up the gold-bearing gravels.

The Fairbanks Exploration Company pioneered the use of dredges in the Fairbanks district. One that combined excavating and concentrating plants was Dredge No. 8, with a steel hull 99 feet (30 m) long, 50 feet (15 m) wide, and a draft of 7 feet, 9 inches (2.4 m). The dredge was manufactured by the shipbuilding division of Bethlehem Steel and assembled on Goldstream Creek, 14 miles (22.5 km) north of Fairbanks, early in 1928.

In action, the dredge resembled an animated houseboat. Its endless chain of 68 steel buckets, each with a capacity of 6 cubic feet (0.17 m^3), dredged to an average depth of 19 feet (5.8 m), discharging the gravel into the upper end of an inclined, revolving screen that separated the gold-bearing gravel from the coarser rock. The relatively heavy gold fell through the screens and was trapped in the riffles, while waste gravel was sent by conveyor to the tailings pile behind the dredge. After retorting, assaying, and sampling, the gold was shipped to the United States Mint.

An adjustable spud at the stern held the dredge in position. With the spud as a center, electric winches and cables anchored to the shore swung the dredge in an arc of 60°. After each swing, the endless bucket was lowered and another cut taken. When the gravel was dredged to bedrock, the dredge was moved forward—making its own waterway as it chewed through pay dirt—and the process was repeated. The company supplied its own power at 4,000 volts (stepped up to 33,000 volts for transmission then down to 2,300 volts in the field) from a turbogenerator plant at Fairbanks.

The dredge crew consisted of a skilled winch operator, who maneuvered the dredge; two oilers; and one or two roustabouts, or general-purpose laborers. A dredgemaster supervised the operation from a control room on the upper deck. Work stopped in the fall, when the dredge pond froze solid, and resumed in the spring, when crews excavated ice from the pond to start the dredges again.

In 1931, working three shifts during the eight-month season (May to December), Dredge No. 8 advanced a distance of 5,057 feet (1,541 m), removing and processing an average of 5,040 cubic yards (3,853 m^3) per day. The dredge operated from 1928 until 1959.

Location/Access

Located at Mile 9 on the Old Steese Highway (P.O. Box 81941, Fairbanks, AK 99708), the dredge is open from late May to early September for tours. Admission fee includes gold panning; keep what you find. Nearby, a former miners' bunkhouse and dining hall now serves as a bar, restaurant, and hotel. Phone (907) 457-6058.

FURTHER READING

Guy R. Plumb, "Washing Gold at Fairbanks," *Mines Magazine* 22 (June 1932): 9–10

Hanford B Reactor

Richland, Washington

Built from 1943 to 1944 as part of the Manhattan Project to produce the atomic bomb, the Hanford B Reactor was the world's first plutonium production reactor.* Its history began with the selection of a vast area of flat, arid scrubland in south-central Washington State as the site of the "Hanford Engineer Works." The Hanford site was chosen because of its proximity to the Bonneville and Grand Coulee dams, reliable sources of plentiful electricity, and the Columbia River, which would provide abundant water for cooling. There were plenty of aggregates for the extensive concrete the project would require, and the remote site offered secrecy and security.

Hanford's primary purpose would be the production of plutonium by the irradiation of natural uranium in large water-cooled, graphite-moderated reactors. Nine production reactors eventually were built at Hanford, of which three (B, D, and F) were built during World War II. (Simultaneously, uranium-enrichment facilities were built at Oak Ridge, Tennessee, since no one was sure which material would produce the best weapon.) Prime contractor for the design, construction, and operation of the Hanford facilities was E. I. du Pont de Nemours & Company of Wilmington, Delaware. In 1946 the General Electric Company succeeded DuPont as operating contractor.

* The X-10 reactor at Oak Ridge, Tennessee, which served as a pilot plant for Hanford, operated as a plutonium production reactor between February 1944 and January 1945. When Hanford went on line, the X-10 was idled until after World War II, when it was used to produce a variety of isotopes. Thus, Hanford B was the first reactor built and operated solely for the production of plutonium.

The Hanford B complex during construction, 1944.

Ground for Camp Hanford, which would house as many as sixty thousand construction workers and their families at the peak of activity, was broken on April 6, 1943. Work on the first of the Hanford production piles began on June 7, 1943, and was completed in just more than fifteen months, as the United States worked to beat Germany in the race for the atomic bomb. On September 13, 1944, the day the construction team left Hanford, nuclear physicist Enrico Fermi (1901–54) inserted the first aluminum-canned uranium slug to begin loading the reactor. (Only twenty months earlier, Fermi had first demonstrated in Chicago that a nuclear chain reaction could be sustained and controlled.) Loading was completed on September 26, and the reactor went critical (i.e., achieved a sustained chain reaction) at a few minutes past midnight. By December 28, all three Hanford reactors had gone critical: plutonium production in quantity had begun.

The Hanford nuclear reservation included fuel element fabrication facilities, production reactors (or "piles"), and chemical separation facilities. Irradiated slugs ejected from a production reactor were temporarily stored in pools of water, then moved in shielded casks on railroad cars to one of three chemical separation buildings, where they were dissolved in hot nitric acid. Precipitation and centrifugal processes separated out radioactive wastes and small quantities of highly purified plutonium nitrate. The plutonium nitrate was shipped by Army convoy to Los Alamos, New Mexico, for final purification and assembly into the world's first atomic weapons.

The reactor required 2,000 tons (1,814 t) of machined graphite bars laid to a tolerance of ±0.005 inch (0.127 mm) and bored with 2,004 channels to hold the uranium slugs. Two channel-flow ribs allowed filtered water to circulate around the canned uranium slugs at a rate of 30,000 gallons (113,550 l) a minute. The graphite bars formed a 36-by-36-by-28-foot (10.97-by-10.97-by-8.53-m) block, surrounded by a 10-inch-thick (254-mm) envelope of cast iron and a 4-foot- (1,220-mm) thick biological shield of steel and concrete. Supplementing the reactor shields, the room walls were solid concrete, 3 to 5 feet (914 mm to 1,524 mm) thick. With the exception of two years when it was idled, B reactor operated continuously, making plutonium for military use until it was deactivated in 1968.

By 1960, eight production reactors, designed solely for defense production, were at work to meet the plutonium needs of the Cold War. All were shut down between 1964 and 1971. N reactor, built to produce both electricity and plutonium for weapons, went on line in 1963 but was shut down in 1987 for safety repairs following the Chernobyl accident. Today, the 560-square-mile (1,450 km^2) Hanford reservation is a ghost town of the Atomic Age, a temporary burial ground for radioactive waste. Plutonium reprocessing and finishing plants are the only operating facilities.

Location/Access

The control room of B reactor is open for tours by special arrangement. Information is available at the Hanford Science Center, phone (509) 376-0557.

FURTHER READING

Richard G. Hewlett and Oscar E. Anderson, Jr., *A History of the United States Atomic Energy Commission*, vol. 1, *The New World, 1939–1946* (University Park, Pa.: The Pennsylvania State University Press, 1962).

Vincent C. Jones, *Manhattan: The Army and the Atomic Bomb* (Washington, D.C.: Center of Military History, U.S. Army, 1985).

Richard Rhodes, *The Making of the Atomic Bomb* (New York: Simon and Schuster, 1986).

First Basic-Oxygen Steelmaking Vessel

Trenton, Michigan

In 1954 McLouth Steel Corporation introduced the basic-oxygen process of steelmaking to the United States. Borrowing from a process used on a smaller scale by two plants in Austria, McLouth purchased three top-blown oxygen converters and built a high-purity oxygen plant capable of producing 3.5 million cubic feet (99,110 m^3) of 99.5-percent pure oxygen per day. By December 1954, the Detroit-area steelmaker was producing 40-ton (36-t) heats of closely controlled steel in eighteen to twenty-three minutes' blowing time.

The basic-oxygen process quickly proved a practical and economical method for making high-quality steel. With a capital investment of just $7 million, McLouth's three vessels were soon each producing up to 66 tons (60 t) of steel every forty-five minutes—a tons-per-hour rate nearly three times the open-hearth record. By early 1955, McLouth had produced 600,000 ingot tons (544,200 t) using the basic-oxygen process. The steel had exceptional drawing quality as a result of the low nitrogen content (as low as 0.0013 percent, well below the usual open-hearth minimum) and the metallurgists' improved ability to control chemical composition—carbon, manganese, phosphorous, and sulfur—within very close limits.

Basic-oxygen steelmaking is not only faster than the open-hearth process, but because it is exothermic (i.e., the reaction produces heat), it requires no fuel. At McLouth, molten iron from the company's 1,350-ton (1,225-t) blast furnace was rolled into the melt shop in 200-ton (181-t) bottle cars and charged into one of three oxygen vessels, each approximately 13 feet (3,962 mm) in diameter and 22 feet (6,705 mm) high, lined with refractory brick. The vessels, fabricated by the Pennsylvania Engineering Corporation of New Castle, Pennsylvania, were suspended from trunnions at ground level to eliminate lifting and pouring the hot metal by crane. Next, scrap (representing approximately 20 percent of the charge) and flux (burned lime, limestone, mill scale) were added.

Finally, a water-cooled lance supported by a jib crane was lowered into the vessel, with its tip just above the molten metal, and oxygen was blown in at high pressure. The oxygen rapidly reacted with the iron to form iron oxide. The turbulence caused by the oxygen jet resulted in rapid mixing of the oxide with the rest of the metal, oxidizing out impurities (principally sulfur and phosphorous) in the

One of the three original vessels, no longer used, is displayed outside McLouth's Trenton, Michigan, plant.

form of slag. About twenty minutes later, the lance was withdrawn, the slag poured off, and the molten steel teemed (poured) into heavy cast-iron molds, where it solidified into ingots for later rolling into plates and sheets.

Operating experience quickly proved the worth of the top-blown oxygen converter. During the period 1949–52, the oxygen process accounted for the production of just 12,000 tons (10,880 t) of steel worldwide; by 1955, it accounted for almost 1.7 million tons (1.5 million t). Today, the basic-oxygen process pioneered in the United States by McLouth (and simultaneously in Canada by Dominion Foundries & Steel of Hamilton, Ontario) accounts for the major part of the world's steel production, and the proportion is still growing.

Location/Access

One of the three original oxygen vessels, no longer used, is on display at McLouth Steel Products Corporation's Trenton Plant, 1650 W. Jefferson, Trenton, MI 48183; phone (313) 285-1200.

FURTHER READING

C. R. Austin, "Oxygen Steel in the United States," *Iron and Steel Engineer* 33 (May 1956): 64–68.

Thomas Hruby, "Oxygen Steelmaking Arrives," *Steel* 136 (4 April 1955): 80–84.

William T. Lankford, et. al., *The Making, Shaping and Treating of Steel,* 10th ed. (Pittsburgh: Association of Iron and Steel Engineers, 1985).

"Big Brutus" Mine Shovel

near West Mineral, Kansas

Built by the Bucyrus-Erie Company in 1962, "Big Brutus" was the second-largest surface-mining shovel in the world. (Eclipsing it was a 115-cubic-yard-[88-m^3-] capacity Bucyrus-Erie shovel that began operations at the Sinclair Mine near Paradise, Kentucky, the same year.) The Pittsburg & Midway Coal Mining Company commissioned the shovel, which was built at Bucyrus-Erie's South Milwaukee factory and shipped to Hallowell, Kansas. There, its assembly occupied fifty-two Pittsburg & Midway employees for eleven months. When the shovel was finally put to work at P&M Mine 19 on June 6, 1963, mine superintendent Emil Sandeen dubbed the mechanical giant—160 feet (49 m) tall and weighing 11 million pounds (5 million kg)—"Big Brutus." The name stuck.

The model 1850B shovel, the only one of its kind ever built, was designed to remove the mine's thick overburden faster and more efficiently than the 950B unit it replaced. A ground-cable system supplied 7,200-volt power to the shovel from a new General Electric transformer station. With a bucket capacity of 90

Bucyrus-Erie's "Big Brutus" mine shovel takes a 90-cubic-yard (69 m^3) bite of Kansas coal.

cubic yards (69 m^3), the shovel averaged 5,000 cubic yards (3,823 m^3) per hour working in normal overburden, handling material averaging 60 feet (18 m) in thickness and doubling production from the mine's 18- and 14-inch (406- and 360-mm) seams. A bulldozer and two smaller shovels working in conjunction with Big Brutus handled the coal-loading job. Electric utilities consumed the bulk of the mine's output.

Big Brutus remained in operation for eleven years, stripping overburden to recover approximately 9 million tons (8.16 million t) of coal during that time. The shovel was retired in 1974. Too big to relocate, Pittsburg & Midway donated the shovel, a site, and funds for its restoration to Big Brutus, Inc., a nonprofit organization whose members have donated thousands of hours refurbishing it as a museum.

Location/Access

The Big Brutus mine shovel is located 6 miles (10 km) west of the junction of Kansas routes 7 and 102 in southeastern Kansas near West Mineral (P.O. Box 25, West Mineral, KS 66782); phone (316) 827-6177. Hours: daily, generally from 9 A.M. till sunset, but it varies with the season. Admission fee.

FURTHER READING

"Second Largest Shovel Ups Tonnage and Life at Mine 19," *Coal Age* 69 (February 1964): 96–100.

Manufacturing Facilities and Processes

INTRODUCTION by Euan F. C. Somerscales

Mechanical engineers are directly involved with all phases of manufacturing, so it is appropriate that a number of the landmarks celebrate their achievements in this field. This chapter includes landmarks that are examples of the basic processes of manufacturing, and it also includes landmarks associated with mass production, that quintessential American contribution to the organization of manufacturing.

Since prehistoric times most manufactured parts have started life by being cast from molten metal, but such materials generally do not have the hardness, strength, or ductility that is required for further processing or use. As an alternative to casting, parts can be formed by an equally ancient process, namely that of forging. In forging, a block of metal, usually heated to a high temperature, is squeezed or hammered into shape. The material that results from this process is particularly useful for applications involving suddenly applied forces, such as those experienced by aircraft landing gear at the moment of touchdown, because it has an inherent strength and ductility not found in the same metal when cast.

In the earliest times, the smith used a handheld hammer for forging, but this limited the amount of metal that could be handled. However, the invention in the fourteenth century of the tilt hammer, which was driven by a waterwheel, greatly increased the smith's capabilities. Nevertheless, by the early years of the nineteenth century, even the tilt hammer was inadequate. This led to the invention of the steam hammer, as described in this chapter. It was a story of simultaneous invention, possibly industrial espionage, and the transfer of technology across international borders.

Modern developments in forging avoid the impulsive blow of the steam hammer by employing presses, which apply the load in a controlled manner, often under the supervision of a computer. The beneficial results on the metal are then much more evenly distributed throughout the material. What are probably the two largest presses in existence can exert forces up to 50,000 tons (45,000 t), and these have been designated as landmarks in recognition of their capacity. As so frequently is the case today, the incentive for producing such massive pieces of machinery has come from the requirements of modern aircraft.

Not all metal parts originate from a casting or a forging. Nowadays, many articles are first formed from finely divided (150 μm to 1 μm diameter, or 0.006 inch to 0.00004 inch diameter) metal powder by compression and by heating to a sufficiently high temperature that the metal particles are fused, or sintered, into one solid mass. Powder metallurgy, as it is now called, entered into the mainstream of manufacturing in 1909, when W. D. Coolidge (1873–1975) of the General Electric Company applied it to the manufacture of tungsten lamp filaments. For the modern engineer, powder metallurgy is of interest not so much because of its ability to form refractory metals, such as platinum and tungsten, but because it allows the rapid, low-cost manufacture of articles in a sequence of fully automated and continuous processes of compacting and sintering. Isostatic compression was introduced in 1930 to eliminate internal stresses in the formed part that led to cracking. The Battelle Memorial Institute extended the concept by devising hot isostatic pressing (HIP) between 1959 and 1964, and this development was recognized by the designation of the first HIP vessel as a Historic Mechanical Engineering Landmark.

The permanent joining of two pieces of the same metal to form one homogeneous piece has always been feasible by forge welding, where the two parts are softened by heating and then hammered together. The process, which is expensive and time consuming, has been replaced by electric arc welding. This dates from about 1900, but, as might be expected, it was only gradually introduced into everyday engineering practice. Probably its most important test came when it was applied in 1930 to the steam drums of high pressure boilers. This accomplishment is recognized by the forty-second landmark, which is a sample welded drum that successfully withstood the application of a pressure six times higher than its calculated safe working pressure. The key was automation and inspection; the objective was to eliminate weld variability by using automatic welding machines, followed by X-ray examination of the welds to detect flaws that might weaken the weld.

Cast and forged parts generally do not have a smooth enough finish for many applications, so the metal has to be formed to its final dimensions by cutting with a sharp tool. The development of such machining processes represents possibly the most important forward step in mechanical engineering. Previously precision and accuracy were limited to parts that could be finished by hand. After the introduction of machine tools, large parts for steam engines, for example, could be machined accurately and to close tolerances, thus allowing machines with a predictable performance, such as the steam engine, to be manufactured.

The earliest machine tools have not survived. In the heart of rural Vermont, however, a truly outstanding collection of historically significant examples has been assembled at the American Precision Museum in Windsor. By a stroke of imaginative genius, this museum, the first of the Mechanical Engineering Heritage Collections, has been established in the former shops of Robbins & Lawrence, one of the "shrines" of early American manufacturing technology.

It was in New England factories, like that of Robbins & Lawrence, that the American Industrial Revolution had its beginnings, and probably the most important of these was the U.S. Government Armory in Springfield, Massachusetts. Here, between 1794, the date of its founding, and about 1850, two of the fundamental techniques of what we would now call mass production were developed. Historians of technology normally refer to the "American system of manufacturing" when discussing the innovations introduced at Springfield. This had two significant characteristics: interchangeability of the parts produced and the use of machine tools, as opposed to hand methods. Interchangeability meant that any part of a rifle or musket produced at the Springfield Armory could be replaced by any one of the corresponding parts produced in the same armory, without any adjustment or machining being necessary.

The manufacture of rifles and muskets, even in very large quantities, for a single customer does not represent what we would today call mass production. Mass production involves a large number of customers as well as production in great quantities. It was the textile industry that really started mass production. The place of textile machinery in the history of mechanical engineering is as important as the often noted role of the steam engine. This has been recognized by the landmark designation of the Watkins Woolen Mill in Lawson, Missouri, and indirectly, in the Slater Mill, which contains a remarkable collection of operating textile machinery.

The application of automatic machinery was extended from the textile industry to many other areas of manufacturing. Striking examples of this are seen in three of the landmarks described in this chapter; namely, the Owens-Corning bottle machine of 1903, the Corning ribbon machine of 1926, which produced electric light bulbs, and the automatic plant for automobile underframe manufacture that was designed and built by the A. O. Smith Company of Milwaukee, Wisconsin, which went into operation in 1921. The last of these probably represents one of the first applications of robots as a replacement for human workers in automobile manufacture.

Mass production, as we see it, depends on interchangeability, machine tools, and, now, robots, but organization is also important. Organization in the context of manufacturing means the factory system. This combines the workers into disciplined groups, working regular hours. It also places large amounts of machine power at the workers' hands, thereby increasing worker productivity. The New England armories and textile mills such as the Watkins Woolen Mill are early examples of factories. However, a much earlier example probably is the shipyard, as exemplified by the Portsmouth-Kittery Naval Shipyard, dating from 1774. At the other end of the time scale we have Joshua Hendy's Ironworks, which is representative of the many late-nineteenth- and early twentieth-century general engineering workshops that have been an important factor in regional and national development.

In reviewing manufacturing, we obtain probably the clearest sense of the change that has occurred in the short history of mechanical engineering, which dates formally from about 1750, although humans have practiced engineering in various guises far longer. The earliest engineers were able to form metal by casting and forging, but when it came to cutting metal, their machine tools were of a crudity that is difficult to comprehend today. Machine tools, combined with factory organization and the development of interchangeable manufacture, has resulted in our being able to produce goods in abundance and of a quality that would be considered luxurious in an earlier age. The mechanical engineer has played the most important part in all this.

Portsmouth-Kittery Naval Shipbuilding Activity

Portsmouth, New Hampshire, and Kittery, Maine

Portsmouth, named after Portsmouth, England, is New Hampshire's oldest settlement and only seaport. There, the mouth of the Piscataqua River forms a good and deep harbor whose singular advantages were recognized and exploited as early as the seventeenth century. In 1603 Martin Pring, the first European to explore the river, described it as "a noble sheet of water, and of great depth, with beautiful islands and heavy forests along its banks." The British government commissioned a survey of the harbor at Portsmouth, and by 1650, timber for masts was being selected, marked, and harvested for use by the Royal Navy. Naval shipbuilding on the Piscataqua began in 1690 with construction of the first of three frigates for the Royal Navy, the 54-gun *Falkland.* The 32-gun *Bedford* followed in 1696; and the 60-gun *America,* in 1749.

In 1774 Fort William and Mary, which had commanded the harbor entrance since 1690, was seized by local colonists. With the Piscataqua free of British forces, shipbuilding operations to support the colonial revolt got under way on Langdon's (now Badger's) Island, marking the origin of U.S. naval shipbuilding activity at Portsmouth-Kittery. Between 1775 and 1800, seven vessels were built here for the Colonial Navy. The first of these was the 32-gun frigate *Raleigh,* launched from Portsmouth on May 21, 1776, six weeks before the Declaration of Independence. Captain John Paul Jones was appointed to command the next ship built here, *Ranger,* a swift sloop of war of 18 guns and a crew of 150, launched in May 1777.

In 1800 the Secretary of the Navy recommended the purchase of the 58-acre (23-ha) Dennett's Island for a new navy yard. The island was gradually cleared, and a blacksmith shop, saw pits, shiphouse, and shed for timber storage were built. In March 1814 the keel of the 74-gun ship of the line *Washington* was laid; the vessel was commissioned on August 26, 1815, following the declaration of peace with England. In 1866 the government purchased the adjacent Seavey's Island, containing 105 acres (42 ha). The two islands gradually were joined by accretions and filling. With time, Portsmouth became a fully integrated shipbuilding operation, with its own foundry, forge, and blacksmith shops; carpenter, tinsmith, and coppersmith shops; rope walk; mast shop; rigging and sail lofts; and floating dry dock. The shipyard even grew some of its own food and kept a small herd of cattle.

Of the thirty-three buildings constructed before 1900 that are still extant, about half predate the Civil War. The oldest is the "Mast and Boat Shop, Rigger and Sail Loft" of 1837. Of granite-and-timber construction, it originally was an open building; a canal ran through the center, into which logs could be floated, then winched up for conversion into masts and spars. Two floors were added in later years, and the building is still used as the riggers' shop.

Portsmouth built and refit dozens of ships for the navy in the nineteenth

USS *Tennessee* in dry dock, Portsmouth Naval Shipyard, 1913. *Official photograph, U.S. Navy.*

century, gradually shifting from sail to steam. The first steam-powered vessel built here was the side-wheel frigate *Saranac*, launched in 1848. The years 1862–63 saw the Portsmouth Navy Yard working to capacity. According to one historian, the yard then employed more than two thousand men, who built seven ships or gunboats and repaired and refit three others.

Portsmouth became the first naval shipyard to build a submarine, the L-8, laying its keel in November 1914. Today, the shipyard is a center for the design, construction, and repair of submarines. Since 1963, its official name has been Portsmouth Naval Shipyard, even though the islands it occupies are part of Kittery, Maine.

Location/Access

The Portsmouth Naval Shipyard Museum is open to the public by appointment. It is located at Portsmouth Naval Shipyard, Code 100H, Portsmouth, NH 03804-5000; phone (207) 438-3550. Nearby, the Kittery Historical and Naval Museum interprets Kittery's and the nation's naval shipbuilding history, as well as local history. It is on Rodgers Road, near routes 1 and 236; phone (207) 439-3080.

FURTHER READING

Walter E. H. Fentress, Centennial History of the United States Navy Yard at Portsmouth, New Hampshire (Portsmouth: O. M. Knight, 1876).

Springfield Armory

Springfield, Massachusetts

Established in 1794, the Springfield Armory was the birthplace of the small-arms industry in the United States and an early example of large-scale manufacture. Here, in the first decades of the nineteenth century, arms-making was transformed from a craft to an industry, muskets from a shop to a factory product. Ordnance from the Springfield Armory has figured in every American war; its name became synonymous with the world's finest military arms. Today, the Springfield Armory National Historic Site houses some twenty thousand rifles—including the first of each weapon manufactured here—making it one of the largest collections of small arms in the world.

Spurred by the U.S. Army's demand for small arms of consistently high quality, a system of national armories was authorized by Congress in 1794. Springfield, on the east bank of the Connecticut River, had served as a federal storage and supply depot during the Revolutionary War. President George Washington selected this hilltop town as the site of the first United States arsenal for a variety of reasons: its location far enough inland to prevent attack by sea; abundant water for power and transportation; the presence of skilled gunsmiths and other artisans; good roads; and its proximity to the northern department of the Continental Army. Armory Square became the site of a cluster of workshops and storage buildings (called the Hill Shops), while operations needing water power (the Mill Shops) occupied scattered sites on the Mill River, a mile (1.6 km) to the south.

"Lock, stock, and barrel" define the principal elements—and thus manufacturing functions—of shoulder arms. The lock, or firing mechanism, requires many small, precision-made metal components strong enough to withstand powerful mechanical stresses. The stock, traditionally made of hardwood, requires the cutting of irregular, curved surfaces for the external form and to accommodate the lock and barrel. The barrel, made of iron or steel capable of withstanding the explosive force and heat of fired ammunition, requires precision shaping, rolling, and welding, then boring or drilling, rifling, and finishing. The army's demand for reliable military weapons of high quality and with interchangeable parts led to Springfield's pioneering advances in large-scale manufacture, including mechanization, milling, and quality control.

In the 1820s, Thomas Blanchard (1788–1864) designed stock-making machinery—a battery of fourteen special-purpose, water-powered woodworking machines that completely mechanized the process—unlike anything seen in America up to that time. Blanchard's biggest contribution was a copying lathe for turning gun stocks or any other irregularly shaped objects. The lathe was widely applied to the manufacture of shoe lasts, handles for axes and agricultural implements, and carriage parts. Together with Blanchard's other special-purpose

Blanchard lathe at the Springfield Armory in the 1920s. *Courtesy National Museum of American History.*

machines, it eliminated skilled labor and set American manufacturing on the road toward mechanized production.

In the 1840s, chief mechanic Cyrus Buckland (1799–1891) designed the arsenal's second generation of gun-stocking machinery, which refined the tasks of the earlier machinery and took full advantage of steam power. Together with mechanic Thomas Warner, Buckland developed new milling and cutting tools, and filing jigs for forming metal parts. By the eve of the Civil War, the Springfield Armory had achieved the U.S. War Department's longtime goal: weapons constructed with uniform parts that could be easily repaired in the field.

During its long career, the Springfield Armory manufactured five major types of shoulder arms, beginning with the French single-shot, smoothbore flintlock muskets produced until 1842. Smoothbore muskets, with locks adapted to percussion ignition of ammunition, were produced from 1844 until 1865; rifled muskets, from 1857 to 1865. Breechloading rifles, the third major weapon type, were made here from 1865 until 1893. Beginning in 1893, the armory concentrated on bolt-action, or repeating, rifles, standard infantry issue until 1931. These included the Krag-Jorgensen rifle, based on a Danish design, and the Model 1903 Springfield rifle, unsurpassed among military small arms during World War I, when a quarter million Springfields were produced. The fifth type of rifle was the semi-automatic M-1, made here from 1937 to 1957, and the fully automatic M-14, made from 1959 to 1963.

Inventor John C. Garand's M-1 did away with the time-consuming, manual operations of unlocking, withdrawing, closing, and locking the bolt between each shot. "So far as fire power is concerned," Garand said in 1943, "one man with this weapon is equivalent to five with the conventional type rifle."

As a result of Garand's success with the M-1, the armory's most modern mass-production machinery was installed, setting the standard for ordnance manufacture. More than 4.5 million M-1 rifles were produced; the weapon served around the world during World War II and the Korean conflict.

By World War II, when the Springfield Armory delivered more than 3.1 million rifles, public arsenals already were a vanishing breed; private industry now made most American military products. Springfield's focus shifted to research and development, but protracted bureaucratic battles led to the Defense Department's controversial decision to close the armory in 1968.

Location/Access

The Springfield Armory National Historic Site, at One Armory Square (off Federal Street), Springfield, MA 01105, encompasses approximately 55 acres (22 ha) and several buildings of the original armory complex. The Main Arsenal building (1840s) houses the world's largest collections of small arms, as well as the original Blanchard lathe and many of John Garand's prototypes for the M-1. Hours: Wednesday–Sunday, 10 A.M. to 5 P.M. Phone (413) 734-8551. Other former armory buildings today house Springfield Technical Community College.

FURTHER READING

David A. Hounshell, *From the American System to Mass Production, 1800–1932* (Baltimore and London: The Johns Hopkins University Press, 1984).

IA, *The Journal of the Society for Industrial Archeology*, Special Theme Issue: Springfield Armory, vol. 14, no. 1 (1988).

Jackson Ferry Shot Tower

near Austinville, Virginia

Erected between 1808 and 1812, the Jackson Ferry Shot Tower is one of six nineteenth-century lead shot towers that survive in the United States. (The others are at Baltimore, Boston, Dubuque, Philadelphia, and Spring Green, Wisconsin.) Its history is inextricably tied with that of the lead mines along the New River in southwestern Virginia, first developed by Colonel John Chiswell in the mid-eighteenth century. Numerous small industrial establishments sprung up around them, and the mines proved important during the Revolutionary War, when Fort Chiswell was garrisoned for their protection. Following Chiswell's

The Jackson Ferry Shot Tower is one of only six extant lead shot towers in the United States.

death in 1776, Chiswell's heirs sided with the Crown. The Commonwealth of Virginia confiscated the mines, and the state assembly empowered the governor to engage "slaves, servants or others" to work the mines "to greater advantage." Later, the mines passed through several owners until Thomas Jackson, an English immigrant who had worked at the mines as a smith, bought them at public auction in 1806. Jackson, who already owned land in the vicinity of the New River and had established a ferry crossing (whence the tower's name), used slave labor to build the tower; a small cemetery nearby containing the bodies of seven blacks is testament to the fatalities that occurred during construction.

The drop method of producing shot—"solid throughout, perfectly globular in form, and without . . . dimples, scratches and imperfections"—was patented in England in 1782 by William Watts of Bristol. His method was simple: molten lead was poured through a sieve at the top of a tower (the height of the fall varying in size according to the size of shot desired), producing droplets. The droplets assumed a spherical shape and solidified as they fell through the air to the bottom of the tower, where the shot was quenched in a pool of water. The new technology spread quickly across Europe, then to the United States following the ban of imported shot in 1808.

The Jackson Ferry Shot Tower consists of a 75-foot (23-m) tower of local limestone, with a 75-foot (23-m) vertical shaft extending below ground. The tower is 20 feet (6,096 mm) square at the base, tapering to 15 feet (4,572 mm) square at the top, with walls 2½ feet (762 mm) thick. A single door at ground level permitted access to a winding wooden stairway leading to the top of the tower. There, another door opened onto a small roofed porch, to which ladles of molten lead were hoisted by rope. Arsenic was usually added to the lead to increase surface tension and improve sphericity.

A tunnel connected the bottom of the 150-foot (46-m) shaft with the riverbank, providing for easy removal of the cooled shot and delivery of fresh water from the New River. After it was removed from the cooling vessel, the shot was bagged for transport by wagon to commercial markets in the South, where it was sold for use in fowling pieces and other shotguns.

Jackson produced lead shot until his death in 1824. His nephew, Robert Raper, continued operations until 1839, when production ceased. The tower has been restored to its historic appearance by the Commonwealth of Virginia Division of State Parks.

Location/Access

The shot tower—part of Shot Tower & New River Trail State Park, Route 1, Box 81X, Austinville, VA 24312—is located on U.S. 52 at the New River, 2 miles (3 km) north of the Poplar Camp exit of Interstate 77. Hiking trails and picnic facilities. Phone (540) 699-6778.

FURTHER READING

Walter Minchinton, "The Shot Tower," *American Heritage of Invention & Technology*, Spring/Summer 1990, 52.

Wilkinson Mill

Pawtucket, Rhode Island

The Wilkinson Mill on the west bank of the Blackstone River in Pawtucket was built between 1810 and 1811 by machinist Oziel Wilkinson. The three-story, rubble-stone mill is significant for its association with his son, David Wilkinson (1771–1852), who played a critical role in the history of textile technology, in steam power generation, and in the development of the machine tool industry.

Oziel Wilkinson, a skilled blacksmith, migrated with his family to Pawtucket from nearby Smithfield, Rhode Island, about 1783. He opened a shop powered by water from the Pawtucket Falls and, aided by his three sons, manufactured farm tools, domestic utensils, and cut nails. Later, he forged anchors for the local shipbuilding trade. The emerging textile industry needed the skills of talented mechanics such as the Wilkinsons, and when Samuel Slater, a master mechanic who had emigrated from Nottingham, built the first successful water-powered textile machinery in North America, he turned to the Wilkinsons for help. From 1788 to 1789, David Wilkinson furnished the iron forgings and castings for Slater's first carding and spinning machines; Slater's success led to construction of the Slater Mill in 1793 and the introduction of mass production technology to a formerly hand-powered home industry.

With their creative ironwork, the Wilkinsons contributed to Pawtucket's reputation as the most important industrial village in America during the period 1790 to 1820. In 1791 Oziel built a reverberatory air furnace, with which he cast iron gudgeons (journals) for Slater's waterwheel, believed to be the first ever made in this country. In 1793 David Wilkinson produced one of the first steam engines for propelling a boat and successfully tested it on the Providence River—fourteen years before Robert Fulton's famous demonstration of the *Clermont*.

From 1794 to 1795, Oziel Wilkinson constructed a water-powered rolling and slitting mill, just south of the Slater Mill, to produce iron plate and nail rods. He began making screws for clothiers' and oil presses, sparking David Wilkinson's interest in cutting and finishing screw threads. By 1796, David Wilkinson had devised a machine for cutting screw threads that incorporated a slide-rest. This machine, which he patented in 1798, featured a heavy carriage supported on three rollers. Later, in 1846, Wilkinson wrote that his screw machine

> . . . was on the principle of the gauge or sliding lathe now in every workshop almost throughout the world; the perfection of which consists in that most faithful agent *gravity*, making the joint, and that almighty perfect number *three*, which is harmony itself. I was young when I learnt that principle. I had never seen my grandmother putting a chip under a three legged milking stool; but she always had to put a chip under a four legged table, to keep it steady. I cut screws of all dimensions by this machine, and did them perfectly.*

Wilkinson's industrial lathe marked a major advance and earned him distinction as the founder of the American machine tool industry.

From 1810 to 1811, Oziel built the Wilkinson Mill, originally designed for cotton spinning, just across from his rolling and slitting mill. It was powered by water and (for backup) steam. David, who was responsible for running the mill, operated a machine shop on the first floor where he manufactured and repaired textile machinery. Working with his brother Daniel, David Wilkinson achieved a national reputation as a master builder of textile machinery; Wilkinson machines

---

* From "David Wilkinson's Reminiscences," *Transactions of the Rhode Island Society for the Encouragement of Domestic Industry in the Year 1861* (Providence, 1862), 100–11.

A portion of the machine shop at Wilkinson Mill.

were sold throughout New England and as far south as Georgia. Wilkinson also perfected a mill to bore cannons and, about 1817, built the first successful power loom in Rhode Island.

In 1829, following a serious depression in the textile industry, Wilkinson lost his business and left Pawtucket. He worked at a succession of jobs in New York, New Jersey, Ohio, and Canada. In 1848 he petitioned Congress for remuneration for his invention of the slide-rest, which by then had been widely adopted; Congress voted to pay Wilkinson $10,000 "for benefits accruing to the public service for the use of the principle of the gauge and sliding lathe, of which he was the inventor." Following Wilkinson's departure from Pawtucket, the Wilkinson Mill was used in the manufacture of woolen goods and cotton braid but was never again at the center of textile innovation as it had been during David Wilkinson's tenure.

Today, the Wilkinson Mill houses the offices of the Slater Mill Historic Site, a museum devoted to American industrial and social history, which includes the Old Slater Mill (1793) and the Sylvanus Brown House (1758), the home of a skilled artisan. The Wilkinson Mill contains a working machine shop on the first floor featuring a nationally significant collection of nineteenth-century machine tools—a tribute to the inventive genius of David Wilkinson. Archeological investigations in the early 1970s revealed the existence of the old breast wheel pit in the basement, and bits and pieces of David Wilkinson's second breast wheel installation (ca. 1826) helped guide the authentic reconstruction of the entire water power system. Today the Wilkinson Mill again operates with power supplied by the Blackstone River.

Location/Access

The Wilkinson Mill, Roosevelt Avenue at Main Street (P.O. Box 696, Pawtucket, RI 02862), demonstrates the operation of a nineteenth-century machine shop run by water power. Phone (401) 725-8638. Hours: Labor Day through November 1, and March through May: Saturday and Sunday, 1 to 5 P.M.; June through Labor Day: Tuesday–Saturday, 10 A.M. to 5 P.M., and Sunday, 1 to 5 P.M. Admission fee.

FURTHER READING

Joseph Wickham Roe, *English and American Tool Builders* (New York: McGraw-Hill Book Company, Inc., 1926).

Robert S. Woodbury, *History of the Lathe to 1850*, Society for the History of Technology Monograph Series, no. 1 (Boston: Nimrod Press, Inc., 1961).

American Precision Museum/ Robbins & Lawrence Armory and Machine Shop

Windsor, Vermont

In fulfilling a contract for the manufacture of 25,000 rifles for the U.S. Government, the firm of Robbins & Lawrence was the first to achieve the interchangeability of machine-made parts on a practical basis, laying the groundwork for mass production. The new manufacturing technology—so novel that British observers called it the "American system"—later spread to the production of a new consumer durable, the sewing machine, and eventually to such products as the bicycle, typewriter, and automobile.

In 1845 the U.S. War Department awarded a contract for 10,000 Model 1841 army rifles to Samuel E. Robbins, Nicanor Kendall, and Richard S. Lawrence. Kendall and Lawrence were both experienced custom gunsmiths, and Robbins was a wealthy retiree from the lumber business. They built a three-story brick armory and machine shop in 1846 and, though their contract called for delivery of the rifles over a period of five years at the rate of 2,000 per year, delivered all 10,000 rifles by 1847. Most important, the quality of the work surpassed that of any other armory, including the national armories.

After completing the first contract, Robbins and Lawrence bought out Kendall. In 1848 the new partnership received a contract for 15,000 more of the same

Robbins & Lawrence Armory and Machine Shop. *Courtesy American Precision Museum.*

rifles, to be delivered at the rate of 3,000 a year over five years. By the early 1850s, Robbins & Lawrence was among the foremost makers of arms and arms-making machinery in the world. The firm's fame grew when it displayed its interchangeable firearms at the London Crystal Palace Exhibition in 1851.

From Robbins & Lawrence, the British ordered 152 rifle-making machines for use at the Enfield Armoury. From the Ames Manufacturing Company in Chicopee, Massachusetts, they ordered twenty-three woodworking machines for stock making; these were improved models of machines first developed at the Springfield Armory by Thomas Blanchard in the 1820s (see "Springfield Armory," p. 146). Ames also furnished numerous small tools such as gauges, jigs, and patterns. These, the committee's major purchases, resulted in the export of American precision, high-production machinery abroad. At home, meanwhile, Robbins & Lawrence mechanics gradually carried their precision manufacturing know-how to the sewing-machine industry when firearms orders collapsed following the close of the Civil War.

The American Precision Museum today occupies the former Robbins & Lawrence armory and machine shop. It contains the largest collection of historically significant machine tools in the nation. Artifacts range from small, handmade machine tools—a lathe, planer, drill press, and gear-cutting machine typical of those used by small mechanics' shops in the early nineteenth century—to the earliest turret lathe known to have been made in the Robbins & Lawrence shop (1861) and the earliest known American-made vernier caliper (1846). The latter tool made it possible to precisely control the dimensions of machined parts. In addition to its large collection of machine tools of all types, the museum also contains various products of machine tools, from dynamos to typewriters.

American Precision Museum.

In 1987 the American Society of Mechanical Engineers recognized the museum's artifacts as a Historic Mechanical Engineering Heritage Collection, at the same time designating the former armory and machine shop as a Mechanical Engineering Heritage Site.

Location/Access

The American Precision Museum is located on the Connecticut River at 196 South Main Street, Windsor, VT 05089; phone (802) 674-5781. Hours: May 30–November 1: Monday–Friday, 9 A.M. to 5 P.M.; Saturday, Sunday, and holidays, 10 A.M. to 4 P.M. Admission fee. The Windsor-Cornish Bridge (1866), the longest covered bridge in the United States, is nearby.

FURTHER READING

E. A. Battison, "The Evolution of Interchangeable Manufacture and Its Dissemination," ASME Paper No. 87-WA/HH-4 (1987).

David A. Hounshell, From the American System to Mass Production: The Development of Manufacturing Technology in the United States (Baltimore: The Johns Hopkins University Press, 1984).

Westmoreland Malleable Iron Works

Westmoreland, New York

In 1826 Seth Boyden (1788–1870) produced the first "blackheart" malleable iron—principally iron and carbon, rendered tough and ductile by a controlled heat-conversion process—in Newark, New Jersey. In the next two decades, malleable iron foundries sprang up in New England and New York for the production of saddlery hardware, carriage and wagon parts, and agricultural implements. The alloy's unique metallurgical structure gave it great strength, remarkable resistance to impact and corrosion, and easy machinability.

The Westmoreland Malleable Iron Works was the oldest malleable iron company in continuous operation in the United States. In 1833 Calvin Adams founded the Oak Hill Malleable Iron Company in New York's Greene County. Within a few years, William Thorpe of Albany joined him as partner and by 1839 took full control of the company. Following Thorpe's death in 1847, the business passed to William Smith and Abel Buell, who brought Erastus W. Clark into the company. In 1850 Buell and Clark moved part of the foundry to Westmoreland. Buell later left the business, which has remained in the Clark family ever since.

The basic process of producing malleable-iron castings has changed very little in the past century, although the techniques of manufacture have advanced tremendously. Small castings of brittle white iron are made malleable

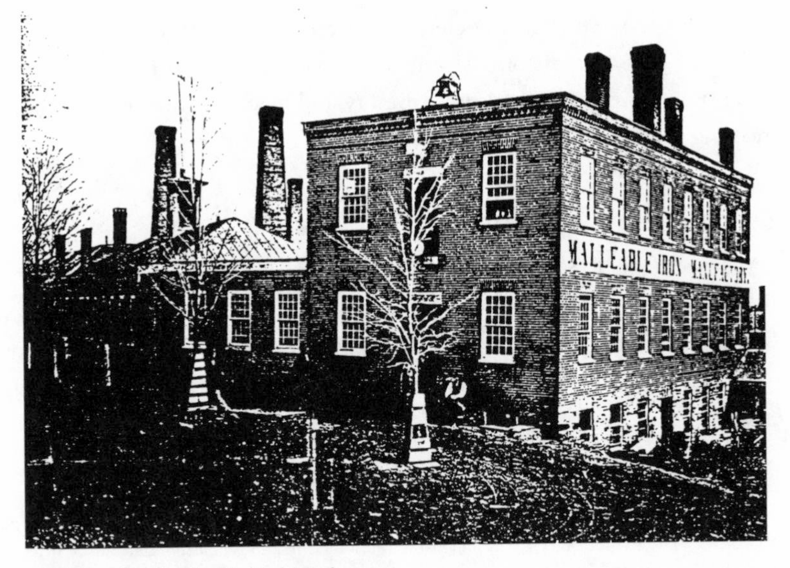

Westmoreland Malleable Iron Works.

by annealing them—applying red heat (1,600°F/870°C)—for several days, then allowing them to slowly cool. The iron is melted by an electric induction furnace instead of the coal-fired cupola of Boyden's day. Charging, melting, casting, and cleaning operations have all been mechanized, speeding up the founding process and helping to assure uniformity in castings. Meanwhile, cores, which were once hand-rammed by gangs of boys seated at long tables, are now formed by core-blowing machines, which force the core mixture in a stream of compressed air into vented core boxes that allow the air to escape but hold the sand in a firm, well-packed mass.

In the early years, the industry was shrouded in secrecy as founders zealously guarded their techniques lest a competitor discover them. Annealing was a matter of guesswork; furnace temperature was judged by eye. Cooperative research followed the formation of the American Malleable Castings Association (succeeded by the Malleable Founders Society) in 1897. (E. C. Metcalf of the Westmoreland Malleable Iron Company was among the group's founders.) Today, independent laboratories analyze the iron for its carbon and silicon content; the sand, for consistency.

Between 1890 and 1910, the railroad industry caused a spectacular demand for malleable iron for such castings as journal boxes and lids, drawbars, brake equipment, boxcar door hangers, and many others. After 1910, automakers also turned to malleable iron for rear-axle housings, differential cases, hubs, steering-gear housings, and other parts requiring a tough and ductile metal in a complex form that would be difficult to produce by forging. Present-day uses of malleable iron demonstrate its versatility and reliability. In addition to agricultural, automotive, railroad, and construction equipment, malleable iron is used in oil-field pumps, chain-hoist assemblies, plumbing parts, valve handwheels, fittings for electric-distribution systems, and hand tools, to name only a few applications.

Location/Access

The Westmoreland Malleable Iron Works was located at Main and Furnace streets (Exit 32 of the New York State Thruway, in the Catskill region). It closed its doors in the early 1990s, and the family contributed the landmark's plaque and other mementos to the local historical society.

FURTHER READING

Malleable Iron Castings (Cleveland: Malleable Founders Society, 1960).

Watkins Woolen Mill

near Lawson, Missouri

The Watkins Woolen Mill contains the finest collection of nineteenth-century textile machinery *in situ* in North America. Built from 1860 to 1861 by Waltus L. Watkins (1806–84), the mill produced yarn and cloth intermittently until about 1900, when it was shut down intact. All of the mill's carding machines,

Third floor of the Watkins Woolen Mill, showing carding and spinning machines. *Photograph by Jet Lowe, Library of Congress Collections.*

spinning jacks, twisters, looms, dyeing vats, and napping and fulling machines remain in place—as used and altered over time—offering an unusually complete picture of the operation of a mid-nineteenth-century woolen mill.

Watkins, a machinist and master weaver from Frankfort, Kentucky, gained experience in textiles by working in the cotton mills of his uncles. In 1832 he moved to Missouri, where he built and for several years operated a small cotton mill before converting it to wool. The mill was destroyed by fire, and Watkins temporarily turned his attention to farming. In 1839 Watkins purchased land in Clay County, in northwestern Missouri, and ten years later built an ox-driven flour and woolen mill. In May 1861 he completed construction of a three-and-one-half-story, brick, steam-driven woolen mill.

Watkins prospered—consumption of woolen goods more than doubled during the Civil War—and the Watkins complex grew to include a flour mill, general store, and broom factory. Mill workers lived in small houses on the grounds. Following Watkins's death in 1884, his three sons carried on the business, turning from cloth to yarn sales in 1886, when cheaper Eastern woolens usurped the market for finished goods. About 1900, mill operatives left their stations. Half-used sacks of dyed wool remained in place; so did the journals, pencils, desks, and chairs of the foremen. The mill remained untouched for the next half century.

The Watkins farm remained in the Watkins family until 1945, and the mill and its machinery were preserved intact. It was opened to tourists as efforts were made to interest the state of Missouri in purchasing the farm and maintaining it as a museum. The Watkins Mill Association, organized by a small group of executives of the Allis-Chalmers Company in nearby Independence, purchased the mill building and its contents. In 1963 Clay County voters approved a $184,000 bond issue for purchase of almost 800 acres (324 ha)—about half the acreage of the original Watkins farm—as a state park.

The Watkins Mill State Historic Site, which includes the woolen mill, the mill owner's house (1854), and the church (1871–76) and school (1856) given to the community by Waltus Watkins, opened in 1965. In 1966 the Watkins Woolen Mill was designated a National Historic Landmark.

Location/Access

The Watkins Mill State Historic Site is located 4 miles (6.4 km) southwest of Lawson in Clay County, at 26600 Park Road North, Lawson, MO 64062; phone (816) 296-3357. Hours: daily, May through October; hours vary November through April. Admission fee.

FURTHER READING

Laurence F. Gross, "The Importance of Research Outside the Library: Watkins Mill, A Case Study," *IA, The Journal of the Society for Industrial Archeology* 7 (1981): 15–26.

Creusot Steam Hammer

Le Creusot, Burgundy, France

James Nasmyth invented the steam hammer in 1838–39 to forge the 30-inch- (75-cm-) diameter paddle-wheel shafts for I. K. Brunel's *Great Britain* (see p. 216), though Brunel later rejected paddle wheels in favor of screw propulsion. Ingeniously simple, Nasmyth's device consisted of an anvil on which to rest the forging; a block of iron, constituting the hammer itself, which would deliver the blow; and an inverted steam cylinder, to whose piston rod the block was attached. Steam admitted to the cylinder would raise the hammer, which would then fall of its own weight upon the forging on the anvil. The hammer, with its wider opening between hammer and anvil, allowed much larger forgings to be worked than was possible with the tilt hammers then in use.

French ironmaster Eugène Schneider (1805–75) and his chief engineer, François Bourdon, reputedly saw Nasmyth's sketch of the hammer during a visit to Nasmyth's ironworks west of Manchester at Patricroft. The two built the first steam hammer at their works near the small town of Le Creusot about 1840. (Nasmyth, meanwhile, built his own hammer and patented the device in 1842.) By the 1850s, Schneider Brothers & Company (later known as Schneider & Company) had earned a worldwide reputation as builders of the largest classes of engines, steamships, and ordnance then known. To keep pace with the growing size of cannon, armor plate, and marine-engine shafts, in 1877 the firm began building a hammer of colossal proportions that would eclipse all others.

The Creusot hammer consists of four distinct parts: the foundation, or substructure, including the anvil; the legs, with their entablature; the steam

Early twentieth-century view of the Creusot steam hammer, attended by two of the four cranes that served it.

cylinder, with its valves and linkages; and, finally, the active mass made up of the piston, piston rod, hammerhead, and die. The foundation consists of solid masonry resting on bedrock 36 feet (11 m) below the soil. The hollow-cast legs are bolted to plates embedded in the masonry; each stands 33½ feet (10 m) high and is joined to the other by wrought-iron plates. One leg supports the pulpit, or operator's platform. Atop the legs, a 30-ton (27-t) table binds the whole into a rigid A-frame that both guided the path of the hammerhead and absorbed the shocks of its blows.

The steam cylinder (actually a stack of two cylinders) is 19 feet, 8 inches (5,890 mm) high, with an inside diameter of 6 feet, 3 inches (1,905 mm). Steam averaging 71 psig (489.5 kPa) was distributed and exhausted through two balanced, single-acting slide valves that admitted steam only beneath the piston to drive it up, not above the piston to force the hammer down; the hammer used gravity to do its work. The piston rod, measuring 14 inches (355 mm) in diameter, was itself an impressive forging. Together with the hammerhead, it delivered a formidable striking force of between 80 and 100 tons (72 and 91 t), depending on the hammer and the length of stroke.

Four stationary, steam-powered, swan-neck cranes and four heating furnaces served the Creusot hammer, providing steel ingots weighing up to 120 tons (109 t). The Creusot hammer was used to work massive iron and steel shafts, piston rods, and other forgings that helped increase world industrial capacity. It stood unchallenged until Bethlehem Iron Company purchased the patent rights from Schneider & Company and built one of nearly identical design in 1891. Bethlehem demolished its hammer in 1902, while that at Le Creusot was retired in 1930. Both hammers ultimately fell victim to hydraulic and mechanical presses, which could apply force slowly and evenly to produce large forgings of uniform internal structure—something not always possible with hammers, which often altered only the outer surfaces, leaving internal stresses.

Though stripped of its life-giving steam, the Creusot hammer continued to impress visitors to the Schneider works. In 1969 it was disassembled and rebuilt in the public square. The Creusot hammer is one of only a small handful of large steam hammers extant worldwide.

Location/Access

The Creusot Hammer stands in the public square. Nearby, the Museum of Man and Industry, Château de la Verrerie (open by appointment), interprets local history, including the mining, iron, and steel industries.

FURTHER READING

T. S. Rowlandson, *History of the Steam Hammer,* a lecture delivered at the Mechanics' Institution, Patricroft, on December 14, 1864 (Eccles, Manchester: A. Shuttleworth, Stationer, & C., 1864).

Joshua Hendy Iron Works

Sunnyvale, California

The Joshua Hendy Iron Works pioneered the manufacture of large machinery in the American West, earning a worldwide reputation for its hydraulic mining machinery, which became an industry standard. The firm was founded by English machinist Joshua Hendy (1822–91), who emigrated to the United States in the mid-nineteenth century, settling in New York and Texas before arriving in San Francisco in September 1849, at the peak of the Gold Rush. Within two months, Hendy started California's first redwood lumber mill. Milled lumber was an elemental commodity expensive to import from the East, and Hendy's business venture was a success.

Hendy expanded his interests into mining. Observing the evolution from manual placer mining with pan and pick to more efficient power machinery, in 1856 Hendy founded the Joshua Hendy Iron Works. Hendy became one of the principal suppliers of gold- and silver-mining machinery in the American West, expanding his San Francisco works from one shop to three. Hendy hydraulic mining equipment—including the Hydraulic Giant, the Challenge Ore Feeder, and the Hendy Ore Concentrator—became the world standard.

The San Francisco earthquake and fire of 1906 leveled the Hendy shops. Company directors (Hendy had died in 1891) decided to relocate to Sunnyvale, a quiet ranchers' trading center 40 miles (64 km) to the south, where promoters offered 32 acres (13 ha) on the Southern Pacific Railroad main line at no cost. In the sprawling new plant, the company branched out into gate valves for

Joshua Hendy Iron Works machine shop, 1919.

flood-control, irrigation, and power projects worldwide. During the First World War, Hendy built reciprocating steam engines for cargo ships; during the Depression, Hendy produced the many huge gate valves for the Boulder and Grand Coulee dams. Other Hendy products included ornamental street lamps, among them the distinctive lampposts of San Francisco's Chinatown district.

In 1940 the Hendy Iron Works was acquired by a consortium led by (among others) Charles E. Moore, K. K. Bechtel, and Henry J. Kaiser. This group, which had teamed up to build Boulder Dam, saw the latent possibilities of the company. Within two years, they expanded the plant from 65,000 square feet (6,039 m^2) to nearly a million (92,903 m^2), and from 60 employees to 11,500. Production was expanded to include precision parts, propulsion steam turbines and reduction gears, corvette engines, and torpedo mounts. Hendy manufactured more than a quarter of all triple-expansion engines for the 2,700 Liberty ships (cargo vessels) built between 1941 and 1945, producing 773 EC2 engines (as they were designated) in just three and a half years.

In 1947 the Westinghouse Electric Corporation purchased the sprawling Sunnyvale plant to provide a western source of equipment for electric utilities. The plant was soon producing steam turbines for power generation, transformers, switch gear, and motors. Since the 1950s, the plant has been engaged in the design, development, and manufacture of missile launch and handling systems for the U.S. Navy.

Location/Access

Some original buildings of the Joshua Hendy Iron Works are still in use as part of Northrop-Grumman. Visitors can contact the Iron Man Museum, 401 East Hendy Avenue, Sunnyvale, CA 94086; phone (408) 735-2020.

Owens AR Bottle Machine

Toledo, Ohio

In 1900 all bottles and many jars manufactured in the United States were still produced by human skill and lung power. Assisted by young helpers, glass blowers used a blowpipe and hand tools to create glasses, jars, bottles, bowls, and vases. To produce relatively uniform containers for beverages, food, and drugs, glassworkers gathered a gob of molten glass on the end of a blowpipe and lowered the glowing mass into a hinged, two-part metal mold into which the glass was blown to form a hollow vessel. After removing the glass from the mold, the team finished the neck and shoulder of the container by hand.

In the late nineteenth century, the increasing demand for bottles by packaged-goods manufacturers was a strong stimulus to the development of a mechani-

The Owens AR automatic bottle machine of 1914. *Courtesy Walbridge & Bellg Productions.*

cal means of producing glassware. A number of British and American inventors patented semiautomatic bottle machines, but these still required the glass gob to be gathered and fed by hand.

The Owens automatic bottle machine, the brainchild of an unorthodox inventor with no technical training, was first placed in commercial production in 1903. Michael Joseph Owens (1859–1923), a skilled glassblower, spent eighteen years working in glass factories before joining Edward Drummond Libbey in 1888 at the New England Glass Company (later Toledo Glass Company) in Toledo, Ohio. With Libbey's financial backing, Owens developed semiautomatic machines to manufacture light bulbs, drinking glasses, and lamp chimneys.

In 1895 Owens turned his attention to designing a fully automatic bottle machine. The greatest obstacle was finding a way to machine-gather the glass in precise, uniform quantities. Owens's ingenious solution was a suction device resembling a bicycle pump in form and function. Withdrawing the piston rod on the pump created a vacuum that sucked up a charge of glass into a mold, forming the neck of the bottle. Suspended by the neck, the gather of glass was next placed in a body mold, where the return stroke of the piston blew the glass into the mold, taking its shape. The first attempts to blow a bottle with the pump yielded distorted "freaks," but successive tries produced a perfect 4-ounce (118-ml) petroleum-jelly jar. With the principle proven, Owens turned his attention to building a complete bottle-making machine.

Owens's first commercial model had six arms, or separate working units, mounted on a circular rotating frame. Each carried a blank mold, a neck mold, and a plunger for forming the neck. The arm dipped to suck up its gather of glass as it passed over the pot. Owens's machine, patented in 1904 (Nos. 766,768 and 774,690), not only made a satisfactory bottle, it made a *narrow-necked* bottle (the semiautomatics had been confined to the production of wide-mouthed ware) and it made it quickly, turning out twelve 1-pint (0.4732-l) bottles per minute, 17,280 bottles every twenty-four hours. The machine reduced manual labor to a minimum, requiring but a single operator.

In 1903 the Owens Bottle Machine Company was incorporated to license established producers. (The Kent Machine Company of Toledo supplied the Owens machines.) By 1914, fourteen domestic licenses had been issued in the United States covering nearly every important kind of bottle and jar. The European Bottle Machine Company was formed to handle operations in Europe, and by 1920, Owens bottle-making machines were at work in England, Germany, Holland, Austria, Sweden, France, Denmark, Italy, Norway, Hungary, Scotland, and Ireland.

Between 1905 and 1926, 317 Owens bottle-making machines, comprising nine different models, were put into production worldwide. Introduced about 1912, Model AR, with ten arms, was the most versatile, manufacturing bottles ranging in size from prescription ware to beer and catsup bottles to gallon-size packers, and producing, on average, 140 bottles per minute.

The Owens bottle machine revolutionized an industry. In 1905 the majority of bottles and containers were still produced by the hands and lungs of skilled craftsmen; by 1922–23, 80 percent of production was machine-made. By 1929, American bottle production had been wholly transformed from a handicraft to a machine process, and the once-dominant hand industry had been relegated to a narrow and quantitatively unimportant field. Uniform containers produced at lower cost meant that glass containers were now readily available for packaging and preserving food and beverages, pharmaceuticals, household cleaners, and other products, while standard height and capacity made high-speed packing and filling lines possible. Finally, the automatic bottle machine put an end to the industry's notorious exploitation of children. (In 1880 children between the ages of ten and fifteen constituted a quarter of the total work force, working ten-hour days for as little as thirty cents a day.) In a letter to the Owens Company in 1913, the National Child Labor Committee of New York wrote that the automatic machine had done more to eliminate child labor than the committee had been able to accomplish through legislation.

The AR bottle machine is no longer extant.

FURTHER READING

Pearce Davis, *The Development of the American Glass Industry* (Cambridge, Mass.: Harvard University Press, 1949).

Warren C. Scoville, Revolution in Glassmaking: Entrepreneurship and Technological Change in the American Industry, 1880–1920 (Cambridge, Mass.: Harvard University Press, 1948).

A. O. Smith Automatic Frame Plant

Milwaukee, Wisconsin

"An automatic frame plant that would run without men." That was the goal of Lloyd R. Smith, president of the A. O. Smith Corporation of Milwaukee, in 1916, when he envisioned the automatic production of automobile frames as a way to corner a larger market share and boost revenues. The nation then produced 1.5 million automobiles annually, with several makers dividing the existing market for frames. Smith gave the order to his large staff of skilled engineers, and ground was broken two years later. Following delays due to World War I, when the company produced wheel-hub flanges, casings for bombs, and other war materiel, Smith engineers completed plans for a plant performing 552 separate mechanical operations on every frame.

The automatic frame plant represented the fulfillment of Lloyd Smith's dream. His grandfather, Charles Jeremiah Smith, had emigrated from England and started a small machine shop in 1874 in Milwaukee, where he built baby buggies, then in the 1890s graduated to bicycles. By 1898, C. J. Smith & Sons was the largest producer of bicycle parts in the world. In 1904 Arthur O. Smith, C. J.'s youngest son, organized the A. O. Smith Company for the production of automobile structural parts. Two years later, the company produced the first pressed-steel automobile frame in the nation for Cleveland's Peerless Motor Car Company. Orders came in from Cadillac, Packard, Elmore, and others.

A. O. Smith automatic frame plant. Shown is a special machine for finishing spring hangers.

With the start-up of its automatic frame assembly plant on May 23, 1921, A. O. Smith became the world's largest manufacturer of automobile frames. The plant did not, as Lloyd Smith first idealized it, run "without men," but required a staff of 180 at supervisory, visual inspection, and control stations. The production line, nearly two city blocks long, consisted of nine units, each unit comprising several stations performing the same operation:

Unit No. 1: Picked up the raw steel strips and examined them for defects, throwing out those that did not meet the required standards of length, breadth, and thickness.

No. 2: Doused the strips in baths of acid pickle to clean them.

No. 3: Fabricated the longer strips into right- and left-side members, bending them, turning up their edges, and punching holes for rivets.

No. 4: Fabricated the shorter strips into cross members.

No. 5: Assembled the side members.

No. 6: Assembled the whole frame, inserting, driving home, and heading the rivets.

No. 7: Inspected the assembled frame, a partly human job.

No. 8: Washed, painted, and dried the frame.

No. 9: Transmitted the finished frame to overhead storage, where it hung with others in carload lots until a worker, using a small crane, delivered it into a waiting freight car.

An hour and a half elapsed from raw steel to finished frame. Every eight seconds a frame was completed and swung into storage—420 an hour, 10,000 a day—for eventual delivery to Pontiac, Chrysler, Chevrolet, and Buick.

As automobiles, including frames, were redesigned, A. O. Smith made corresponding changes in the production line. Eventually, however, automobile designers began to change the models of each make of car each year, requiring costly and time-consuming changes in the frame production line. That, plus the fact that riveting had given way to welding in the manufacture of automobile frames, caused the A. O. Smith frame plant to close. The last frame came off the assembly line on June 24, 1958.

Location/Access

The A. O. Smith automobile frame assembly plant is no longer extant.

FURTHER READING

Stuart Chase, "Danger at the A. O. Smith Corporation," *Fortune*, November 1930, 62–67.

"Making Automobile Frames Automatically," *Iron Trade Review* 83 (23 August 1928): 441–43.

Corning Ribbon Machine

Dearborn, Michigan

Following his invention of a successful incandescent lamp in 1879, Thomas Edison chose the Corning Glass Works to manufacture the glass bulbs for his first lamps. The bulbs had to be individually blown by skilled glassblowers. Working at top speed in the red-orange glow of a glass-melting tank, a gaffer (as this master craftsman was called) and an assistant could produce up to two bulbs per minute.

As electric lightbulbs began to assume commercial importance, a faster, cheaper method of producing them became imperative. In 1915 the Empire Machine Company, a Corning subsidiary, brought out a semiautomatic bulb machine operated by electricity. Although it still required the molten glass to be gathered and fed by hand, the Empire semiautomatic more than doubled worker productivity and reduced labor costs by 70 percent. The race for the production of a fully automatic bulb machine began in earnest, pitting Corning's Empire against Libbey's Westlake Machine Company.

In the spring of 1921, Corning engineer William J. "Will" Woods (1879–1937), who had helped develop the Empire semiautomatic, conceived the simple but revolutionary idea of blowing bulb blanks through a hole in a metal plate. His theory was that if a gather of molten glass were flattened and then placed on a plate with a hole of the proper size, the glass would sag, by its own weight, through the hole to form a globular bag. Air could then be forced into this bag to form the shape of the bulb blank; to perfect the shape, a mold could be closed around it, and the air pressure continued. If a series of such plates were hinged to form an endless chain, and a flat stream or "ribbon" of molten glass laid on the belt while in motion, perfect bulbs might be made in continuous—and rapid—succession.

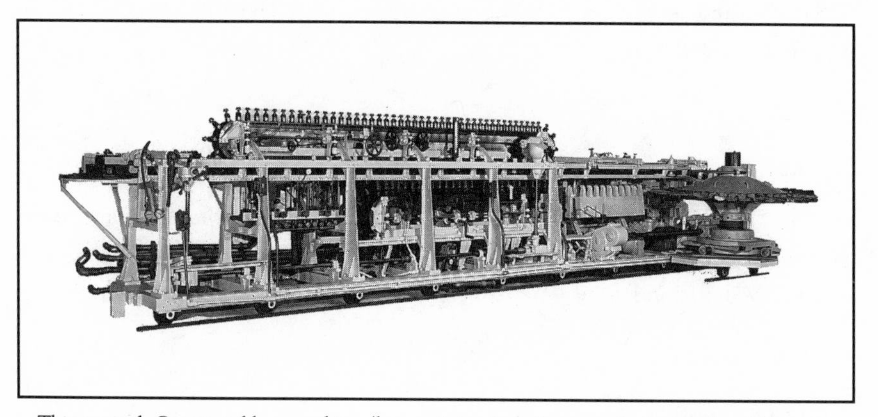

This 3.9-inch Corning ribbon machine (here, missing its bulb-removal wheel and molds) saw service at the Corning plant in Wellsboro, Pennsylvania. *Photograph from the Collections of Henry Ford Museum & Greenfield Village.*

Woods began to experiment with a single plate and a plunger, or blowhead, by which he could introduce air into the bag. He then designed a mechanism that would first form the desired blanks, then conduct them with properly maintained temperatures and predetermined speed through the elongating and blowing operations.

Working with Corning chief engineer David E. Gray, Woods developed the first successful ribbon machine in 1926. A glass-melting tank sat above one end of the machine, feeding a stream of molten glass down between two rollers. The rollers squeezed the hot glass into a thick, glowing ribbon, which next met up with an endless belt made up of flat sections and driven, like a bicycle chain, by sprockets. Each section contained a hole approximately 1 inch (25.4 mm) in diameter.

As the glass sagged through the holes, taking on a bulbous shape, a series of blowheads descended on the hot ribbon, blowing air into the partially formed bulbs. Meanwhile, a third chain below and inside the first thrust up a series of split molds, the latter snapping together around the glass to give final shape to the bulb. The entire process lasted only ten or twelve seconds, resulting in a shower of finished bulbs—almost 300 each minute, 400,000 per day— as they were cut off by a rotating knife and deposited onto a conveyor for the trip to the annealing lehr.

By 1927, the automatic production of incandescent bulbs was firmly established, with two machines, the Empire and the Westlake, accounting for 95 percent of U.S. production. By 1930, the perfected Corning ribbon machine reached a production rate of 600 to 800 bulbs per minute—up to 1 million per day. By decade's end it had eclipsed its competition.

More than sixty years later, the Corning ribbon machine remains the state of the art. With the exception of some handmade specialty bulbs, fewer than fifteen Corning ribbon machines supply world demand for incandescent bulbs. Today, Corning Engineering, a subsidiary of Corning Glass, licenses its ribbon machine worldwide.

Location/Access

A 1928 Corning ribbon machine is exhibited at the Henry Ford Museum, 20900 Oakwood Boulevard, Dearborn, MI; phone (313) 271-1620. Hours: daily, 9 A.M. to 5 P.M. Admission fee.

FURTHER READING

Pearce Davis, *The Development of the American Glass Industry* (Cambridge, Mass.: Harvard University Press, 1949).

Fusion-welded Test Boiler Drum

Chattanooga, Tennessee

When the Charles L. Edgar Station of the Edison Electric Illuminating Company (see p. 90) went on line in Weymouth, Massachusetts, in 1925, it used steam generated at the unprecedented pressure of 1,200 psig (8,274 kPa). This was about double the highest pressure then used in generating stations. Owing to uncertainty about the safety of using a standard boiler drum of riveted construction, the Edgar Station drum—32 feet (9.8 m) long by 4 feet (1,219 mm) in diameter, with walls 4 inches (102 mm) thick—was forged from a single steel plate. The production of such a forging was a remarkable demonstration of the blacksmith's art, but it was also extremely costly.

Forge welding of high-pressure boiler drums had been tried in Germany as early as 1913. The steel plate was rolled into a cylinder, then the longitudinal seam was welded by hammering it while heating the metal locally with gas flames. A simpler, less expensive method appeared with the advent of electric-arc welding on a large-scale commercial basis in the 1920s.

In the late 1920s, the Hedges-Walsh-Weidner Company of Chattanooga, Tennessee, along with other boiler manufacturers, began welding and testing boilerplate. As it developed and perfected coated electrodes for electric-arc fusion welding—a process of welding metals in the molten state without applying mechanical pressure or blows—the company began an experimental program to hydrostatically test welded boiler drums. On May 2, 1930, it tested a welded boiler drum to destruction, with significant results.

The drum had been fabricated from rolled shell plate one inch (25.4 mm) thick, manufactured to American Society for Testing and Materials (ASTM)

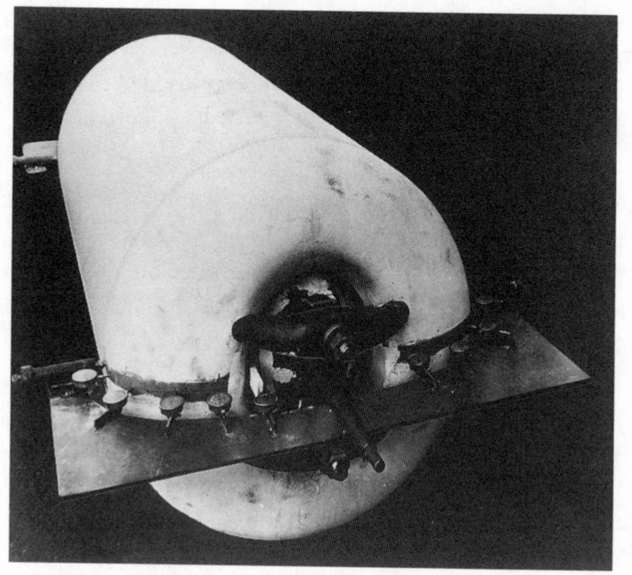

Fusion-welded test boiler drum with test instrumentation installed to measure deformation.

standards of 55,000 psig (379,170 kPa) tensile strength for firebox boilerplate. The drum was 98 inches (2,489 mm) long, with an inside diameter of 34 inches (863 mm). A single longitudinal seam weld joined the edges of the rolled shell. Two girth seam welds joined the dished heads to the shell. One head was blank; the other was pierced by a 12-inch-by-16-inch (304-mm-by-406-mm) oval manhole cover. There was a distance of 72 inches (1,829 mm) between the head seam welds. The test drum was assembled by through-welding using the flux-coated electrodes developed by Hedges-Walsh-Weidner.

The vessel was mounted on a laboratory test stand. Dial indicators measured the extent of two-dimensional strains as hydrostatic pressure was applied in 250 psig (1,724 kPa) increments. Based on tentative design calculations, the safe working pressure for the drum was estimated to be 517 psig (3,565 kPa). A small group of engineers watched, but none could have predicted that the test pressure would reach 3,250 psig (22,408 kPa)! As the vessel expanded under hydraulic pressure, the flanged manhole finally bulged, causing a leak that prevented further testing. Nevertheless, the test was a success, proving conclusively that the welded joints were 100-percent efficient and could withstand stresses more than six times those considered safe.

Following the first successful test, the company's experiments with welded boiler drums continued in earnest. In June 1930, plant superintendent A. J. Moses presented the test results to the annual meeting of the National Board of Boiler and Pressure Vessel Inspectors. Later that year, Moses wrote a paper describing the details of the test work, concluding: "The process of metallic arc welding developed by the Hedges-Walsh-Weidner Company is safely applicable to power boilers and pressure vessels."

Fusion welding rapidly gained recognition. By 1931, the entire boiler industry was engaged in the development of welding processes for pressure vessels. That year, the Boiler Code Committee of the American Society of Mechanical Engineers adopted rules for the fusion-welded construction of boilers and pressure vessels, and established requirements for X-ray examination of welded seams for stress, a practical, nondestructive test that proved to be even more rigorous than physical testing.

In replacing riveted construction, fusion welding resulted in increased working efficiencies for steam power plants by allowing higher working pressures and temperatures, and the fabrication of larger units of improved safety. It also stimulated new interest in the welding art.

Location/Access

The pioneer fusion-welded test boiler drum is displayed outside the metallurgical and materials laboratory of ABB Combustion Engineering, Inc., 911 West Main, Chattanooga, TN 37402; phone (423) 752-2100.

FURTHER READING

W. Cross, The Code: An Authorized History of the ASME Boiler and Pressure Vessel Code (New York: American Society of Mechanical Engineers, 1990).

A. J. Moses, "X-ray Examination of Welded Pressure Vessel Seams," *Combustion* 3 (September 1931): 17–20, 35.

———, "Practical Application of the A.S.M.E. Welding Code," *Journal of the American Welding Society* 11 (February 1932): 13–15.

"Results of Tests on Welded Drums," *Power* 72 (15 July 1930): 112.

Alcoa 50,000-ton Hydraulic Forging Press

Cleveland, Ohio

On May 5, 1955, U.S. Air Force Secretary Harold E. Talbott put into production the largest machine tools in the world: towering 35,000- and 50,000-ton (31,751- and 45,350-t) hydraulic die-forging presses. The $40 million installation, at Aluminum Company of America's Cleveland plant, marked the halfway point in the Air Force's $179 million heavy-press program to build ten presses—three forging and six extrusion presses—to turn out structural members for high-speed military aircraft.

The Air Force heavy-press program grew out of the Cold War and a desire to strengthen Air Force capabilities in supersonic aviation by building stronger and lighter aircraft of fewer components. The program was spurred by the discovery, early in World War II, that the Germans were using large forging and extru-

Mesta-built Alcoa 50,000-ton hydraulic forging press, Cleveland, Ohio, in raised position. *Photograph by Jet Lowe, Library of Congress Collections.*

sion presses—larger than any then known—to fabricate aircraft parts in a single piece. Larger presses would greatly reduce expensive and time-consuming machining and subassembly operations; instead of bolting, riveting, and welding many small units to form a structural member, large forgings would be pressed out between closed dies. The resulting one-piece sections would be stronger and lighter, with superior aerodynamic surfaces.

The 50,000-ton forging press, built by the Mesta Machine Company, is 87 feet (26 m) high, extending 36 feet (11 m) below ground and 51 feet (15 m) above. It weighs about 8,000 tons (7,000 t). Sixteen huge steel castings poured in the Mesta foundries at West Homestead, Pennsylvania, comprise the major elements of the press.

A moveable die table, or platen, holds the lower forging die; the upper die is clamped to the upper platen, which in turn is attached to the lower moving crosshead. The entire moving crosshead assembly, the upper "jaw" of the press, consists of eight steel castings totaling almost 1,150 tons (1,043 t). A manipulator moving on rails inserts ingots between the dies and removes the forged parts.

The press force is generated by a hydro-pneumatic pressure system consisting of four pre-filler bottles, two horizontal reciprocating pumps driven by 1,500-horsepower (1,118-kW) motors, and four forged alloy-steel, pressure-accumulator bottles. A pressure of 4,500 psig (31,027 kPa) is built up in each accumulator and released to the eight pressure cylinders housed in the stationary crossheads at the top of the press. The combined effort of these cylinders produces the 50,000-ton forging capacity.

Fluid flows are staggering, ranging from 11,750-gallons- (44,478-l-) per-minute high pressure to 26,4380-gallons- (100,078-l-) per-minute pre-fill pressure—enough to fill a good-sized house to the rafters in less than sixty seconds. At the end of the press cycle, the hydraulic force is reversed and directed to eight pull-back and balancing cylinders, which lift the moving crosshead assembly to its raised position.

Aluminum die forgings are used at key structural points in all modern aircraft. Die-forged members provide strength and can be shaped in complex forms with relatively little machining. Traditional methods of fabricating parts of this type required costly machining from bigger pieces of metal or else building up from smaller components.

Location/Access

The 50,000-ton hydraulic press is located at Aluminum Company of America, Forging Division, 1600 Harvard Avenue, Cleveland, OH 44105. It is not open to the public.

FURTHER READING

"Press Plant Specially Built for Large Aircraft Forgings," *Steel Processing* 41 (June 1955): 350–60.

Wyman-Gordon 50,000-ton Hydraulic Forging Press

North Grafton, Massachusetts

In 1944 the War Production Board selected the Wyman-Gordon Company to operate a Mesta-built 18,000-ton (16,329-t) hydraulic forging press, then the largest in the United States, at a new government-built plant in North Grafton, Massachusetts. Six years later, Wyman-Gordon was again selected, along with the Aluminum Corporation of America (see "Alcoa 50,000-ton Hydraulic Forging Press," p. 171), to operate two even larger presses as part of the U.S. Air Force heavy press program.

Two hydraulic forging presses were built at North Grafton: one of 50,000 tons (45,350 t), dubbed by its builder, Loewy Construction Company, "Major"; the other of 35,000 tons (31,751 t), called "Minor." Along with a similar press in Cleveland, Wyman-Gordon's 50,000-ton press was the largest machine tool in the world. These statistics suggest its size: it weighed approximately 10,000 tons (9,000 t); its foundations went 100 feet (30 m) into bedrock; above ground, it soared ten stories high; and the production floor covered six city blocks.

Both Wyman-Gordon presses were of the "pull-down" type—i.e., the cylinders were located below the lower press bed, and the upper entablature and upper platen were pulled down against the work in the dies between the two platens. The 50,000-ton press consisted of nine hydraulic cylinders and six columns. The

Loewy-built Wyman-Gordon 50,000-ton hydraulic forging press, North Grafton, Massachusetts, in raised position. *Photograph by Jet Lowe, Library of Congress Collections.*

columns were so large—each weighed close to 300 tons (272 t) —that they had to be built up of three rectangular laminations and secured with tie-rods.

The hydraulic cylinders were designed to put the structural members of the press in nearly pure compression. A single operator commanded over a million pounds of pressure, while automatic safety controls monitored strain and guarded against damage from eccentric loads. The 50,000-ton Loewy press turned out its first forgings in October 1955.

The Air Force heavy press program revolutionized plane making, just as its champion, Lt. Gen. K. B. Wolfe, had predicted in 1951. A dramatic example was the development of the Boeing 747 in the 1960s, when Wyman-Gordon produced the massive support beam for the main landing gear. Twenty feet (6,096 mm) long by 4 feet (1,219 mm) wide, and weighing 4,000 pounds (1,814 kg), this was the largest closed-die titanium forging in the world.

Today, the 50,000-ton press forges a variety of alloys, stainless steels, refractory metals, and titanium, turning out airframe and structural components in a variety of shapes and sizes, including fuselage bulkheads, wing spars, and rotor hubs for helicopters. Wyman-Gordon purchased the North Grafton plant, including the three heavy presses, from the federal government in 1982.

Location/Access

Open by application to Wyman-Gordon Company, 105 Madison Street, Worcester, MA 01615; phone (617) 756-5111.

FURTHER READING

F. T. Morrison and R. G. Sturm, "World's Largest Forging Press," *Mechanical Engineering* 75 (March 1953): 191–93.

H. C. Hood, "Some Problems in the Development of a 50,000 Ton Press," *Steel Processing* 39 (December 1953): 642–46.

First Hot Isostatic Processing Vessels

Columbus, Ohio

In only twenty-five years, hot isostatic processing (HIP) has grown from a laboratory curiosity to a manufacturing technique having broad commercial application. Initially conceived as a relatively low-volume process for cladding nuclear fuel elements, HIP today is widely used to fabricate parts made from high-temperature superalloys, ceramics, and composite materials.

In 1955 the Atomic Energy Commission asked researchers at the Battelle Memorial Institute's Columbus Laboratories to develop a process to bond components of small Zircaloy-clad, pin-type nuclear fuel elements for the Shippingport,

HIP vessel.

Pennsylvania, pressurized-water reactor (see "Shippingport Atomic Power Station," p. 103). Four scientists—Russell Dayton, Edwin Hodge, Stan Paprocki, and Henry Saller—decided to try a novel diffusion-bonding technique. At elevated temperatures, they would apply isostatic gas pressure—that is, equal pressure from all directions—to the material.

The researchers fabricated a pressure vessel using a 3-foot-long (914-mm) stainless steel tube by plugging one end and welding it closed, and threading the other end to accept a high-pressure valve. They inserted a sample pin, then attached the valve, which in turn was attached to a feeder line connected to a helium cylinder. The researchers pressurized the vessel to approximately 2,000 psig (13,788 kPa) and inserted the closed end into a heat-treat furnace at a temperature of about 1,500°F (815°C).

Though the process was too slow, taking up to thirty-six hours, the hot-wall experiments achieved excellent Zircaloy bonding, as well as Zircaloy-to-core bonding, with the desired dimensional control. Thus was born the technique of gas-pressure bonding, or hot isostatic processing (HIP) as it is known today. With the principle proven, the researchers replaced the tube vessel with large hot-wall laboratory vessels and conducted similar experiments at higher pressures.

Limitations—in size, temperature, and pressure capabilities—eventually led to the use of a resistance furnace located inside a water-cooled pressure vessel. The Battelle team demonstrated that by applying pressure they could improve the properties of most materials and produce complex shapes unattainable by other methods.

In the 1960s, the application of HIP to the production of high-speed tool steel from powdered metals helped prove its commercial viability. Using HIP to consolidate powdered metals was a natural outgrowth of the fabrication of nuclear materials. Battelle's demonstration that HIP-processed powders enjoyed properties equivalent to forged metals set off a flurry of government- and industry-funded research. Manufacturers of aircraft components, especially turbine buckets and blades, began replacing forgings with HIP-cast parts, resulting in substantial cost savings and improved tensile and fatigue strength. HIP-production equipment, meanwhile, evolved from small, slow, and unreliable furnaces to the 4-foot-diameter (1,219-mm) autoclave (a heated pressure vessel) installed at Battelle beginning in 1972.

Today, HIP is used to perform six distinctly different processes: (1) powdered metal consolidation, called hot isostatic processing, which is particularly useful for forming parts with complex shapes (for example, tool steel for machine tools and superalloy parts for jet engines); (2) diffusion bonding, called gas-pressure bonding, isostatic diffusion bonding, or HIP welding, used for forming complex nuclear elements and complex shapes from wrought materials that cannot be fabricated by conventional means; (3) densification of cemented carbides, to improve the properties of tool bits and remove flaws from steelmaking rolls; (4) healing defects in castings to improve their properties and enhance their resistance to fatigue; (5) healing creep damage in used parts (for example, extending the life of turbine blades in jet engines); and (6) pressure infiltration of molten materials into porous solids to obtain the combined properties of both materials.

There are now more than three hundred research and production HIP systems in the United States and others throughout the world. HIP development continues at Battelle and elsewhere, and its applications continue to expand.

Location/Access

The early HIP vessel, once displayed in the lobby at Battelle, has been given to the Smithsonian Institution in Washington, D.C., but the plaque remains at Battelle Memorial Institute Communications Department, 505 King Avenue, Columbus, OH 43201-2693.

FURTHER READING

H. D. Hanes, D. A. Seifert, and C. R. Watts, *Hot Isostatic Processing* (Columbus, Ohio: Battelle Press, 1979).

"HIP Makes Stronger, Cheaper Turbine Parts," *American Machinist* 119 (November 1975): 126–27.

Food Processing

INTRODUCTION by Euan F. C. Somerscales

Mechanical engineering plays, and has played, an important but probably unrecognized role in bringing food to the table. This has been true from early times, but as the scale of food handling has increased with urbanization, many food preparation processes that were originally confined to the family are now carried out on a very large scale in an industrial setting, which has led to the growing involvement of the mechanical engineer in this vital social task.

The story starts with the milling of grain by mechanical means, which dates from antiquity. Originally, the production of flour was done by one person grinding the grains between two stones (quern). The mechanization of this process, so that the stones were moved by water power, resulted in a very simple water mill now known as the Greek or Norse mill. A horizontal circular stone was rotated by a horizontal waterwheel that was turned by a jet of water obtained from a dammed-up stream. The grain to be ground was placed between the rotating stone and a stationary stone, with a hole pierced in the latter that allowed the driving shaft from the mill wheel to be connected to the upper, rotating stone. This extremely simple device, invented by an early but anonymous mechanical engineer, involves the elements of mechanical engineering in the conversion of energy and the transmission of power.

The vertical waterwheel, which replaced the horizontal wheel, is thought to have been invented by the Romans. Because the wheel rotated in the vertical plane and the millstones rotated in the horizontal plane, bevel gear wheels were used to connect the waterwheel shaft and the millstone shaft. Food processing thereby led to the introduction of another mechanical device, the gear wheel, which subsequently has been applied in a vast range of situations, and which has been developed into a device of the very highest technical sophistication.

The water-driven mill represents the origins of mechanical engineering, and it also represents its early history, up until about the eighteenth century, when the introduction of the steam engine changed this branch of engineering. However, the water mill using a waterwheel was only slightly affected by the advance of technology, chiefly by the introduction of iron into the construction of wheels, gears, and shafts. A mill, such as the landmark Graue Mill would be recognizable to even a Roman miller even though it was not built until the nineteenth century.

Just as cereal grains must be crushed to produce the flour, the orange must be squeezed to extract its juice. We all know this is a tiresome chore, despite the pleasure of consuming the end result. The large-scale marketing of orange juice had to eliminate the labor-intensive squeezing process to be commercially successful. The mechanical engineer's talents are as applicable to this process as were the skills of that proto-mechanical engineer who devised the Greek or Norse mill in the dim recesses of antiquity. Today, the engineer has available materials, sources of power, and a body of knowledge undreamed of by the early mill engineer. Nevertheless, there is a clear link encompassing the whole of mechanical-engineering history that joins the Graue mill and the landmark FMC Citrus Juice Extractor.

Animals, bacteria, fungi, insects, and plants all can extract nutrition from food that is intended for human consumption. To discourage this loss, various preservation methods have been devised. Methods such as drying, smoking, pickling, and salting are so old that their origins are unknown. Newer methods involve high-temperature heating, vacuum packing, refrigeration, and freezing. Mechanical engineers have been particularly involved in these latter preservation methods. Canning, which combines high-temperature heating and vacuum packing, dates from the close of the eighteenth century. It is practiced today in the home, but that experience demonstrates the need for an automated process if a mass market is to be served with canned foodstuffs. Considerable ingenuity is required to do this, but it was accomplished in the FMC Rotary Pressure Sterilizer in the first two decades of this century. It has been recognized as a Historic Mechanical Engineering Landmark because it involves basic elements of mechanical engineering, in the automatic handling of the cans and in the control of the heating process.

The landmarks in food processing serve to illustrate the extent of the mechanical engineer's contribution to society. These are not only in the obvious areas of, say, manufacturing and power production but involve even the food we eat.

Graue Mill

Oak Brook, Illinois

Built by German immigrant Frederick Graue (1819–81) in 1852, the Graue Mill was operated by three generations of Graues until 1920. In the mid-nineteenth century, thousands of such water-powered gristmills dotted the American landscape, grinding grain for local farmers and serving as the economic mainstay of their communities. Today, they are a vanishing breed.

In 1849 Frederick Graue, then thirty-one years old, together with William Asche, purchased a site on Salt Creek and erected a sawmill. Three years later, Graue bought out his partner's interest and erected a three-story brick grist mill, 45 by 28 feet (13.7 by 8.5 m) in size. A New York millwright installed the mill machinery, which is believed to have included an undershot waterwheel and "two runs of buhrs" (two pairs of millstones). The mill ground wheat, corn, oats, and buckwheat for the farmers of Brush Hill (today's communities of Hinsdale and Oak Brook).

Plans of the structure drawn by the Historic American Buildings Survey in 1934 record the mill as it then stood. (No original plans survive.) The millrace, diverting water from Salt Creek, led east from the south side of the mill pond to an undershot wheel and emptied under an arched opening at the lower end of the race, rejoining the stream well below the dam.

Records show that a more efficient vertical-shaft Leffel turbine replaced the undershot wheel in 1868. A steam power plant was added sometime before 1874; it was destroyed by an explosion in 1880 and rebuilt in 1884. (There are no data for either engine.)

After 1916 the mill operated only occasionally. In 1931 the property was

Graue Mill.

added to the DuPage County Forest Preserve District. Beginning in the 1930s, the Graue Mill was reconstructed; its undershot waterwheel, wooden gearing, belt power-transmission system, and stone millrace were rebuilt to reflect its presumed appearance and operation during the period 1852–68. Today, an electric motor powers a single pair of stones producing corn meal, while another motor turns the gearing in the cellar. The reconstructed waterwheel, meanwhile, is not in use.

Location/Access

Graue Mill and Museum, York and Spring roads, P.O. Box 4533, Oak Brook, IL 60521; phone (708) 655-2090. Hours: daily, mid-April to mid-November, 10 A.M. to 5 P.M. Admission fee. Stone-ground cornmeal and recipes for sale.

FURTHER READING

Oliver Evans, *The Young Mill-Wright and Miller's Guide* (Philadelphia: Blanchard and Lea, 1860).

Anderson-Barngrover Continuous Rotary Pressure Sterilizer

Santa Clara, California

Stimulated by the offer of a prize of twelve thousand francs from the French government for better methods of preserving food for Napoleon's army and navy, Parisian confectioner Nicolas Appert began his studies of food preservation in 1795. In 1809 he succeeded in preserving food in specially made glass bottles that he kept in boiling water for varying periods of time. He published his results in a book, *The Art of Preserving Foods*, the following year. But while Appert is considered the "father" of canning—the preservation of foods in hermetically sealed containers by sterilization by heat—not until a half century later, as a result of Pasteur's work, were the causes of food spoilage understood.

Microorganisms are present in all natural foods, which must be processed at high temperatures—212°F (100°C) to 240°F (116°C) or higher, depending on their acidity—to destroy them. In 1874 A. L. Shriver of Baltimore was granted a patent on a steam-pressure retort (similar to a large domestic pressure cooker) that was subsequently widely adopted by the canning industry for sterilizing canned foods. The filled and sealed cans were loaded by hand into mesh baskets and lowered into the retort; the canned product was cooked under pressure to resist the steam pressure buildup within the cans, cooled, and the baskets were removed. The start-and-stop of batch operation was slow and labor intensive. It also took a long time for the heat to penetrate to the center of the immobile cans.

The continuous rotary pressure sterilizer, developed between 1913 and 1920,

Albert R. Thompson's continuous rotary pressure sterilizer brought automation and uniformity to the processing of canned goods.

brought automation and product uniformity to the processing of canned goods, and vast labor and energy savings to the canning industry. It solved a problem that had baffled engineers for years: how to introduce filled, sealed cans into a pressurized chamber full of steam, heat and cook the contents uniformly, remove the cans without affecting the steam pressure, then cool them, all in a continuous stream.

Albert R. Thompson (1879–1947), chief engineer of the Anderson-Barngrover Manufacturing Company (later FMC Corporation) of San Jose, California, supplied the solution. The Anderson-Barngrover continuous rotary pressure sterilizer introduced in 1920 was a massive cylinder of riveted boilerplate, about 20 feet (6,000 mm) long and 5 feet (1,524 mm) in diameter. With precise synchronization, a rotating pocket valve admitted cans at one end and propelled them, gradually and continuously, through the tank on a reel and spiral running the length of the cooker. As the reel turned, the cans rode against the spiral, gradually moving forward in their channels. The constant agitation of the cans' contents allowed rapid heat penetration and reduced processing time. At the discharge end, a pressure cooler employing the same mechanical handling system was joined to the cooker. Another pocket valve transferred the cans from the cooker to the cooler.

The Anderson-Barngrover continuous rotary pressure sterilizer was an immediate success. It was continuous and automatic, it cooked cans of food quickly and evenly and immediately cooled them, and it was fast—processing up to four hundred cans per minute. The machine reduced cook-room labor as much as 15 to 1 and reduced steam consumption by 50 percent while turning out canned goods with better color, flavor, and texture.

The continuous rotary pressure sterilizer was refined over the years. Welding replaced riveting; the drive pulley gave way to an electric motor and gear reducer; in the 1940s, the American Society of Mechanical Engineers' pressure-tank standards were adopted for the shells, and working pressures rose from 20 psig (137 kPa) to 33 psig (227 kPa) and beyond. Today's units can process two thousand or more cans per minute. But after more than seventy years the basic machine remains unchanged, testimony to the quality of its engineering.

Anderson-Barngrover and the John Bean Spray Company, also of San Jose, merged in 1928 to form the Food Machinery Corporation (later FMC Corporation). FMC and its predecessors have built more than fifteen hundred continuous rotary pressure sterilizers and coolers, which are used to process about half of the world's canned food.

Location/Access

This landmark served as a laboratory machine until 1989, when it was replaced with a simulator. A display at FMC demonstrates the principles of operation. The plaque is mounted at FMC Corporate Technology Center, 1205 Coleman Avenue, Santa Clara, CA 95052.

FURTHER READING

W. V. Cruess, *Commercial Fruit and Vegetable Products* (New York: McGraw-Hill Book Company, Inc., 1924).

FMC Citrus Juice Extractor

Lakeland, Florida

FMC Corporation introduced the first rotary whole-fruit juice extractor in 1946, operating it experimentally on grapefruit at the Sunkist Exchange Plant in Tempe, Arizona. Model 402X, with 24 heads, operated at a rate of 20 strokes/480 fruit per minute. Despite some problems, the machine's overall performance was encouraging. The company manufactured three more machines and operated them commercially on oranges at the Sunkist plant in Ontario, California, the following year. By the 1947–48 season, FMC extractors were also at work in Florida and Texas.

Early citrus juice extractors suffered from maintenance problems. They also mixed core, membrane, pulp, and seeds with the juice stream. With these deficiencies in mind, FMC designed the Inline extractor in late 1947 and tested a

The FMC citrus juice extractor solved the problem of separating "undesirables"—the membrane, pulp, and seeds—from the juice stream.

prototype, Model 659, in Florida the following year. In addition to using the whole-fruit extraction principle (see sidebar), the new machine incorporated a unique prefinishing system to remove undesirables from the juice during the extraction process. The juice stream now contained only juice and juice sacs, making it possible to employ a completely enclosed juice-handling system. This feature allowed improved sanitation and more efficient cleanup.

Two additional units joined the original prototype in tests during the 1948–49 season. Based on these results, FMC designed limited tooling for the manufacture and installation of thirty units during the 1949–50 season. Full-scale commercial production followed. More than four hundred units were manufactured and installed in citrus plants for the 1950–51 season.

Since then, the Inline extractor has undergone several major model changes. The latest model, with an improved feed hopper and a peel-oil recovery system that reduces water requirements and waste, operates at a speed of 100 strokes/500 fruit per minute. While it bears only a distant resemblance to the prototype model 402X, in principle it is a direct descendant of the revolutionary machine that brought orange and grapefruit juice to breakfast tables around the world. Today, FMC citrus fruit juice extractors squeeze and prefinish 70 percent of the world's citrus juice.

Location/Access

The earliest extractors, Model 402X, were destroyed following introduction of subsequent models. Some Model 718 extractors, built in the early 1950s, remain in service. One is displayed at FMC Corporation, Citrus Machinery Division, Fairway Avenue, Lakeland, FL 33801; phone (941) 683-5411. It may be viewed by appointment.

Whole-Citrus Fruit Juice Extraction: How It Works

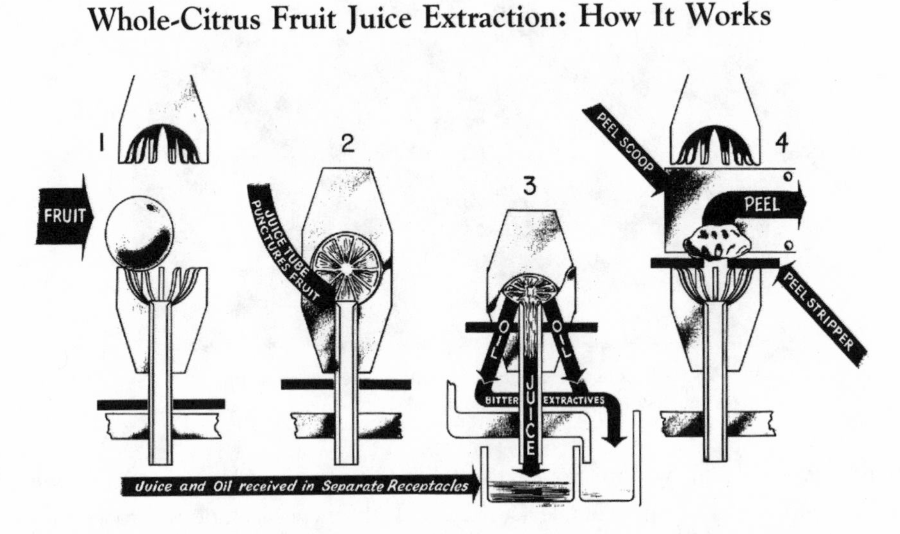

Sources: *The FMC Whole Citrus Juice Extractor: The Story of Its Conception and Evolution*, commemorative brochure (Lakeland, Fla.: FMC Corporation, n.d.).

Materials Handling and Excavation

INTRODUCTION by Robert M. Vogel

No single realm of the mechanical engineer touches more areas of technology, industry, and commerce than that of materials handling. Manufacturing, transportation, mining, logging, construction, and on and on—all involve at one point or another, continually or occasionally, the conveying, lifting, digging, or otherwise handling of materials, either in bulk or by the unit.

Closely paralleling the other branches of mechanical engineering, the design of materials-handling and excavating machinery has gone through an evolutionary development that saw construction change from timber to iron to steel, and propulsion systems change from muscle to steam to diesel and electricity. Each change was accompanied by improvements in refinement of control and mechanical detail as well as capacity and operating efficiency.

Until nearly the nineteenth century, most such equipment was directed to the sinking of mine shafts and the subsequent hoisting of the mineral or its ore; to the erection of buildings, structures, and public works; and to the handling of the goods of commerce at wharfside and in the warehouse. Mine hoists were simple affairs of timber, powered by men or animals or at the deepest pits by waterwheels. In erecting even the largest structures—the great cathedrals—stone blocks, timbers, and other massive components were raised into place by relatively simple machinery, invariably muscle powered. Ships and warehouses were loaded and unloaded similarly. The equipment used was rooted in antiquity, based on the classical "simple machines" known to the Romans: the pulley (in the form of multiplying tackle), the inclined plane, the lever, the screw, and the wheel and axle.

The range of time and technology covered by the landmarks in this category is surprisingly broad, reaching from the water-powered, timber-built man-engine at the Grube Samson to the great hoisting engine at the Quincy Mine, which can be considered to represent the highest expression of steam-powered stationary machinery, to the PACECO container crane, prototype of the single machine that today handles the vast bulk of ocean freight.

As exemplified by the PACECO crane, electricity has totally displaced steam in all fixed materials-handling equipment, a trend evident as early as 1905 in the form of the "Pit-cast" jib crane and typical as well in the factory traveling cranes of the period. In mobile machinery for handling materials and excavating, the diesel engine has superseded the steam engine, and, except in the very lightest service, muscle power has all but disappeared as small gasoline engines have reached a point of nearly absolute reliability.

Samson Mine Reversible Waterwheel and Man-Engine

St. Andreasberg, Germany

Mining is one of Germany's oldest industrial activities. The Samson Silver Mine, opened in 1521 and productive by 1533, was one of the first in the Harz, a region blessed by abundant water power. There, two remarkable survivals of early nineteenth-century technology are preserved: a reversible overshot wheel, used for hoisting ore out of the mine and believed to be the only survivor of its kind; and a man-engine, powered by an even larger overshot wheel, used to transport miners to and from the mine's lower levels.

The reversible waterwheel first appeared in the sixteenth century in the Erzegebirge, where it was used for dewatering mines by hoisting large buckets of water. Flooding presented no problem in the Samson shaft, where water was easily drained through adits, or horizontal shafts. The first reversible waterwheel at Samson was installed in 1556 to hoist ore from a depth of 200 feet (60 m). The present wheel, 30 feet (9 m) in diameter, was installed in 1824 and initially hoisted from a depth of 2,300 feet (700 m).

As mines became deeper, the problem of getting miners to the lower levels became acute. Imagine descending to the 600-foot (180-m) level—equivalent to the height of a sixty-story building—by ladder. Even worse, imagine climbing out at the end of a grueling day's work. A mine warden named Doerell is credited with having installed the first man-engine, at Zellerfeld, in 1833. The device allowed miners to descend and ascend the deepest shafts with a minimum of effort.

The man-engine has been described as an adaptation of an ordinary pump to pump men instead of water. The man-engine at Samson consisted of two reciprocating rods equipped at intervals of 10½ feet (3.2 m) with small platforms for the miner to stand on. As one rod moved up, the other moved down. Between strokes, when the platforms of the two rods were at the same level and momentarily at rest, the miner stepped from one platform to the other for the ride to the next station (up or down) on the adjacent rod.

An overshot waterwheel 40 feet (12 m) in diameter, which was attached to a 75-foot (23-m) connecting rod, powered the Samson man-engine. The horizontal reciprocating motion was turned into vertical motion by two interconnected bell cranks 180 degrees out of phase; thus, one man-engine rod ascended while the other descended after the common rest, during which the miner stepped from one platform to the other.

The Samson man-engine was installed in 1837, when the mine was 1,970 feet (600 m) deep. Whereas formerly it had taken a miner 90 minutes to climb down and an exhausting 150 minutes to climb out, with the man-engine the miner rode, relatively without effort, for 45 minutes each way. The man-engine increased the life expectancy of the average miner and, at the same time, increased productivity by increasing the amount of time the miner spent at work below ground.

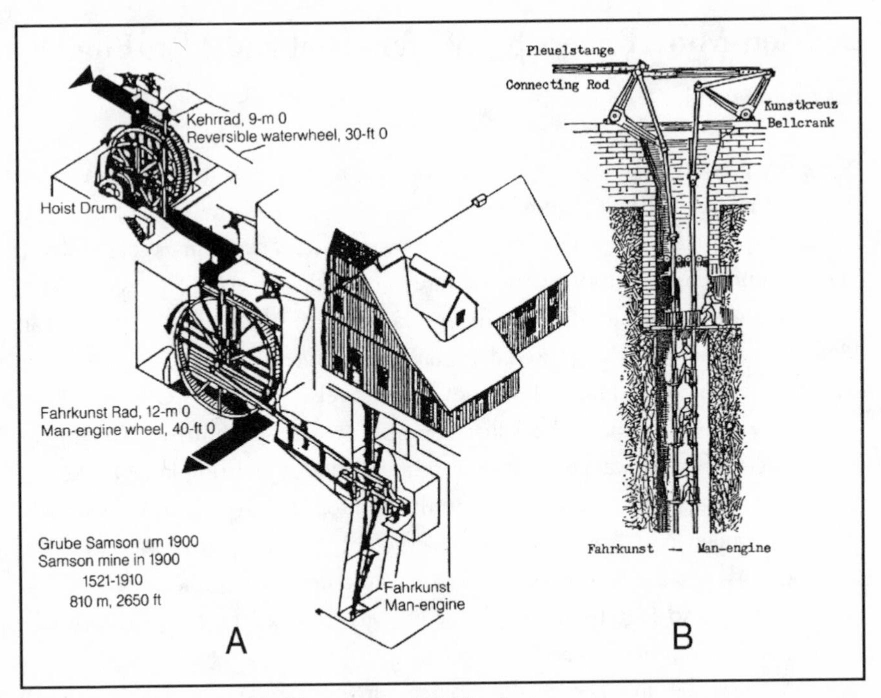

A) Samson Mine Reversible Waterwheel and B) Man-Engine.

By 1845, ten man-engines, or *fahrkunsten*, were at work in the Clausthal mining area (then in Prussia), raising and lowering men at speeds between 49 and 72 feet (15 and 22 m) per minute. The idea spread to Belgium, France, and Austria, then to Cornwall and the Isle of Man. Their reign was short—only about fifty years—for they were soon succeeded by elevators, which offered greater safety, comfort, capacity, and speed.

The Samson Mine was closed in 1910, having attained an ultimate depth of 2,656 feet (810 m). The axle and driving crank of the man-engine are original, but the waterwheel was rebuilt in 1954. The man-engine was fitted with electric drive in 1922 and today is used only by service personnel to reach the 623-foot (190-m) level and the lower of two electrical generating plants that continue to use water power at the site.

Location/Access

In addition to the silver mine and man-engine, Grube Samson, 3424 St. Andreasberg, Niedarsachsen, Germany, also features a museum of local history and the history of silver mining in the Upper Harz.

FURTHER READING

David H. Tew, "The Continental Origins of the Man-Engine and Its Development in Cornwall and the Isle of Man," *Transactions of the Newcomen Society* 30 (1955–56): 49–62.

Buckeye Steam Traction Ditcher

Findlay, Ohio

The northwest corner of Ohio, which today yields bumper crops of corn and tomatoes, was once known as the Black Swamp. The pear-shaped wasteland, 120 miles (193 km) long and 20 to 40 miles (32 to 64 km) wide, was thickly forested and covered by malarial bogs and pools of water. Beginning in the mid-nineteenth century, however, the land was extensively cleared and reclaimed. Today, the only reminders of the former swamp are the parallel rows of tile-lined drainage ditches that cross the fields like strings on a harp.

To pipe water away from croplands, farmers dug open ditches along gradients following the fall of the land. Later, they laid underdrainage tiles, a technique introduced in America in 1821 by John Johnston of Geneva, New York, a native of Scotland, and carried westward by settlers. Underdrainage lowered the water table and loosened the soil, allowing it to "breathe." Together with crop rotation and the planting of deep-rooted legumes such as clover, underdrainage unlocked the fertility of the Black Swamp soils. By 1920, Ohio had some 25,000 miles (40,000 km) of drainage ditches, of which 15,000 miles (24,000 km) were located in the Lake Erie drainage basin of northwest Ohio.

Ditches initially were dug by hand, then by horse and plow. But in 1893, James B. Hill (1856–1945), working in a Bowling Green, Ohio, machine shop, built a steam-driven mechanical ditcher. He was granted a patent for his traction ditching machine (No. 523,790) the following year. Hill's steam- and (after 1908) gasoline-powered ditchers enabled any farmer, regardless of skill, to dig ditches quickly and accurately.

Buckeye steam traction ditcher.

First, surveyors and engineers laid out the direction, depth, and grade of the ditch. The Buckeye ditcher was properly aligned, and the adjustable digging wheel, attached to a wood frame, was engaged to rotate. As the machine moved forward, the digging wheel was gradually lowered to the desired depth. A support shoe was set and locked in place behind the digging wheel, and the cables supporting the back end of the wheel frame were slackened. The operator, standing or sitting on a platform, sighted over a guide to the grade stakes, making sure to keep the digging wheel on the proper grade.

Buckets on the digging wheel scooped the dirt, carried it to the top of the wheel, and dumped it onto a transverse belt-conveyor, which deposited it to one side of the trench. The digging wheel had neither spokes nor axle, allowing it to dig to a depth nearly equal to its diameter; the width of the ditch could be altered by changing the size of the digging buckets. The ditcher excavated the full depth of the trench in a single pass, digging 3 lineal feet (914 mm) to a depth of 3 feet (914 mm) per minute in ordinary soil, or 1,800 feet (550 m) on an average working day.

The Buckeye ditcher revolutionized underdrainage ditching. Eventually, the machines could be found in almost every farm community in northwest Ohio and southern Ontario. The Buckeye Traction Ditcher Company (Hill sold the company in 1902) became the world's largest builder of ditching and trenching machinery. By 1910, some seven hundred Buckeye traction ditchers had been shipped; ultimately, more than two thousand were sold in northwestern Ohio and southern Ontario alone. Although designed to dig ditches for agricultural drainage tile, the Buckeye traction ditcher could dig trenches for pipelines or for open drainage ditches as well. Buckeye ditchers helped dig the lacework of canals that drained the Florida Everglades starting at the turn of the century and dug thousands of miles of ditches for oil pipelines around the world.

In 1936 the Buckeye Company began looking for an early ditcher to use in its advertising. They found one, No. 88, in Oklahoma and refurbished it. Buckeye displayed the ditcher at county fairs and in parades, then exhibited it in front of its plant. The steam-powered ditcher, built in 1902, has a single-cylinder engine, with a piston 5½ inches (140 mm) in diameter with a 7-inch (180-mm) stroke. The vertical boiler is 5 feet (1,524 mm) tall and 3 feet (914 mm) in diameter. The ditcher's drive wheel is 4 feet (1,219 mm) in diameter; and the ditching wheel is 7½ feet (2,286 mm) in diameter.

Hill's unique labor-saving device was the forerunner of traction ditchers used worldwide. A modified version of his machine is still manufactured by the Ohio Locomotive Crane Company in Bucyrus, Ohio.

Location/Access

The Buckeye steam traction ditcher is unassembled and in storage at the Hancock Historical Museum, 422 West Sandusky Street, Findlay, OH 45840; phone (419) 423-4433.

FURTHER READING

Frank C. Perkins, "The Buckeye Traction Ditcher," *Scientific American*, September 10, 1904, 177–78.

Peter W. Wilhelm, "Draining the Black Swamp: Henry and Wood Counties, Ohio, 1870–1920," *Northwest Ohio Quarterly* 56 (Summer 1984): 79–95.

"Pit-cast" Jib Crane

Birmingham, Alabama

This is the only survivor of six jib cranes fabricated for the American Cast Iron Pipe Company in 1905 for the manufacture of cast-iron pipe by the pit-cast method. The first recorded use of cast-iron pipe was the system that delivered water from the River Seine to the Palace of Versailles near Paris in 1664. In the United States, by as early as 1817, the Watering Committee in Philadelphia laid 9-foot (2.7-m) lengths of cast-iron pipe imported from England. Cast iron rapidly became the standard for both water and gas pipe, and its success spurred a growing demand for pipe manufactured domestically. In 1819 the first cast-iron pipe was made at the Weymouth Furnace on the Great Egg Harbor River in New Jersey, while in 1834 the first foundry devoted exclusively to making cast-iron pipe was built in Millville, New Jersey.

Initially, cast-iron pipe was made in horizontal molds in lengths of 4 or 5 feet (1.2 or 1.5 m). Two half-molds were closed around a reinforced core of baked sand whose diameter was that of the pipe bore. Pipe length was limited to the

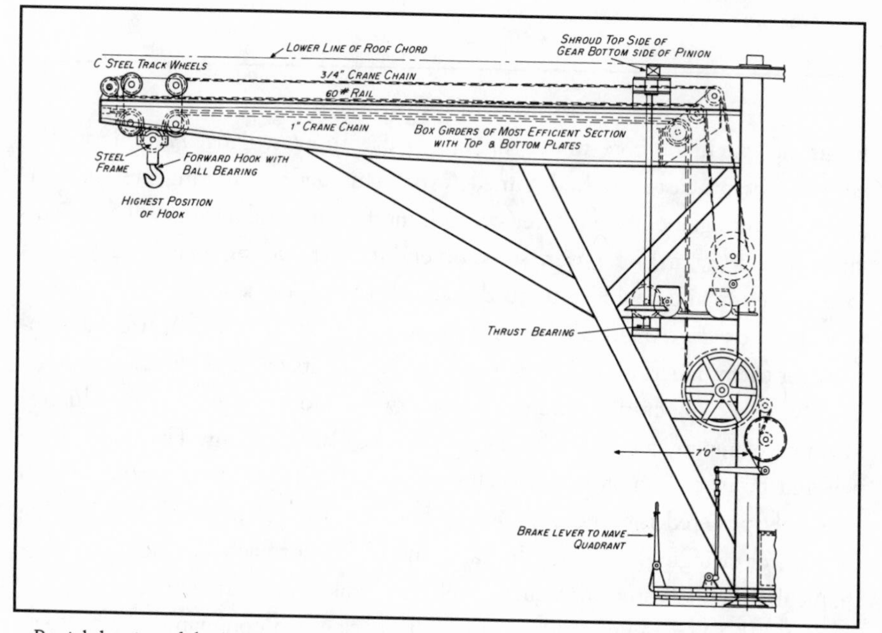

Partial drawing of the "pit-cast" jib crane.

The Manufacture of Pit-cast Pipe

The casting floor of the pit cast department of a pipe foundry is a series of pits in which the molds are rammed and poured. The molds are made in cylindrical containers, called flasks. The barrel pattern is a metal cylinder with handling rings at one end.

Empty flasks and molding sand are brought to the pits to be rammed. Damp sand is thrown in at the top between the pattern and the flask and rammed, or compacted, to form a separating wall. For pipe to be made with bell up, a bell pattern is then placed over the barrel pattern and more sand rammed around it until the mold is full. The barrel pattern is withdrawn by the crane, and the complete mold is carried to a drying oven. Hot gases bake the mold until it is thoroughly dry.

Meanwhile, cores are being prepared in another department. Both barrel and head cores are made of a mixture of sand and clay, and after being formed, are baked. When mold and cores are dry they are ready for assembly. The barrel core is lowered through the mold and seated, and the bell core is placed over it. When a group of molds and cores are assembled, molten iron is brought from the cupola in a ladle and poured into the molds. The iron solidifies; the core bar is withdrawn; the flask is lifted out of the pit and suspended horizontally over a rail runway leading to the cleaning floor; clamps are knocked off, and the pipes roll out. After cleaning, inspection, and coating, each pipe is subjected to the final hydrostatic test. In the testing process, the pipe is filled with water and must withstand a pressure considerably in excess of what it will encounter in actual service.

Source: *Handbook of Cast Iron Pipe*, 2d ed. (Chicago: Cast Iron Pipe Research Association, 1952), 25–26.

length at which the core would support itself without sagging. After 1850, the vertical—or pit-cast—method gained favor. This vertical casting method, imported from England, gave greater strength, uniformity, and accuracy. The molds, up to 16 feet (4.9 m) long, were stood on end in a pit. During pouring, impurities rose to the top and could be cropped once the casting cooled.

In 1905 the American Cast Iron Pipe Company (ACIPCO) was incorporated for the production of pit-cast pipe. It equipped its original plant at Birmingham with six jib cranes purchased from two Ohio companies: the Alliance Machine Company and the Cleveland Crane & Car Company. The cranes were powered by electric motors, a novelty in 1906. Each had three motors, which separately powered the crane's hoisting, booming (in-out motion), and slewing (turning or swinging) motions. Initially, the brakes for each function were mechanical; air brakes and, subsequently, electric brakes were added later.

ACIPCO's six cranes operated back-to-back, each commanding a 25-foot-deep (7.5-m) pit. Vertical pipe molds, lined with sand and fitted with a core to form the hollow interior of the pipe, stood in the pits. The jib crane supported the ladle from which molten iron was poured into the molds; following solidification, the mold was opened and the jib crane extracted and removed the pipe. The crane itself is made of structural steel. The large, braced box-girder jib is about 36 inches (914 mm) deep.

In the 1920s, the pit-cast method for making pipe gave way to centrifugal casting in which molten iron is poured into horizontal metal or sand-lined molds that are rotated at high speed. The metal is flung against the mold by centrifugal force, eliminating the need for a core. The resulting pipe is denser, stronger, and of more uniform wall thickness than pit-cast pipe. Following the introduction of centrifugal casting at ACIPCO, the old pits were filled in and the jib cranes removed. Only one crane, still used for general lifting, survives.

Location/Access

The jib crane is now unassembled and in storage at the Sloss Furnace Museum, 1st Avenue North and 32nd Street, Birmingham, Alabama 35202; phone (205) 324-1911. Hours: Tuesday–Saturday, 10 A.M. to 4 P.M., and Sunday, noon to 4 P.M.

FURTHER READING

Handbook of Cast Iron Pipe, 2d ed. (Chicago: Cast Iron Pipe Research Association, 1952).

Quincy Mining Company No. 2 Mine Hoist

Hancock, Michigan

Michigan's Keweenaw Peninsula on the southern shore of Lake Superior was a repository of abundant native copper. The district was equally favored by its location only a few miles from the Great Lakes waterway to the east. By 1940, Keweenaw had yielded more than 8 billion pounds (3.6 billion kg) of copper. Two mining companies dominated that output: Quincy, formed in 1846, and Calumet & Hecla, formed in 1864.

High on a hill above Hancock, sheltered by the 150-foot- (47-m-) tall Quincy No. 2 shaft-rockhouse, stands a giant among mine hoists, now silent but judged "a magnificent piece of machinery" by *Power* in its day. Built by the Nordberg Manufacturing Company of Milwaukee, the compound-condensing hoisting engine, with two high-pressure cylinders and two low-pressure cylinders, 32 and 60 by 66 inches (810 and 1,520 by 1,680 mm), had an ultimate winding capacity of 13,300 feet (4,054 m) of 1⅝-inch (41.2-mm) wire rope in a single

Quincy Mining Company No. 2 mine hoist, ca. 1925. *Courtesy L. G. Koepel, Library of Congress Collections.*

layer, the greatest on record. Quincy looked to the giant hoist, built at a cost of $371,000, to raise larger loads faster and to consume less fuel, thereby helping ensure the company's economic survival.

The hoist operated in balance—one skip car rose as the other descended—raising a load of 20,000 pounds (9,072 kg) of rock per trip at a rope speed of 3,200 feet (975 m) per minute, or 36 miles per hour (58 km/hr). It is enormous; together with condensing equipment, the unit weighs 1,765,000 pounds (800,586 kg), covers a floor space of 60 by 40 feet (18 by 12 m), and stands 60 feet (18 m) high. Containing more than 3,000 cubic yards (2,294 m^3) of concrete, the foundation for the hoist and its condensing equipment was the largest ever poured for an engine.

The cross-compound hoisting engine is really four engines in one. Arranged on an inverted V frame, the engine's two high-pressure cylinders are inclined at a 45-degree angle and connected to a common crankpin turning the drum shaft at the apex of the triangle. The two low-pressure cylinders are similarly placed at the opposite end of the drum.

The hoist's grooved, cylindro-conical drum, 30 feet (9 m) long and 30 feet (9 m) in diameter at the middle, was designed as a truss bridge of 48 cast-iron sections and drawn together by steel tension rods. The conical ends helped equalize torque on the engine by winding on a smaller drum diameter when the skip was at the bottom and the entire weight of the hoisting cable was being raised.

The new hoist greatly improved the Quincy Mine's hoisting efficiency, outperforming the duplex-noncondensing hoist it replaced. During 1921, its first year of operation, the new hoist consumed 2,400 fewer tons (2,177 t) of coal than the hoist it replaced to perform the same amount of work.

Unfortunately, by the time the hoist was installed, declining copper prices and the rise of the great Western copper regions already had reduced Michigan's role as a copper producer, and Quincy suffered one unprofitable year after another. The Depression dashed any hope of an upturn in the copper market, and the nation's deepest copper mine closed in 1931.

Location/Access

The Quincy No. 2 shaft-rockhouse is located on Route 41 north of Hancock on Michigan's Keweenaw Peninsula. The Quincy Mine Hoist Association, Inc., 201 Royce Reed Road, Hancock, MI 49930, has restored the Nordberg hoist and opens it for tours during the summer months. For information, contact: phone (906) 482-3101.

FURTHER READING

Charles K. Hyde and Larry D. Lankton, *Old Reliable: An Illustrated History of the Quincy Mining Company* (Hancock, Mich.: The Quincy Mine Hoist Association, Inc., 1982).

Thomas Wilson, "Quincy Hoist-Largest in World," *Power* 53 (18 January 1921): 90–95.

PACECO Container Crane

Alameda, California, and Nanjing, China

By 1950, the handling of ship cargo had not changed markedly from the methods used in antiquity. Even mechanized lifting facilities, capable of lifting large, heavy cargo and swinging it between ship and shore, had done little to improve dock efficiency. Vessels, meanwhile, had dramatically increased in size, and so had turnaround time (the time required to unload and load cargo), resulting in costly delays to ship and cargo owners.

In the mid-1950s, Malcolm McLean, founder of Sea-Land Service, Inc., pioneered the concept of shipping goods in intermodal containers—containers that detach from the truck chassis for loading on ships or railroad cars and vice versa. Containerization drastically reduced labor costs and turnaround time. But most ports were not equipped to handle the heavy containers except by mobile-type revolving cranes, which were cumbersome and lacking in stability.

In 1956 the Matson Navigation Company embarked on a two-year, multi-million-dollar study of containerization. Among other things, Matson set out to determine the most efficient crane for loading containers between ship and shore.

PACECO container crane at Encinal Terminals, Alameda, California, the site of its original installation.

The company concluded that an ore-unloading type, having a horizontal boom and through-leg trolley, came closest. Early in 1958, Matson finalized performance specifications for a new crane and put the project out for bid. Pacific Coast Engineering Company, Inc. (PACECO), of Alameda, California, won the contract.

PACECO engineers, led by PACECO president Dean Ramsden, chief engineer Chuck Zweifel, and assistant chief engineer Murray Montgomery, analyzed each component of the "hook cycle"—that is, the steps in the process of loading or unloading, from "hook on" to "unhook." Following the philosophy that the best design has the fewest number of pieces, PACECO designed a simple A-frame crane of 260 tons (236 t) deadweight, replacing the usual trussed construction with all-welded box girders wherever possible. Controlled by switches, the traveling crane could handle 25-ton (23-t) loads with ease. Multiple-cable rigging, meanwhile, gave excellent load stability.

On January 7, 1959, Matson put the world's first high-speed container crane into service as part of its West Coast—Hawaii trade at Encinal Terminals in Alameda. The PACECO container crane revolutionized the handling of cargo, cutting turnaround time from as much as three weeks to as little as eighteen hours.

Whereas a longshore gang using a ship's burtoning gear could handle approximately 9 tons (8.2 t) of cargo per hour, a container crane operator, working on a three-minute hook cycle per 20-ton (18-t) container, could handle 400 tons (363 t) of cargo per hour. Containerization also reduced damage and pilferage—after loading at the warehouse, the container remained sealed until arrival at the consignee's warehouse—and allowed a closer and more efficient alliance among rail, road, and water transport.

Matson installed two more PACECO cranes, at Los Angeles and Honolulu, in 1960. In the 1960s, the International Standards Organization adopted a uniform corner fitting and guidelines for container dimensions. Meanwhile, shipping companies worldwide, increasingly aware of the advantages of containerization, commissioned new vessels specially designed to handle containers.

The PACECO container crane at Encinal Terminals, modified to enable it to serve larger ships, remained in regular use until 1984. In 1988 it was dismantled and shipped to the Port of Nanjing, China. Today's ship-to-shore container cranes are direct descendants of the first one, and their basic design remains unchanged.

Location/Access

The original PACECO crane is now located at the Port of Nanjing, China, a government-operated port near Shanghai.

FURTHER READING

"Systems Approach Puts Matson Cargo in Containers," *Modern Materials Handling* 14 (April 1959): 93–96.

Environmental Control

INTRODUCTION by Robert B. Gaither

Among the many differences that distinguish people from other animals is one that many regard as among the most prominent, namely, the fact that people have controlled much of their own evolution. We did not survive by adapting to changes in the environment. We made adjustments to our immediate surroundings to an extent that we could not only survive but could live and work in comfort while surrounded by the most severe climatic conditions on earth.

The conditioning of air to match people's desire to be comfortable has its beginnings in prehistory, when our earliest ancestors devised schemes for keeping warm when the weather became cold. There are numerous accounts of people in early civilizations using natural ice, running water, and heated stones to alter the temperature and humidity of spaces. However, the existence of systematic efforts to heat or cool and control the humidity of air is not known to us before the early nineteenth century. During the first millennium, the Romans heated a number of structures using heated stones and cleverly arranged stone ducts to carry flue gases from fires under the rooms of these structures. Nevertheless, the field of environmental control saw no major advances until 1740 to 1745, when Benjamin Franklin invented the Pennsylvania Fireplace, which probably was the first true heating system. This system took in cool outside air and, by natural convection, passed it over plates that were being heated on their other sides by fire. Then with carefully arranged ducts, the air was distributed into the rooms of a house. The so-called Franklin Stove was only a part of the Pennsylvania Fireplace.

Prior to the advent of prime movers, air movement or ventilation could be accomplished only by natural convection or by human or animal power. In the years after 1800, the beginnings of the Industrial Revolution and the attendant interest in the development of novel mechanical systems driven by engines were making their appearance. In 1815, the Marquis de Chabannes was granted a

British patent for ". . . a method for conducting air and regulating the temperature in houses and buildings." The method used for cooling the air involved the use of a fan to pass it through an evaporative cooling tower. Other schemes developed shortly thereafter called for passing air over metal plates that could either be heated or cooled with running water or ice.

Jacob Perkins is credited with being the first person to design a closed vapor-compression refrigeration system in 1834 (British patent No. 6,662). Although he built a working model of the system, it received little attention. It took another twenty years before others built mechanical refrigeration machines that were used in industry. In the 1850s, vapor-compression machines were developed in the United States and other countries. At this same time, Ferdinand Carré, in France, developed the ammonia absorption refrigeration system.

For the next twenty years, experimentation continued alongside early efforts to manufacture commercial systems that would wash, heat, and refrigerate air. In July 1869, Benjamin F. Sturtevant took out a U.S. patent on a system comprised of a fan, duct work, and a heat exchanger that could be used to heat or refrigerate air. With that patent and the know-how, Sturtevant began a prosperous business. In 1873, a Frenchman, A. Jouglet, wrote a detailed account of various methods for cooling air using ice, underground tunnels, and refrigeration machines. In 1894, Herman Rietschel, a professor at the Berlin Institute of Technology, began describing the heating, cooling, and humidity control of air as a recognized science in a series of publications entitled "Guide to Calculating and Design of Ventilating and Heating Installations."

Rietschel's scientific approach to the design of systems to condition air was first introduced to the United States by Alfred Wolff, a consulting engineer who designed several heating and cooling systems for medical colleges and hospitals. Wolff crowned his career in 1901 by using waste heat to operate a cooling system to air condition the New York Stock Exchange. Other engineers followed with a series of innovations and set the stage for a visionary who possessed a sharp understanding of business and what it takes to mold an infant enterprise into a strong and prosperous industry. The person who accomplished this was Willis H. Carrier.

In 1901, Carrier graduated from Cornell University with a degree in electrical engineering and immediately took a position with the Buffalo Forge Company. After successfully completing the design of several air-handling systems, Carrier began a series of experiments to better understand the complex mechanisms taking place when lowering the humidity in an airstream. During the first decade of the twentieth century, Carrier established the fundamental principles of psychrometry and wrote a handbook containing all of the formulae needed to design systems that could condition air. In 1907, he persuaded Buffalo Forge to establish a subsidiary company with himself at the helm and proceeded to take steps in the development of the air-conditioning industry. At the Winter Annual

Meeting of the American Society of Mechanical Engineers in December 1911, Carrier presented a paper that has become the single most fundamental document in the air-conditioning industry. In this paper, he clearly explained the basic precepts of psychrometry and offered a psychrometric chart for making calculations. The formulae and chart contained in Carrier's paper have since been reproduced in virtually every textbook on ventilation and air-conditioning in use today.

In 1915, Carrier left Buffalo Forge and established his own company. In the 1920s, he introduced small and reliable air-conditioning units, and in 1930 placed air conditioners on railroad cars and in theaters throughout the nation. Later in that decade, he installed air conditioners in the chambers of the Senate and House of Representatives in the U.S. Capitol.

Since that time, mechanical engineers have designed air-conditioning systems of incredible size and impressive sophistication. Today, the air-conditioning industry holds a strong position in the economy of many nations and is given credit for the conversion of tropical areas into areas of industrial productivity as well as locations for comfortable living.

Holly System of District Heating

Lockport, New York

While Birdsill Holly (1822–94) is best known for his waterworks machinery (see "Holly System of Fire Protection and Water Supply," p. 10), the last years of his life were devoted largely to the development of district steam heating. In 1876 Holly improvised a boiler in the basement of his home and laid a 700-foot (213-m) steam line around his yard and an adjoining property. The small-scale installation functioned perfectly, convincing Holly that buildings over a wide area could be heated by steam from a central plant.

In 1877 Holly organized the Holly Steam Combination Company and served as its chief engineer. The company laid steam pipes that supplied residences, churches, hotels, and other buildings in Lockport, New York, with heat. Holly's system was designed to overcome the inefficiency of heating buildings with small, individual boilers. From a large central boiler plant, Holly furnished steam under moderate pressure to a group of closely located buildings through a loop of insulated supply and return mains. Each customer was charged for the steam consumed, measured by a meter of Holly's own design.

Following the system's successful demonstration in Lockport during the winter of 1878, Holly was hired to install similar systems in other cities. The size of the installations ranged from 1½ to 16 miles (2 to 26 km) of underground pipe. The company was reorganized as the American District Steam Company in 1880. Holly continued to serve as engineer until his retirement in 1888, then continued as consulting engineer until his death six years later.

Holly's first district heating plant, at Elm and South streets, operated continuously from 1877 until 1970, when the boiler house was shut down and demolished. Holly's ideas have proved more durable: district heating today is enjoying a rebirth as a practical and economical way to heat buildings in compact urban districts. (See "Detroit Edison District Heating System," p. 203.)

In 1987 the American Society of Mechanical Engineers designated the Holly System of Fire Protection and Water Supply and the Holly System of District Heating as Mechanical Engineering Heritage Sites. The designations, the first of their kind, recognize important developments in the history of mechanical engineering even though a structure or object may no longer be extant.

Location/Access

The American Society of Mechanical Engineers plaque is located at the Erie Canal Museum, New York State Canal Corporation, 80 Richmond Avenue, Lockport, NY 14094; phone (716) 434-3140.

FURTHER READING

Morris A. Pierce, "The Introduction of Direct Pressure Water Supply, Cogeneration, and District Heating in Urban and Institutional Communities, 1863–1882" (Ph.D. diss., University of Rochester, 1993).

Stirling Water-tube Boilers

Dalton, Georgia

When they were installed in the Elk Cotton Mills in 1906, the Stirling water-tube boilers represented the state of the art in steam boiler design. Today the coal-fired, hand-fed boilers, which supplied steam to power mill machinery until 1975, are among the oldest extant steam boilers in a cotton mill.

As demand for greater amounts of power grew in the late nineteenth century, it was necessary to build ever larger boilers operating at higher pressures. Fire-tube boilers then in use, built of small plates riveted together, were limited in capacity and pressure, and explosions were commonplace.

The water-tube boiler was developed gradually, with dozens of engineers tackling the problem. Stephen Wilcox (1830–93) together with the Scottish-born mechanical engineer Allan Stirling (1844–1927) are among those credited with its substantive development. The Stirling water-tube boiler, with its bent tubes (Stirling himself patented a machine for bending steel and wrought-iron tubes in 1893) was a superior design.

In the water-tube boiler, the water and steam are contained inside the tubes, while the hot gases are in contact with the outer tube surfaces. The water-tube boiler offered quicker steaming, greater ease of cleaning, improved fuel efficiency, and more efficient use of space; it was also able to operate at higher steam pressures than fire-tube boilers. But perhaps its greatest contribution was improved safety. Rapid circulation reduced temperature stresses and the unequal expansion and contraction that were the common cause of fatal explosions, while the smaller drums of the water-tube boiler could be made of thinner metal rolled uniformly, which was less likely to rupture. In fact, the highest temperatures were in the tubes; were a rupture to occur, the damage would be localized.

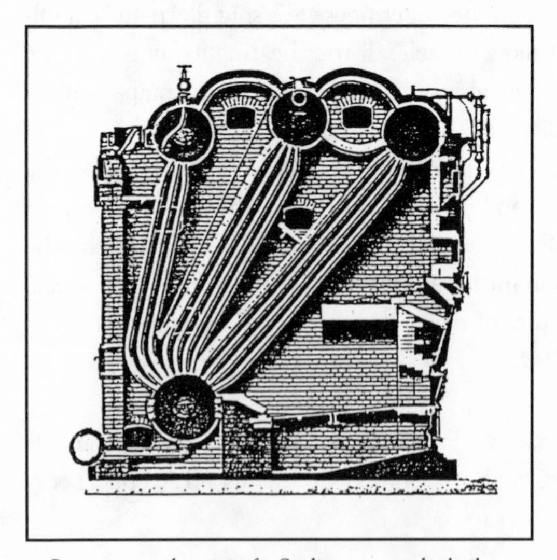

Cross-section drawing of a Stirling water-tube boiler showing its characteristic bent-tube construction.

The Stirling water-tube boiler was built in a number of different classes to meet varying conditions of floor space and headroom. It consists of three steam-and-water drums, 36 to 54 inches (914 to 1,371 mm) in diameter, set parallel and each connected by a bank of curved water tubes to a lower mud drum. Shorter tubes connect the steam spaces of the upper drums and the water spaces of front and middle drums. The boiler is supported on a structural steel framework and surrounded by a brick housing to contain the combustion and minimize heat loss.

The Stirling water-tube boiler was first manufactured commercially by the International Boiler Company of New York in 1889. In 1890 the Stirling Boiler Company was established and purchased the assets of the International Boiler Company. In 1906 the Babcock & Wilcox Company acquired the Stirling Consolidated Boiler Company with its 65-acre (26-ha) plant in Barberton, Ohio. By this date, the Stirling water-tube boiler was in widespread use. It still forms the basis of most modern boilers, particularly in ships.

The Stirling boilers at the Elk (later Crown) Cotton Mills supplied steam at 180 psig (1,241 kPa) and 2,500 horsepower (1,864 kW) to operate a Hamilton compound engine powering the mill's line shafting, a Fleming high-speed engine driving a generator, and a fire pump. They remained in continuous use until 1975, when the company switched to commercial electric power. The boilers continued to heat the mill until 1986, when they were relegated to standby status.

Location/Access

The Stirling boilers are open upon application to CrownAmerica, Inc., 714 Chattanooga Avenue, Dalton, GA 30720; phone (706) 278-1422.

FURTHER READING

Glenn R. Fryling, ed., *Combustion Engineering: A Reference Book on Fuel Burning and Steam Generation*, rev. ed. (New York: Combustion Engineering, Inc., 1966).

Steam: Its Generation and Use (New York: The Babcock & Wilcox Co., various editions).

Allan Stirling, "Shell and Water-Tube Boilers," *Transactions of the American Society of Mechanical Engineers* 6 (November 1884 and May 1885): 566–618.

The Stirling Company, *Stirling: A Book on Steam for Engineers* (New York: The Stirling Company, 1905).

Detroit Edison District Heating System, Beacon Street Plant

Detroit, Michigan

In an effort to cut energy costs and attract businesses back downtown, many cities today are returning to an old method of energy distribution: district heating. In district heating, steam turbine exhaust from electrical generating plants or steam from dedicated boilers is distributed through underground pipes to homes and businesses located in densely built downtown districts. District heating does away with the need for boilers in individual buildings, saving space and reducing both start-up and operating costs.

Detroit-Edison Beacon Street Plant in 1926, shortly after its completion.

Birdsill Holly introduced district heating at Lockport, New York, as early as 1877, demonstrating how a single, large steam plant could operate at higher overall thermal efficiency than a series of small, isolated boilers, especially in the commercial districts of cities (see "Holly System of District Heating," p. 201). But until the early twentieth century, engineers doubted the commercial practicability of district heating in conjunction with electric lighting.

By the basic laws of thermodynamics, electric power plants waste thermal energy, since their maximum thermal efficiency—the difference between the thermal content of the fossil fuel they use and the thermal energy contained in the electricity generated—can never exceed about 35 percent. The remaining energy is transferred to the environment, mostly in the condenser cooling water and stack gasses. Little by little, experience showed that district heating could be a profitable service wherever the power plant was located near a closely built part of the city, where large loads could be served with a minimum investment in distribution mains. By 1928, twenty-six U.S. companies were providing district heating to just more than ninety-three hundred customers. One of the largest of these was the Detroit Edison Company.

In 1903 the Central Heating Company was organized to distribute surplus exhaust steam from Detroit Edison's Willis Avenue station to buildings in what was then an affluent residential district of Detroit. There, the peak of the heating load—early morning, when residents wanted to warm up houses that had been allowed to cool down during the night—neatly corresponded with the peak electrical demand of the Detroit United Railway Company. By burning softer, cheaper coal under the station's boilers than would be burned by private homeowners, Central Heating could enjoy a profitable rate for supplying heat.

Detroit's district heating system was soon expanded with the addition of two new plants—Farmer Street in 1904 and Park Place in 1912—and the purchase, in 1914, of Murphy Power Company, which had been supplying steam heat in the south end of the central business district. Detroit Edison purchased all of the assets of Central Heating in 1915.

To meet increased demand for heating service, from 1925 to 1926 Detroit Edison erected the Beacon Street plant, equipped with two 4,155-horsepower (3,098-kW) boilers. A third, 4,237-horsepower (3,159-kW) boiler was added the following year, and a fourth, of 4,155-horsepower (3,098-kW), in 1929. A single turbine, installed to act as a pressure-reducing valve on the most heavily loaded feeder, produced by-product electricity, which was delivered to Detroit Edison's electrical system.

Detroit Edison provided heating service by the feeder method, distributing steam through high-pressure mains that connected the heating plant with street mains that distributed it to individual customers. The street mains carried a nominal pressure of 35 psig (241 kPa), from which individual service was offered at a guaranteed pressure of 10 psig (69 kPa). Sold on a metered basis, district heating was especially popular for office and other commercial buildings. As new buildings rose in the district served by Detroit Edison, they invariably were connected to the district heating system, precluding the need for chimneys or space devoted to boilers, fuel, and auxiliary devices.

In 1959 Detroit Edison installed a new boiler and turbine-generator at Beacon Street, along with a new 24-inch (609-mm) main steam line. The boiler is the largest in Detroit Edison's central heating system, producing up to 500,000 pounds (226,000 kg) of steam per hour at 900 psig (6,205 kPa) and 700°F (371°C). The steam produces up to 19.5 megawatts of electricity through the new turbine-generator before being exhausted into the steam mains for customer use. This "cogeneration" results in high thermal efficiency.

Now gas-fired, the Beacon Street plant still serves southern Detroit. Detroit Edison's three district heating plants (Beacon, Willis, and Boulevard) supply steam through 53.6 miles (86 km) of mains carrying from 30 to 135 psig pressure (207 to 931 kPa), depending on customer demand. Since less than 5 percent of the steam is returned to the plant in the form of condensate, the system requires approximately 240,000 gallons (908,400 l) of water per hour to produce steam.

Location/Access

Contact the Detroit Edison Company, General Office, 2000 2nd Street, Detroit, MI 48226; phone (313) 237-8000.

FURTHER READING

Thomas C. Elliott, "District Heating and Cooling: Renewed Interest in Old Concept," *Power* 131 (February 1987): 15–22.

E. E. Dubry, "Central Heating in Detroit," *Heating and Ventilating Magazine* 26 (April 1929): 67–70; (May 1929): 73–75.

J. H. Walker and A. R. Mumford, "Present Status of District Heating," *Power Plant Engineering* 34 (1 September 1930): 994–96.

Holland Tunnel Ventilation System

Jersey City, New Jersey, and New York City, New York

The 1.6-mile-long (2.6-km) Holland Tunnel, connecting Jersey City with lower Manhattan, was the first tunnel under the Hudson River designed for motor vehicles and, upon its completion in 1927, the longest subaqueous tunnel in the world. It was also the first mechanically ventilated vehicular tunnel in the world. (London's Blackwall and Rotherhithe tunnels were ventilated by the natural movement of air through the shafts and portals.) Built by the states of New Jersey and New York, the Holland Tunnel pioneered solutions to novel civil and mechanical engineering problems, especially the problem of ventilation, and served as a model for the subsequent construction of the Lincoln, Queens-Midtown, Brooklyn-Battery, and other vehicular tunnels throughout the world.

Since 1906, both New Jersey and New York State had sought some way to supplement the ferries plying between Jersey City and Manhattan and relieve traffic congestion. In 1919 the bridge and tunnel commissions of the two states, formed to devise a solution to the problem, received authorization to build a tunnel. A triumverate of engineers planned and supervised the project: chief engineer Clifford Milburn Holland (1883–1924), who died during construction and for whom the tunnel is named; Milton H. Freeman (1871–1925), who succeeded Holland but died five months later; and Ole Singstad (1882–1969), who supervised completion of the tunnel.

Holland evaluated numerous tunnel cross sections and roadway widths before deciding on twin tubes, each 9,250 feet (2,819 m) long and 29.5 feet (8,991 mm) in diameter, with a two-lane roadway 20 feet (6,096 mm) wide. He considered trench, caisson, and shield methods of construction. The great volume of river traffic and the soft, silty river bottom were decisive factors in selecting the shield method, invented and first employed by Marc Isambard Brunel for excavating a tunnel under the Thames in London in 1825.

The modern tunnel shield is a steel cylinder whose forward edge acts as a cutting edge and whose rear end overlaps the tunnel lining and provides protection for the work. As hydraulic jacks push the shield forward, sandhogs (laborers working under compressed air inside the tunnel) bolt the tunnel's cast-iron rings together. An atmosphere of compressed air in the working area counterbalances the pressure of the water and prevents it from entering the tunnel.

A paramount challenge was to design a ventilation system to clear the tunnel of noxious automobile and truck exhaust fumes. Following physiological and mechanical tests conducted at Yale University, the University of Illinois, and by the United States Bureau of Mines, engineers devised a transverse-flow system of ventilation that served as a model for all subsequent vehicular tunnels.

The air is moved by 84 giant fans of 6,000 total horsepower (4,474 kW)—42

One of eighty-four fans at the Holland Tunnel. *Courtesy Port of New York Authority.*

blower units and 42 exhaust units, the "lungs" of the tunnel—arranged in four ventilating buildings (two on each side of the river). Fresh air is drawn into the ventilation buildings and blown by fans into a fresh-air duct running the length of the tunnel beneath the roadway; the fresh air enters the tunnel through narrow flues, spaced 15 feet (4,572 mm) apart, in the roadway curb. Meanwhile, exhaust fans pull the vitiated air through ports in the ceiling into exhaust ducts running the length of the tunnel and discharge it into the atmosphere through stacks in the ventilation buildings. A tunnel operator at a central control board monitors the carbon monoxide generated by tunnel traffic and changes the rate of ventilation as needed. The ventilation system, manufactured and installed by the B. F. Sturtevant Company, is capable of completely changing the tunnel air every ninety seconds.

Construction of the Holland Tunnel began on October 12, 1920. It was opened to traffic on November 13, 1927, eliminating the time-consuming trip by ferry and strengthening the economy of the New York–New Jersey metropolitan region.

In 1984 the tunnel was jointly recognized by the American Society of Civil Engineers and the American Society of Mechanical Engineers as a historic civil and mechanical engineering landmark.

Location/Access

The Holland Tunnel links 12th and 14th streets in Jersey City with Canal and Spring streets in lower Manhattan. It is operated by the Port Authority of New York & New Jersey. There is a toll for eastbound vehicles.

FURTHER READING

B. F. Sturtevant Company, *The Eighth Wonder* (Boston: B. F. Sturtevant Company, 1927).

Magma Copper Mine Air-Conditioning System

Superior, Arizona

In 1937 the Magma Copper Company installed an underground refrigeration plant to air-condition the 3,400- and 3,600-foot (1,036- and 1,097-m) levels of the Magma Mine at Superior, Arizona. Dr. Willis H. Carrier (1876–1950), celebrated air-conditioning pioneer and chairman of the Carrier Corporation of Syracuse, New York, personally designed the Magma Mine installation and supervised the manufacture of the equipment for what became the first mine in North America to be cooled by mechanical refrigeration. (Carrier previously had designed the air conditioning installations at the Morro-Velho Mine in Brazil about 1914 and at the Robinson Deep Mine near Johannesburg, South Africa, in 1934.)

The sulfides in copper oxidize when the ore comes into contact with air, generating heat. Rock temperatures as high as 140°F (60°C) on the Magma's 4,000-foot (1,219-m) level, combined with humidity caused by groundwater inside the mine, made air-cooling necessary. The company's usual practice was to open up a level and let it stand for several years to dry out and cool off through the use of ventilating fans; even after the level had been ventilated, production efficiency was hampered by the heat and humidity. By installing underground air-conditioning, Magma hoped to hasten its mining operations and improve miners' efficiency.

To cool the Magma Mine, Carrier furnished two centrifugal refrigeration units, each powered by a 200-horsepower (149-kW) induction motor, on the 3,600-foot (1,097-m) level. (The size of the air-conditioning equipment was limited by the size of the shaft compartments—40 inches by 60 inches [1,016 mm by

Dr. Willis H. Carrier (second from the right) inspects the rotor of a centrifugal refrigeration unit of the type installed at the 3,600-foot (1,097-m) level of the Magma Copper Mine.

1,520 mm]. For this reason, Magma installed two units, rather than a single, larger one.) Chilled water was pumped to fin coils on the 3,400- and 3,600-foot (1,036- and 1,097-m) levels. Fans powered by 50-horsepower (37-kW) motors drew air over the coils at the rate of about 30,000 cubic feet (850 m^3) per minute. In passing over the coils, the air was cooled below its dew point, resulting in dehumidification as well as cooling. The Magma engineering department worked out the problem of water supply, transporting groundwater from the 2,500-foot (762-m) level in open ditches and pipes to a sump on the 3,600-foot (1,097-m) level; from there, pumps delivered it to the refrigeration units.

The day before the Magma air-conditioning plant started up, temperatures on the 3,600-foot (1,097-m) level averaged 101°F (38°C) dry bulb and 93°F (34°C) wet bulb; after four months of air-conditioned ventilation, the average of all working places on both levels had been lowered to 80°F (27°C) dry bulb and 72°F (22°C) wet bulb. Air-conditioning not only improved the comfort and efficiency of the miners but also accelerated its development.

By 1941, Magma had extended cooling to four additional levels of the mine. That year, Carrier engineer J. F. Kooistra succinctly described the significance of the Magma installation. The air-conditioning of deep-shaft gold and copper mines, he wrote in *The Mining Journal*, "may be considered as one of the greatest steps being taken by mankind in the search for new methods to increase production, safeguard investments, and improve the working conditions of human beings."

Location/Access

The Magma's pioneer refrigeration units were cannibalized and abandoned in place after the ore was removed.

FURTHER READING

William Koerner, C. B. Foraker, and J. F. Kooistra, "Air Conditioning Mines," *The Mining Congress Journal* 23 (November 1937): 21–25.

J. F. Kooistra, "Air Conditioning of Magma Mine," *The Mining Journal* 20 (15 May 1937): 3–4.

———, "Doubling of Magma's Air Conditioning Plant," *The Mining Journal* 25 (15 December 1941): 3–5.

Equitable Building Heat Pump

Portland, Oregon

Considered a benchmark of modern architecture because of its pioneering use of a thin glass-and-metal curtain wall, the twelve-story Equitable Building, designed by Portland architect Pietro Belluschi and completed in 1948, enjoys a second distinction: it was the first large commercial building in the nation to incorporate a heat-pump system for heating and cooling.

The theoretical conception of the heat pump was described as early as 1824 by French physicist Nicolas Léonard Sadi Carnot (1796–1832) in his classic *Réflexions sur la puissance motrice du feu (Reflections on the Motive Power of Heat)*. In 1854 Sir William Thomson (Lord Kelvin), a pioneer of mechanical refrigeration, called it "tomorrow's method of heating." But the potential of the heat pump (also known as reverse-cycle refrigeration) remained unrealized until the 1940s, when the further development of air-conditioning and commercial and industrial refrigeration suggested the practicality of heating applications. Electric utilities saw the heat pump as a way to boost the consumption of kilowatt hours and encouraged further research and development. By 1947, there were more than 150 commercial and residential heat-pump installations nationwide, the majority in California.

Following World War II, consulting engineer J. Donald Kroeker collaborated with Belluschi in planning a new 212,000-square-foot (19,695 m²) headquarters building for the Equitable Savings & Loan Association. Determined to build the most modern office building possible, Equitable officials gave Belluschi the green light to design a sleek glass-and-aluminum-sheathed building with year-round air-conditioning. The Equitable Building became a prototype for new office buildings nationwide.

Portland's moderate winters and warm, humid summers made use of the heat pump attractive—air-conditioning would require cooling capacity greater than heating capacity—while the availability of cheap hydroelectric power made it economically feasible. The heat pump operates as does a household refrigerator, with essentially the same elements and with exactly the same cycle. However, instead of maintaining a building at a lower temperature than the surroundings, it supplies heat. The concept can be demonstrated by touching the condenser coils at the back of a refrigerator; they are warm. This heat is equal to the sum of the

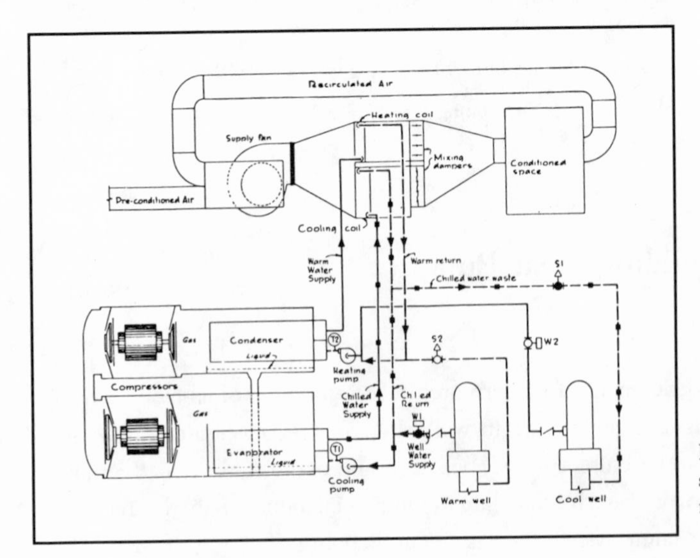

Basic heat-pump system in heating mode.

heat removed from the food compartment and the heat equivalent of the energy supplied to the electric motor driving the refrigerator's compressor.

The heat pump in the Equitable Building has four different modes of operation, each automatically controlled, depending on the outside air temperature. When it is below 50°F (10°C), the heat pump transfers energy from warm (about 63°F, 17°C) well water to the cold air entering the building, thereby raising its temperature. The system in this instance is operating as a heat pump because, in contrast to the household refrigerator, the object is to increase the temperature of the incoming air, not decrease the temperature of the warm well water.

When the outside air temperature is equal to 50°F (10°C), no energy is needed from the warm well water because the energy generated by the building's occupants, office equipment, lights, etc., is sufficient to balance that lost to the outside air through air leakage. For outside air temperatures between 50°F (10°C) and 75°F (24°C), it is necessary to dehumidify the incoming air by lowering its temperature until the moisture in the air condenses. The energy that must be extracted is transferred by the heat pump, either to water being pumped from the cold (57°F, 14°C) well to the two warm wells, or to the dehumidified incoming air. The system is now operating as a refrigerator, rather than as a heat pump, because the object is to cool the incoming air (for dehumidification) rather than to raise the temperature of either the warm well water or the dehumidified incoming air. Once the outside air temperature rises above 75°F (24°C), it is no longer necessary to raise the temperature of the dehumidified air, and all the energy removed from the incoming air in the dehumidifier is transferred by the heat pump (refrigerator) to the well water that is being pumped from the cold to the hot wells.

The heat-pump system incorporates an additional, ingenious feature that conserves energy when the air temperature is below 50°F (10°C). After leaving the evaporator of the heat pump at a temperature between 50°F (10°C) and 53°F (12°C), the water from the warm well passes through a heat exchanger located in the duct carrying warm air leaving the building. The water leaves the heat exchanger at about 53.7°F (12°C), then, before being discharged to the cold well, passes to another heat exchanger placed in the duct that carries the cold air leaving the building. The water is colder than the leaving warm air and warmer than the incoming cold air, thereby transferring energy from the leaving air to the entering air. This saves about 30 percent of the energy that would otherwise have to be supplied to the building and, furthermore, avoids the need for auxiliary heat.

The refrigeration system, which forms the core of the heat pump, consists of four units—two of 200-ton (700-kW) capacity using Freon 11 as refrigerant, and two of 70-ton (250-kW) capacity using Freon 113 as refrigerant. The latter units are used when the system is operating as a heat pump; with an outside air temperature of 10°F (-12°C), the heat pump cools the warm well water, which is being pumped at 600 gallons per minute (2,271 l/m) to the cold well, from 58.7°F

(15°C) to 50°F (10°C). The 200-ton (700-kW) units are used when the system is operating as a conventional air-conditioning system, cooling and dehumidifying the incoming air; these cool the water that is supplied to the heat transfer units (coils) in the air-conditioning system to 40°F (4.4°C).

In more than forty years of operation, some changes have been necessary—all water chillers, for example, have since been replaced—but the heat-pump system continues to provide economical heating and cooling. In 1953 calculations of comparative costs showed that district steam heat would have cost six times as much as the heat pump; oil, four times as much.

Location/Access

The Equitable (now Commonwealth) Building is located at 421 S.W. Sixth Avenue in downtown Portland.

FURTHER READING

"Equitable Builds a Leader," *Architectural Forum* 89 (September 1948): 98–106.

J. Donald Kroeker and Ray C. Chewning, "A Heat Pump in an Office Building," *Heating, Piping & Air Conditioning* 20 (March 1948): 121–28.

Water Transportation

INTRODUCTION by Euan F. C. Somerscales

Robert H. Thurston, the distinguished engineer and engineering historian, wrote in 1878: "The realization of the hopes, the prophecies, and the aspirations of earlier times, in the modern marine steam engine, may be justly regarded as the greatest of all triumphs of mechanical engineering." (*A History of the Growth of the Steam Engine*, 2d ed. [Ithaca, N.Y.: Cornell University Press, 1939], 221). While we might, in some respects, temper that judgment today when we consider, for example, the development of the aircraft turbojet engine, nevertheless it contains substantial truth when we look at marine-propulsion developments subsequent to Thurston's day, namely, the steam turbine, the gas turbine, and the turbo-charged, two-stroke compression ignition (diesel) engine (the most efficient prime mover currently in use). The landmarks associated with water transportation start with the SS *Great Britain*, which was launched in 1843. Although now lacking her engine, since replaced by a wooden replica, this ship is remarkable because it was the first vessel to embody all the elements of the modern ship: metal construction, steam-driven screw propeller, and a large size intended to ensure that it could be operated profitably on long voyages.

In the early days of the application of marine steam engines, those used for naval purposes and those applied in merchant ships followed a different design philosophy. Typically, the engines of naval vessels were of low overall height because it was considered essential for as much of the engine as possible to be below the waterline in order to protect it from enemy fire. The landmark engines of the TV *Emery Rice*, a so-called back-acting engine, are typical of the "folded" arrangements associated with the first application (up until the late 1880s) of steam in naval vessels. From about 1890 onward, the construction of large ships with substantial draught for the navy allowed the use of vertical engines, driving the propeller directly without any intervening gears, chains, or levers. This was a type that had been used since about 1860 in merchant vessels. The USS *Olympia*, one of the landmarks described in this chapter, was an early example of the use of this type of engine in naval vessels.

Two of the most significant advances in marine steam-power-plant design were the invention of the surface condenser and the adoption, in about 1850, of the compound engine,which greatly improved the efficiency of the marine steam power plant. A 6 percent improvement in efficiency, which was typical, represented a saving of 100 tons of coal in one transatlantic voyage. Besides saving in coal costs, it allowed 100 tons more cargo to be carried by a merchant ship and represented an increase in the steaming range of a naval ship.

With further advances in steam pressure, as allowed by improvements in boiler design, triple-expansion engines were introduced in 1874. In these, the steam was expanded successively in three cylinders of increasing diameter. The engines of the USS *Olympia*, USS *Texas*, and the SS *Jeremiah O'Brien*, described in this chapter, are all of the triple-expansion type.

Those of the *Texas* were the ultimate in the design of naval reciprocating engines. Nevertheless, they were an anachronism. The next development in marine propulsion power, the steam turbine, was already in use by the U.S. Navy, but, as the article in this chapter on the *Texas* explains, the early application of the steam turbine to naval vessels was a victim of premature enthusiasm.

Initially, it was the demand for increasing power, without the large bulk of the reciprocating steam engine, that interested ship owners and navies in the steam turbine. Ultimately, it was the possibility of employing a significantly more efficient power plant that led to the turbine superseding the reciprocating engine. Although, in the case of the transatlantic "greyhounds" the feasibility of producing the necessary power from a steam turbine of much smaller dimensions than the comparable reciprocating steam engine must also have been an important consideration. Charles Parsons (1854–1931), the inventor and developer of the turbine that bears his name, demonstrated the capabilities of the steam turbine in marine applications by building the *Turbinia*, which is described in this chapter.

It was inevitable that the nuclear reactor should be considered as a power source for marine applications. In those early, heady days following the demonstration of controlled nuclear fission in the atomic pile built in Chicago in 1942, when electric power produced by atomic fission was going to be "too cheap to meter," the atomic reactor must have seemed to be an ideal source of marine power. This appears to have been true for naval applications, but the outcome has not been so happy for commercial vessels. The NS *Savannah*, one of the landmarks described in this chapter, was a cargo ship that used an atomic reactor. As with land-based nuclear power plants, the practice has not reached the expectations of the dreams. Given the controversy that surrounds the use of nuclear energy for electric-power generation, it is difficult to envision a revival of the idea embodied in the *Savannah*. It is, perhaps, an interesting landmark to the optimism of engineers.

The Evinrude outboard motor is included among the landmarks in this chapter, and it is, at first sight, difficult to reconcile this small device—the prototype weighed 62 pounds, or 28 kg, and developed 1.5 horsepower (1.1 kW)—with the two engines of the USS *Texas*—each of which developed 14,050 horsepower (10,477 kW)—but the link between such disparate engine types is there, nevertheless. Today, the largest vessels, up to 401,554 tons (408,000 t) are powered by two-stroke diesel engines—admittedly of very high power output, typically 4,700 horsepower (3,500 kW) per cylinder. So the two-stroke engine is seen at both the highest and lowest ends of the range of powers used in marine transportation. In both cases, the high ratio of the power to the weight are important. The two-stroke engine probably is also attractive in outboard motorboat engines because of its mechanical simplicity, since it avoids the use of poppet inlet and exhaust valves and their associated valve gear.

Having opened this essay with the words of Robert H. Thurston, it is appropriate to close by paraphrasing some others of his from the same publication: "The landmarks in this chapter exemplify the history of the development of the marine power plant, if not from the earliest days, at least from the time (c. 1840) when steam-powered ocean voyages became an everyday occurrence."

SS *Great Britain*

Bristol, England

Launched in 1843, the SS *Great Britain* was the first vessel to embody all the elements of the modern ship: metal construction, steam-driven screw propeller, and large size aimed at good economy. This pioneer vessel was the creation of Isambard Kingdom Brunel (1806–59), an engineer of courage and foresight who designed and equipped three great ships (the others are the SS *Great Western* of 1837 and the SS *Great Eastern* of 1859). By the time the iron keel plates were laid in July 1839, Brunel had made five design studies for the *Mammoth* (as the vessel originally was to be named). The fifth showed a paddle-propelled, iron-hulled vessel of an unprecedented 3,270 gross tons (2,966 t)—the largest in the world. In 1840 Brunel made the momentous decision to abandon paddle propulsion in favor of the screw propeller.

The *Great Britain*, 322 feet (98 m) long overall with a beam of 51 feet (15.5 m), had no central keel in the manner of a wooden ship. Instead, the riveted double bottom was composed of ten longitudinal girders running along the bottom of the ship for its entire length and angle-iron frames or ribs. Iron plates over the longitudinals formed the lowest deck, while overlapping, double-riveted plates measuring about 3 by 6 feet (910 by 1,820 mm) formed the outer bottom, or skin, of the ship. Five transverse, watertight bulkheads running across the hull added strength.

The engine was an inverted-V type, with two pairs of cylinders each of 88-inch (2,235-mm) bore and 72-inch (1,828-mm) stroke. The Mersey Iron Works fabricated the massive overhead crankshaft, 17 feet (5,181 mm) long and some 28

SS *Great Britain* in its home port at Bristol (Avon), England. *Courtesy South West Picture Agency Ltd.*

inches (711 mm) in diameter, which attracted much attention in its day. Supplied with steam at 15 psig (103 kPa), the engine had an indicated horsepower of 1,800 (1,342 kW) at 18 rpm. The propeller shaft was driven at about three times engine speed by sprockets and chain.

By summer 1843, the great ship was ready. In sea trials the following December, the vessel turned 12½ knots, exceeding expectations. (Later, it would turn nearly 14 knots.) Following five months of exhibition, much of it in London, the *Great Britain* went to Liverpool to embark passengers and cargo for its maiden voyage. In the face of westerly gales and fog, the vessel made passage to New York in 14 days and 21 hours at an average speed of 9¼ knots. The trip was a conspicuous success, with some 21,000 visitors inspecting the vessel during its 19-day layover.

The *Great Britain* made several transatlantic crossings before running aground in Dundrum Bay on the west coast of Ireland, a disaster that taxed the resources of Brunel's Great Western Steamship Company beyond its limits. In 1850 the *Great Britain* was sold and fitted out for the Australian trade. It was later sold again and rebuilt as a sailing vessel. In 1886, carrying coal for Panama, the ship met heavy weather and came to rest in the Falkland Islands, where it was converted to a hulk for the storage of wool and coal. Fifty years later, it was scuttled in Sparrow Cove and declared a crown wreck.

In 1967 Dr. Ewan Corlett wrote to the *London Times* about the ship's plight, instigating an ambitious rescue effort. In 1970 it was towed home on a barge to the same dry dock in which it had been built 127 years earlier. There, the SS *Great Britain* Project Committee has restored (and, in some cases, reconstructed) what Corlett has called "the great-great-great-great-grandmother" of today's ocean-going vessels.

The *Great Britain*'s innovative iron hull is original. The engine, unfortunately, vanished long ago, but a replica is being constructed based on the few published drawings that survive.

Location/Access

Brunel's great iron ship rests in the dry dock where it was built, Great Western Dock, Gas Ferry Road (off Cumberland Road) in Bristol, Avon BS1 6TV Great Britain; phone (117) 926-0680. On the dock is a small museum illustrating its history. Admission fee.

FURTHER READING

Ewan Corlett, The Iron Ship: The History and Significance of Brunel's Great Britain (New York: Arco Publishing Company, Inc., 1975).

L. T. C. Rolt, *Isambard Kingdom Brunel: A Biography* (London: Longmans, Green and Co., 1957).

K. T. Rowland, *The Great Britain* (Newton Abbot, Devon: David & Charles, 1971).

TV *Emery Rice* Engine

Kings Point, New York

The back-acting screw engine of the training vessel *Emery Rice* represents a period of momentous change in U.S. naval history. Sail was giving way to steam power, iron hulls were replacing wood, and the very character of naval engagement was transformed by new guns and armor plate. Thanks to foresight, this typical nineteenth-century marine engine was saved when the vessel it powered was sent to the breakers in 1958.

Constructed in 1873 and commissioned in 1876, the *Emery Rice* began its long career as the USS *Ranger*, an iron gunboat rigged as a three-masted barkentine. One of the last four iron ships to be built (all subsequent naval ships were steel), the *Ranger* was powered by a back-acting ("return connecting-rod" in England) screw engine having all the parts of a conventional reciprocating engine adroitly "folded back" to form a short, compact horizontal compound engine. The novel configuration enabled it to lie athwart the keel, protected below the waterline, out of sight of enemy guns.

The engine was designed and constructed by the Bureau of Steam Engineering of the U.S. Navy. The arrangement of its parts derived from British naval practice beginning in the 1840s. Instead of being beyond the crossheads, the cranks were located between them and their cylinders. The connecting rods reached back, or "returned," from the crossheads to couple to the crankpins. To allow this, the usual single piston rods were replaced by two piston rods under and over the crankshaft on the low-pressure cylinder and by a yoked piston rod on the

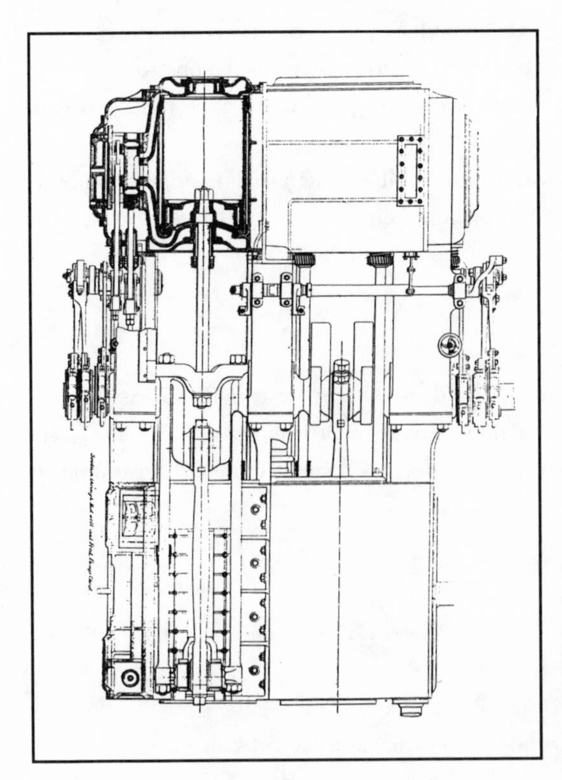

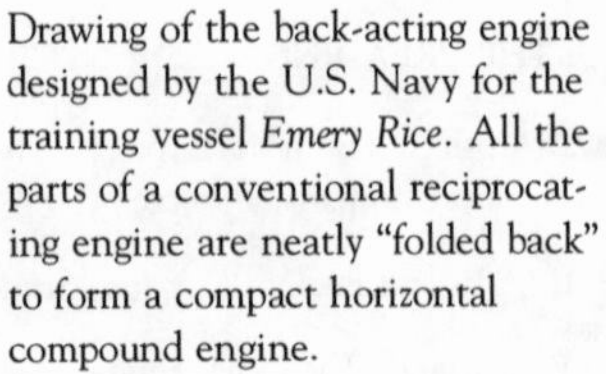

Drawing of the back-acting engine designed by the U.S. Navy for the training vessel *Emery Rice*. All the parts of a conventional reciprocating engine are neatly "folded back" to form a compact horizontal compound engine.

high-pressure cylinder. Although somewhat cramped, the arrangement allowed connecting rods of reasonable length to keep the lateral thrust on the crosshead guides within bounds. The main disadvantage was the need for two stuffing boxes—the seals in the cylinder head through which the piston rod passes—components that were liable to wear and, hence, to leak steam, because of the continual reciprocating motion of the piston rod.

The *Ranger* led an eventful life, serving with the Atlantic and Pacific fleets, performing magnetic survey duty along the western coasts and crossing the equator countless times. In 1909 the vessel was transferred to the Massachusetts Nautical Training School and was successively known as the USS *Rockport, Nantucket*, and *Bay State*. In 1942 she was transferred to the U.S. Merchant Marine Academy at Kings Point, New York, and renamed to honor Captain Emery Rice, a distinguished veteran of the Spanish-American War and World War I. The vessel was retired from sea duty in 1944 and scrapped in 1958. Thanks to the efforts of Karl Kortum, curator of the San Francisco (now National) Maritime Museum, the engine was put into storage. Rear Admiral Thomas J. Patterson, Jr., who earlier had interceded on behalf of the Liberty ship *Jeremiah O'Brien* (see p. 229), succeeded in having the engine returned to Kings Point for display in the academy museum.

Back-acting marine engines disappeared toward the end of the nineteenth century as advances in armor plating of ships' hulls gave protection to conventional multicylinder vertical engines offering vastly greater power.

Location/Access

The engine of the *Emery Rice* is on display at the American Merchant Marine Museum, U.S. Merchant Marine Academy, 60 Cuttermill Road, Great Neck, NY 11021; phone (516) 482-8200, ext. 304. Hours: Saturday and Sunday, 1–4:30 P.M., and Wednesday, by appointment; closed during July.

FURTHER READING

Emory Edwards, *Modern American Marine Engines, Boilers and Screw Propellers* (Philadelphia: Henry Carey Baird & Co., 1881).

Turbinia

Newcastle upon Tyne, Tyne and Wear, England

By 1880, the reciprocating steam engine was fast approaching the practical limits of its development. In 1884, in one of the landmark events in the history of mechanical engineering, Charles A. Parsons (1854–1931) introduced the steam turbine as a practical prime mover. The steam turbine, which derived energy from the velocity of expanding steam rather than from its pressure, did

"Low in the water, long and narrow in the body . . ., sharp as a knife at the bow, speed in every line": thus did one contemporary observer describe Charles Parsons's experimental launch *Turbinia*.

away with the limitations imposed by the mechanics of the piston engine, allowing the power that could be developed with a given weight of machinery in a given space to be substantially multiplied.

The experimental launch *Turbinia*, designed by Parsons in 1894, represents the first application of the steam turbine to marine propulsion. Parsons himself conducted extensive model tests at Ryton, then his home, and at Heaton, home of the C. A. Parsons & Co. turbine works, to determine hull characteristics and power requirements. The experimental launch was 100 feet (30.4 m) long, with a beam of 9 feet (2.7 m) and a total displacement of 44½ tons (40 t). The hull, of steel plate, featured a wedge-shaped bow and rounded body (to decrease drag).

The original turbine engine fitted in the vessel was designed to develop upwards of 1,500 horsepower (1,118-kW) at a speed of 2,500 rpm, with direct drive to a single two-bladed propeller. But early trials proved disappointing; propeller slip was nearly 50 percent and speeds were low. Parsons persevered, however, trying different propeller arrangements. To do this, he devised the world's first propeller testing tank, making it possible to photograph the "vacuous cavities" that seemed to be hindering speed.

The answer appeared to lie in using multiple propellers with larger blade areas. Parsons replaced the single propeller with three propeller shafts, each driven by its own compound turbine and having a combined horsepower of 2,100 (1,566 kW). Each shaft had three screws (propellers) placed at intervals of several feet. The division of the turbines, which applied one-third of the total power to each shaft, greatly increased propeller efficiency and speed. By December 1896, the

The Compound Steam Turbine

In the simplest form of steam turbine, a high-speed jet of steam is directed by a nozzle into a row of buckets, blades, or vanes attached to the periphery of a wheel. By this means, part of the thermal energy of the steam is converted into kinetic energy, then into mechanical energy at the revolving shaft that carries the turbine wheel. This mechanical energy is then available to do work—drive an electrical generator, for example, or turn the propeller of a boat.

But this simple steam turbine has a major drawback: the turbine wheel rotates too fast—from 10,000 to 30,000 rpm. A practical steam turbine must rotate at much lower speeds. The key to reducing turbine speed is to pass the steam through a succession of nozzles and wheels, called stages, with only a small drop in pressure occurring in each stage.

A turbine having multiple stages, called a compound turbine, offered numerous advantages over reciprocating engines. Charles Parsons enumerated them in 1897:

1. *Increased speed.*
2. *Increased economy of steam.*
3. *Increased carrying power of vessel.*
4. *Increased facilities for navigating shallow waters.*
5. *Increased stability of vessel.*
6. *Increased safety to machinery for war purposes.*
7. *Reduced weight of machinery.*
8. *Reduced space occupied by machinery.*
9. *Reduced initial cost.*
10. *Reduced cost of attendance on machinery.*
11. *Diminished cost of upkeep of machinery.*
12. *Largely reduced vibration.*
13. *Reduced size and weight of screw propellers and shafting.*

Source: *Journal of the American Society of Naval Engineers*, May 1897.

Turbinia had reached an average speed of 29.6 knots. Fitted with new propellers of increased pitch ratio, the *Turbinia* attained a record speed of 34.5 knots and was, briefly, the fastest vessel in the world.

With the *Turbinia*, Parsons demonstrated the advantages of the compound steam turbine over the reciprocating engine—among them, increased speed and economy of steam consumption (see sidebar)—and proved the turbine's worth for marine propulsion. In less than a decade, steam turbines would be propelling transatlantic liners and battleships.

The *Turbinia* settled into life as a high-speed demonstration vessel. In 1902 the last change was made in its basic form: single propellers of 28-inch (711-mm) diameter and pitch replaced the triple screws on each shaft. Five years later, the *Turbinia* steamed for what proved to be the last time. An accident cut it in two. In 1961 the two halves were reunited with a reconstructed center section and put on public display.

Location/Access

The *Turbinia,* housed at Exhibition Park, Great North Road, is open by appointment only. Contact Tyne and Wear Museums Service, Blandford House, Blandford Square, Newcastle upon Tyne NE1 4JA, England; phone (091) 232 6789.

FURTHER READING

S. V. Goodall, "Sir Charles Parsons and the Royal Navy," *Transactions of the Institution of Naval Architects* 84 (1942): 1–16.

Cleveland Moffett, "The Fastest Vessel Afloat: The 'Turbinia,' and the New Era She Promises in Ocean Travel," *McClure's Magazine,* July 1898, 243–52.

Charles Parsons, "The Application of the Compound Steam Turbine to the Purpose of Marine Propulsion," *Journal of the American Society of Naval Engineers* 9 (May 1897): 374–84.

R. H. Parsons, *The Development of the Parsons Steam Turbine* (London: Constable and Company Ltd, 1936).

Vertical Reciprocating Steam Engines, USS *Olympia*

Philadelphia, Pennsylvania

The USS *Olympia* is best known as the flagship of Commodore George Dewey, the naval commander who defeated the Spanish fleet at the Battle of Manila Bay in the Philippines in the first action of the Spanish-American War. The date, May 1, 1898, marked the beginning of the United States' reign as a world power. But the protected cruiser, named after the capital of Washington State, is otherwise distinguished as one of the first naval vessels to be fitted with vertical reciprocating steam engines, marking a departure from the usual horizontal cylinders designed to give a low profile and, hence, reduce the vulnerability to gunfire. (See "TV *Emery Rice* Engine," p. 218. Merchant vessels already had adopted vertical engines as the propeller ship displaced the side-paddle steamer.)

Built as part of a program to modernize the U.S. Navy, the *Olympia* was one of the country's first steel ships. Construction was authorized in 1888, and the contract was awarded to the Union Iron Works of San Francisco. Launched on November 5, 1892, and commissioned in February 1895, the *Olympia* was classi-

Vertical reciprocating steam engine of the USS *Olympia*.

fied as a protected cruiser—of moderate size, with a large number of medium-caliber, rapid-fire guns and a curved protective plate of armor over the ship's vitals just above the waterline.

The *Olympia* had twin screws 14.75 feet (4,495 mm) in diameter, each driven by a three-cylinder, triple-expansion engine of 8,425 horsepower (6,283 kW) at 139 rpm, with steam at 160 psig (1,102 kPa) for a maximum speed of 21.6 knots. A stroke of 42 inches (1,067 mm) was common to all cylinders, the bores being 42, 59, and 92 inches (1,067; 1,499; and 2,337 mm). Four double-ended and two single-ended Scotch boilers with a total of forty furnaces under forced draft in a closed stoke-hold system supplied the steam. Trust in steam was not absolute, however; the ship also carried the auxiliary sail rig of a two-masted schooner.

Originally the flagship of the navy's Asiatic Squadron, the *Olympia* became the flagship of the small Caribbean Division in the early twentieth century. In World War I, the ship patrolled the North Atlantic from New York to Nova Scotia. After the war, the *Olympia* served as flagship in the eastern Mediterranean. Her last mission, in 1921, was to bring home the body of America's "unknown soldier" from France for burial in Arlington National Cemetery in Washington, D.C.

The *Olympia* was decommissioned in Philadelphia in 1922 and berthed at the navy yard for the next twenty years. Following Presidential intervention, the vessel was designated a naval relic of the Spanish-American War in 1942 but received no maintenance. In 1954 the navy tried to dispose of all its historical relics (except the USS *Constitution*), spurring the formation of the Cruiser *Olympia* Association in 1957 to raise funds for restoration. A commercial shipyard made some repairs, but the work was slovenly, a large portion of the port engine disappeared, and the yard went bankrupt. A new association was formed, and restoration is proceeding as money becomes available.

Location/Access

The USS *Olympia* and the submarine USS *Becuna* are docked at Penn's Landing, Delaware Avenue and Spruce Street, Philadelphia, Pennsylvania; phone (215) 922-1898. Hours: daily, 10 A.M. to 4:30 P.M., and until 6 P.M. in summer; closed Christmas and New Year's Day. Admission charge.

FURTHER READING

George Dewey, *Autobiography of George Dewey, Admiral of the Navy* (New York: Charles Scribner's Sons, 1913).

Kenneth J. Hagan, *This People's Navy: The Making of American Sea Power* (New York: The Free Press, 1991).

Evinrude Outboard Motor

Milwaukee, Wisconsin

Melted ice cream reputedly led Norwegian-born Ole Evinrude (1877–1934) of Milwaukee to design the first commercially successful outboard motor. Evinrude, so the story goes, was picnicking with his girlfriend, Bess Cary, on Lake Okauchee. Bess expressed a desire for ice cream, and Ole rowed 2 miles (3 km) across the lake to get some. But by the time he had rowed back, the ice cream had turned to soup in the summer heat, inspiring him to design a portable power plant that would eliminate rowing.

The Evinrude outboard motor was quickly accepted by the boating public. It revolutionized recreational boating and stimulated a new industry. Between 1910 and 1920, thirty-eight new companies went into the business of manufacturing outboard motors, with thirteen more following in the next decade. By the 1950s, annual sales topped the half-million mark.

Outboard "motors" to propel boats—in push forms as foot-powered paddle wheels and screw propellers, and as electric propellers powered by bulky storage batteries—had been around for more than forty years when Ole Evinrude designed and built his first prototype in 1907. A practical outboard motor awaited the invention of the internal combustion engine.

The American Motors Company produced a forerunner of the outboard motor in 1896 when it began building its "portable boat motor with reversible propeller." After 1900, the field became more crowded as seven American-made outboards competed. Working with Oliver E. Barthel, Cameron B. Waterman of Detroit developed the first U.S. production model in 1906. The Waterman Porto "outboard"—Waterman is credited with coining the term—was an air-cooled, single-cylinder motor with the flywheel enclosed in the crankcase.

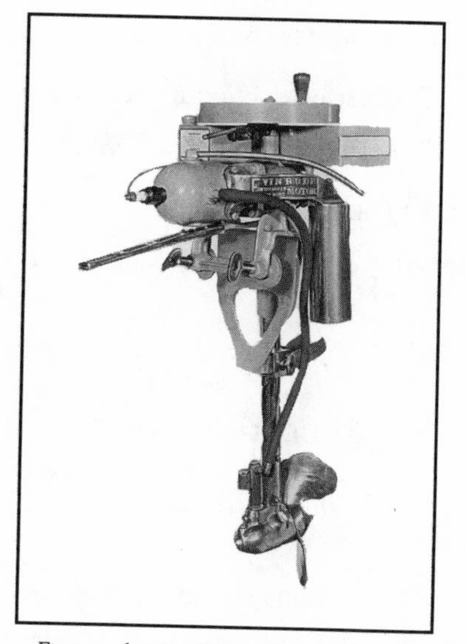

First production Evinrude motor, developed in 1909.

Evinrude's first production motor, developed in 1909, was everything other outboards were not: lightweight, easy to use, dependable, and relatively powerful. The two-stroke motor developed 1.5 horsepower (1.1 kW) at 1,000 rpm and weighed just 62 pounds (28 kg). It used a design that has remained the standard for outboard motors ever since, with a horizontal cylinder, vertical crankshaft, and right-angle gears and propeller shaft housed in an underwater unit.

Evinrude and Bess Cary, now his wife and business partner, formed the Evinrude Motor Company in 1909 and began production. Ole oversaw manufacturing operations, while Bess managed the office and wrote the advertisements that appeared first in the Milwaukee papers, then nationally: "Don't Row. Use the Evinrude Detachable Row Boat Motor."

By 1913, more than three hundred employees were at work in the Evinrude factory to meet demand in the United States and Europe. Late that year, owing to Bess's poor health, Evinrude sold his share of the business and retired, agreeing not to enter the outboard motor business for five years. The Evinrudes toured the country by auto with their young son, but by 1921, both were back in business, this time as the Elto (for "Evinrude Light Twin Outboard") Outboard Motor Company, manufacturing a 3-horsepower (2.2-kW) motor made of aluminum and weighing just 46 pounds (21 kg). Innovation followed innovation: Elto introduced the first exhaust through the underwater propeller hub (for quieter operation), the first waterproof ignition, and the first remote steering.

In 1929 the Evinrude Motor Company merged with Elto and another firm to form the Outboard Motors Corporation. In 1934, on the Silver Jubilee of the Evinrude outboard motor, OMC introduced "hooded power"—a power head enclosed by a streamlined metal hood, now standard.

By World War II, outboard motors were powering native craft all over the world, widening the opportunities for both recreational and occupational pursuits. The Evinrude outboard motor is the first consumer product to be designated a Historic Mechanical Engineering Landmark.

Location/Access

The first production-model Evinrude outboard motor is on display at OMC Milwaukee, 6101 N. 64th Street, Milwaukee, WI 53218; phone (414) 438-5097.

FURTHER READING

W. J. Webb and Robert W. Carrick, *The Pictorial History of Outboard Motors* (New York: Renaissance Editions, Inc., 1967).

Reciprocating Steam Engines, USS *Texas*

San Jacinto Battleground State Park, Texas

The reciprocating steam engine was born, lived, and died in the span of two centuries, but as a motive power for warships, its life was much briefer. Not until the 1880s did steam power replace sail on new battleships. Then, marine power evolved rapidly. By the time construction of the first 14-inch- (355-mm) gun U.S. dreadnoughts—the USS *Texas* and her sister, the USS *New York*—was authorized by Congress in 1910, the U.S. Navy already had five turbine-powered battleships. Why, then, did the navy revert to reciprocating engines?

Poor fuel economy at cruising speeds was the principal defect of early turbines. Further, most repairs to reciprocating engines could be made at sea using the ship's own facilities, whereas turbine problems were likely to be more complex—nozzle erosion, stripped blades, or rotor corrosion, for example—requiring special dockyard facilities or even return to the turbine maker's works. These considerations seemed compelling in 1910, when battleships might cruise for many months at a time thousands of miles from home port. Finally, the decision to use reciprocating engines followed a protracted navy dispute with contractors over turbine standards.

The two engines of the *Texas* followed the standard design for express liners, high-speed channel steamers, and warships: four-cylinder, triple-expansion with four cranks (at 90 degrees) and two low-pressure cylinders. (It was necessary to split the low-pressure stage to avoid cylinders of excessive diameter.) As with most four-cylinder, triple-expansion engines, the two low-pressure cylinders were placed at the ends of the engine to balance the reciprocating forces and reduce vibration, a common problem.

The *Texas* was a twin-screw ship with a combined indicated horsepower of 28,100 (20,954 kW). The ship had an average speed at full power of 21.05 knots, with the shafts turning at 125 rpm, appreciably higher than the 80 rpm that was about tops for merchant ships of the time. Steam was supplied by fourteen Babcock & Wilcox coal-fired, water-tube boilers. The *Texas* was built by the Newport News (Virginia) Shipbuilding & Dry Dock Company at a bid price of $5,830,000.

USS *Texas*, following modernization in 1925. The protected cruiser sports new tripod masts and hull armor for torpedo protection. Launched in 1912, the *Texas* was among the last U.S. dreadnoughts to be powered by reciprocating engines.

The keel was laid down on April 17, 1911, and the ship was launched on May 18, 1912. The *Texas* left Newport News and was commissioned in March 1914. Carrying a crew of 1,300, the ship is 573 feet (175 m) long, with a beam of 95 feet, 2.5 inches (29 m) and a design displacement of 27,000 tons (24,494 t). Fitted on each shaft is a three-blade manganese bronze propeller with a diameter of 18 feet, 7.75 inches (5.68 m) and a pitch of 20 feet (6.1 m).

Major reconditioning from 1925 to 1926 radically changed the appearance of the *Texas*. The vessel's lattice masts were replaced with a single tripod foremast and a short tripod mainmast, blisters were added to the hull for torpedo protection, new oil-burning boilers were installed, and the main deck was strengthened with additional steel plating.

The *Texas* saw service in two world wars. During World War II, the vessel defended convoys in the North Atlantic and supported the invasions of North Africa, Normandy, southern France, Iwo Jima, and Okinawa. Decommissioned in 1948, the *Texas* was turned over to the state of Texas to serve as a memorial and given a permanent berth as part of the San Jacinto Battleground State Park. (Its sibling, *New York*, was exposed to the atomic bombs at Bikini in 1946 and sunk off Pearl Harbor by conventional weapons two years later.)

Although the reciprocating engines of the *Texas* are the largest extant, the most powerful marine reciprocating engines belonged to two ships of the North

SPECIFICATIONS

Reciprocating Steam Engines, USS *Texas*

Boilers, 1914–25:

Description:
Fourteen Babcock & Wilcox straight-tube, sectional header, coal burning, manually stoked; some with superheaters

Pressure: 285 psig (1,965 kPa)*

Temperature: 417°F (214°C)*

Fuel: coal, with supplemental oil

Total heating surface: 65,480 feet^2 (6,083 m^2)

1925 to present:

Description:
Six navy-designed, three-drum, Express type, each with two superheaters (removed after 1931)

Pressure: 285 psig (1,965 kPa)

Temperature: 417°F (214°C)

Fuel: oil

Total heating surface: 40,410 feet^2 (3,754 m^2)

*assumed same as 1925 Express boilers

Engines

Description:
Two vertical, double-acting, four-cylinder, triple-expansion, direct drive—each engine drove one propeller: starboard engine, right-hand rotation; port engine, left-hand rotation.

Builder/Date:
Newport News Shipbuilding & Dry Dock Company, Newport News, Virginia, 1914

High-pressure cylinder, diameter: 39 inches (991 mm)

Intermediate pressure cylinder, diameter: 63 inches (1,600 mm)

Low-pressure cylinders (2), diameter: 83 inches (2,108 mm)

Length of stroke, all cylinders: 48 inches (1,219 mm)

Valve gear: Stevenson open-link, steam-driven reversing gear

Indicated horsepower, each engine: 14,050 (10,477 kW)

Steam inlet conditions (assumed same as boiler conditions):
Pressure: 285 psig (1,965 kPa)
Temperature: 417°F (214°C)

German Lloyd Line. One, the *Kaiser Wilhelm II*, set the Atlantic speed record in 1903 at a clip of 23 knots. Each of its twin screws was driven by two engines coupled in line for a total of eight cylinders and six cranks. The engines were quadruple-expansion, with the high-pressure cylinders over the intermediate-pressure cylinders, the tandem pistons driving the middle cranks. Each engine developed 21,500 horsepower (16,033 kW) at 80 rpm. The German ships were seized by the United States in 1917 and sold for scrap in 1940. The *Kaiser Wilhelm II* was been renamed *Agamemnon*. Regrettably, none of these unique engines was saved.

Steam turbines, meanwhile, gradually improved in efficiency and reliability, and reciprocating engines were abandoned for battleship propulsion. The USS *Oklahoma*, launched in 1914, was the last warship built with reciprocating engines.

Location/Access

The USS *Texas* is moored 22 miles (35 km) east of downtown Houston via Texas 225 at the edge of San Jacinto Battleground State Park, 3523 Highway 134, La Porte, TX 77571; phone (713) 479-2431. It is open for tours daily from 10 A.M. to 5 P.M. year-round. Admission fee.

FURTHER READING

John Kennedy Barton, Naval Reciprocating Engines and Auxiliary Machinery: Text-book for the Instruction of Midshipmen at the U.S. Naval Academy, 3d ed., rev. and rewritten by H. O. Stickney (Annapolis, Md.: United States Naval Institute, 1914).

Norman Friedman, *U.S. Battleships: An Illustrated Design History* (Annapolis, Md.: Naval Institute Press, 1985).

"The Latest United States Battleship," *International Marine Engineering* 19 (January 1914): 1–4.

SS *Jeremiah O'Brien*

San Francisco, California

President Franklin D. Roosevelt called them "ugly ducklings." Admiral Emory Scott Land, chairman of the U.S. Maritime Commission, countered by dubbing them the "Liberty Fleet." The dire necessity of World War II produced these practical, if inelegant, ships, whose purpose was to provide rapid transatlantic cargo service to the war fronts. Between 1941 and 1945, U.S. merchant shipyards built more than twenty-seven hundred EC2 cargo vessels, or "Liberty ships," of which the SS *Jeremiah O'Brien* is the last unaltered survivor.

Faced with massive tonnage requirements and a dearth of steam turbines, the United States Maritime Commission in 1941 decided upon a single-screw ship driven by a triple-expansion steam engine of 2,500 horsepower (1,864 kW). "When the supply of high-powered machinery had been completely earmarked,

The SS *Jeremiah O'Brien*, part of the fleet of U.S. "Liberty ships," survived World War II intact.

any additional ships either had to be slower ships, or empty hulls without engines," was how Rear Admiral Howard L. Vickery, Maritime Commission vice-chairman, explained the choice of the reciprocating engines in 1943.

Of English design, with raked stem and cruiser stern, the Liberty ship had an overall length of 441 feet, 6 inches (134.5 m); a beam of 57 feet (17.4 m); a depth of 37 feet, 4 inches (11.3 m); and a cargo capacity of 9,146 tons (8,297 t). Cylinders of 24.5 inches (622 mm), 37 inches (940 mm), and 70 inches (1,780 mm) in diameter and a stroke of 48 inches (1,220 mm) drove the four-bladed, 18-foot- (5.5-m-) diameter propeller at 76 rpm for an average cruising speed of 11 knots. The Liberty ship had five main cargo holds, three forward and two aft of the propulsion machinery. Steam winches and booms handled the cargo.

The Liberty ship program introduced the techniques of mass production to the shipbuilding industry, with the work spread through eighteen shipyards specially built for the project (see sidebar) and more than five hundred manufacturing plants nationwide. As work progressed, innovations in yard arrangement, equipment, and construction methods transformed the industry.

The EC2s initially were scheduled to be turned out in a period of six months from keel-laying to delivery. Following Pearl Harbor, the rush for tonnage accelerated construction to 105 days, and in January 1942, the 79 emergency cargo ships delivered averaged only 52.6 days, while one yard, Oregon Shipbuilding Corporation in Portland, turned out a ship in just 46 days. This spectacular reduction in building time was made possible by standardization, prefabrication

EC2 Cargo Vessels (Liberty Ships) Delivered in 1943, by Shipyard:

Alabama Dry Dock & Shipbuilding Co., Mobile, Ala.	2
Bethlehem-Fairfield Shipyard, Inc., Fairfield, Baltimore, Md.	192
California Shipbuilding Corp., Wilmington, Calif.	166
Delta Shipbuilding Company, Inc., New Orleans, La.	35
Houston Shipbuilding Corp., Houston, Tex.	74
J. A. Jones Construction Co., Inc., Brunswick, Ga.	21
J. A. Jones Construction Co., Inc., Panama City, Fla.	15
Kaiser Co., Inc., Vancouver, Wash.	8
Marinship Corp., Sausalito, Calif.	10
New England Shipbuilding Corp., South Portland, Maine	91
North Carolina Shipbuilding Co., Wilmington, N.C.	75
Oregon Shipbuilding Corp., Portland, Oreg.	197
Permanente Metals Corp., Richmond, Calif.	279
St. John's River Shipbuilding Co., Jacksonville, Fla.	25
Southeastern Shipbuilding Corp., Savannah, Ga.	36
Walsh-Kaiser Co., Inc. Providence, R.I.	6
Total, 1943	1,232

U.S. Liberty Ship Engine Builders, Number Built:

Alabama Marine Engine Co., Birmingham, Ala.	11
American Ship Building Co., Cleveland, Ohio	40
Clark Brothers Co., Inc., Olean, N.Y.	21
Ellicott Machine Corp., Baltimore, Md.	44
Filer & Stowell Co., Milwaukee, Wis.	140
General Machinery Corp., Hamilton, Ohio	779
Hamilton Engineering Works, Brunswick, Ga.	1
Harrisburg Machinery Corp., Harrisburg, Pa.	91
Iron Fireman Manufacturing Co., Portland, Ore.	309
Joshua Hendy Iron Works, Sunnyvale, Calif.	773
National Transit Pump & Machine Co., Oil City, Pa.	28
Oregon War Industries, Inc., Portland, Ore.	43
Springfield Machine & Foundry Co., Springfield, Mass.	8
Toledo Shipbuilding Co., Inc., Toledo, Ohio	5
Vulcan Iron Works, Wilkes Barre, Pa.	69
Willamette Iron & Steel Corp., Portland, Ore.	211
Worthington Pump & Machinery Corp., Harrison, N.J.	115

(subassembly units—an entire bow section, for example—were fabricated elsewhere, "ahead of the ways"), advances in material-handling facilities (especially larger cranes), and the use of welded instead of riveted construction.

With a normal crew of forty-four, Liberty ships crossed the Atlantic in convoys, calling at nearly every major world port with foodstuffs, coal, oil, locomotives, aircraft, ammunition, motor vehicles and vehicle parts, C-rations, and books. These "shopping baskets of World War II," as one radio announcer described them, sailed bravely, many of them—especially at the beginning of the program—without defensive weapons. Later, most were equipped with armament and carried contingents of the U.S. Navy Armed Guard in addition to the usual merchant marine crew. Fewer than two hundred were lost.

Named after the intrepid Maine sea captain who in 1775 led the first naval action of the Revolutionary War, the SS *Jeremiah O'Brien* was built in fifty-six days by the New England Shipbuilding Corporation, a unit of the Bath Iron Works, and launched in June 1943 from South Portland, Maine. General Machinery Corporation of Hamilton, Ohio, one of fourteen American engine builders participating in the Liberty ship program (see sidebar), manufactured the engine.

In 1966 Commodore Thomas J. Patterson, Jr., initiated an effort to save the *Jeremiah O'Brien* as an example of her class and a memorial to the men and women who built, operated, defended, repaired, and supplied the Liberty ships of World War II. Patterson chose the *O'Brien*, then in mothballs as part of the Reserve Fleet at Suisun Bay, California, because the ship had never been altered. In 1978 the National Liberty Ship Memorial, Inc., was formed to manage the restoration. Countless volunteers donated their time to restoring the vessel, which had been inoperative for more than thirty years. On October 6, 1979, the *Jeremiah O'Brien* sailed out of Suisun Bay under her own power, heading west to San Francisco Bay and, eventually, a permanent berth at Fort Mason.

Location/Access

The SS *Jeremiah O'Brien* is berthed at Pier 2 just off the Bay Bridge. For information, contact the Fort Mason Center, Building A, San Francisco, CA 94123; phone (415) 441-3101. The vessel is open for tours Monday–Friday, 9 A.M. to 3 P.M., and Saturday and Sunday, 9 A.M. to 4 P.M. Admission fee. Each May, the *O'Brien* sails San Francisco Bay, carrying some seven hundred passengers on the annual Seamen's Memorial Cruise.

FURTHER READING

"Building Liberty Ships in 46 Days," *Engineering News-Record* 129 (16 July 1942): 62–67.

"Liberty Ships Built in Basins and on Ways," *Engineering News-Record* 129 (2 July 1942): 64–67.

Howard L. Vickery, "Shipbuilding in World War II," *Marine Engineering and Shipping Review* 48 (April 1943): 182–90.

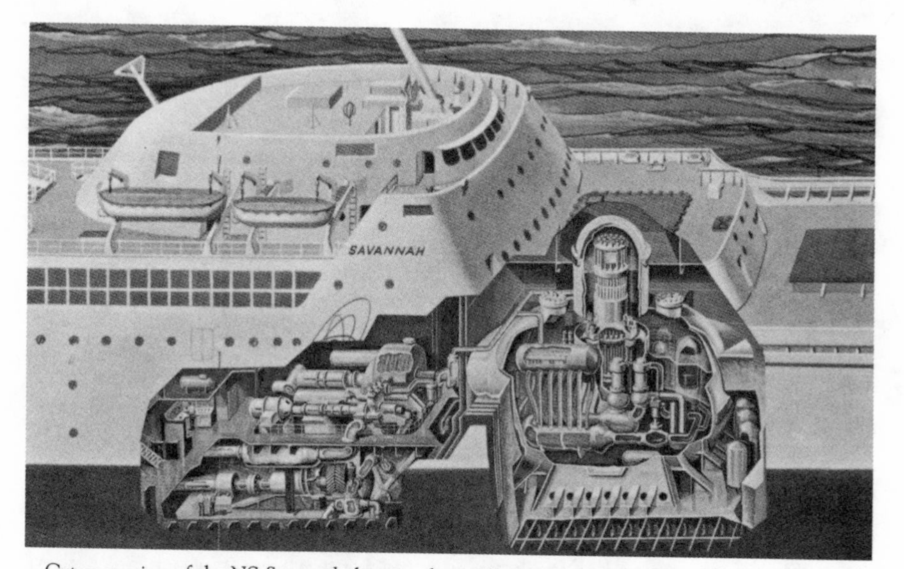

Cut-away view of the NS *Savannah* showing the nuclear propulsion system.

NS *Savannah*

Mt. Pleasant, South Carolina

The sleek, white ship christened by Mamie Eisenhower before slipping down the ways into the Delaware River at Camden, New Jersey, was to be both diplomat and pioneer. The NS *Savannah*, the world's first nuclear-powered merchant vessel, would demonstrate the technical and operational feasibility of nuclear energy as a source of power for commercial vessels and ensure the acceptance of nuclear ships in the world's harbors. Nuclear-powered ships, many hoped, would improve the competitiveness of a merchant marine on the verge of obsolescence.

The *Savannah* was aptly named after the first vessel to use steam power on an Atlantic crossing. A sailing ship fitted with an auxiliary steam engine, the 320-ton (290-t) *Savannah* began its epoch-making voyage from Savannah, Georgia, on May 22, 1819, arriving in Liverpool, England, twenty-nine days later. Just as its namesake had ushered in the Steam Age of ocean travel, the NS (for Nuclear Ship) *Savannah* would usher in the Atomic Age.

President Dwight D. Eisenhower first proposed the construction of a nuclear-powered merchant ship in a 1955 address to the Associated Press. "The new ship, powered with an atomic reactor, will not require refueling for scores of thousands of miles of operation," he said. "Visiting the ports of the world, it will demonstrate to people everywhere this peacetime use of atomic energy, harnessed for the improvement of human living." The following year, Congress authorized $42.5 million for the development and construction of the ship. Construction of the *Savannah*, administered jointly by the Maritime Administration of the Department of Commerce and the Atomic Energy Commission, began in May 1958.

George G. Sharp, Inc., of New York designed the *Savannah*, which was built by the New York Shipbuilding Corporation at its Camden yards. Babcock & Wilcox designed and built the power plant; De Laval Steam Turbine Company supplied the propulsion equipment. The single-screw, combined passenger and cargo ship was designed to carry 60 passengers and a crew of 109. It was 595 feet (181 m) long, with a beam of 78 feet (24 m) and a deadweight tonnage of 9,990 tons (9,063 t). Under normal power, the ship was designed to cruise at a speed of 20 knots.

The propulsion system of a nuclear-powered ship differs from that of conventional ships primarily in the source of heat for generating steam for driving the propulsion turbine, using a nuclear reactor instead of an oil-fired boiler. The *Savannah*'s pressurized-water reactor consisted of a reactor vessel into which was loaded the core of fissionable material—thirty-two fuel elements containing 17,000 pounds (7,711 kg) of uranium dioxide, enough energy to operate the ship for three years. Water was pumped through the reactor core, where it was heated, then through a steam generator (heat exchanger), where it gave up its heat, producing saturated steam in a secondary system. This steam turned the 22,000-horsepower (16,405-kW) main propulsion turbine (driving the single propeller through mechanical reduction gears) and powered the ship's auxiliaries. The training program for engineering officers serving on the *Savannah* included field work at the landmark Vallecitos Boiling Water Reactor at Pleasanton, California (see p. 329).

First operated for the government by States Marine Lines, the *Savannah* made its maiden trip in 1962, stopping at Seattle, where it was shown off to crowds at the World's Fair. But the travels of this goodwill ambassador for nuclear power were abruptly halted when a labor dispute led engineers to shut down the ship's reactor. The *Savannah* sat idle for almost a year before undertaking its first global voyage, with a new crew, under the American Export Isbrandtsen Lines flag.

The *Savannah*, a costly prototype, was never expected to compete economically with conventionally powered ships. But, in the future, nuclear propulsion was expected to offer several economic advantages that would offset higher capital costs. It would eliminate the space required for fuel oil and save on its weight, increase cargo-carrying capacity, allow longer cruising ranges (thereby making nuclear ships virtually independent of fuel supplies outside their home ports), and, finally, operate at higher speeds.

The *Savannah* was retired in 1971. In 1980 Congress chartered the vessel to Patriots Point Development Authority, an agency of the state of South Carolina, for use as a museum. In 1995, the vessel was moved to a U.S. Maritime Administration facility, possibly to be scrapped.

FURTHER READING

U.S. Atomic Energy Commission, Division of Technical Information, *Nuclear Propulsion for Merchant Ships*, by A. W. Kramer (Washington, D.C.: U.S. Government Printing Office, 1962).

Rail Transportation

INTRODUCTION by J. Lawrence Lee

Mechanical engineers have been stimulated by the challenges of railroading from its earliest days. In many ways railroads and engineering have grown up together. The need to travel and transport materials overland goes back to ancient times. No one knows who first moved objects by rolling them on logs, thus making more efficient use of animal and human power, and no one has identified that inspired individual who first conceived the wheel, axle, and bearing combination that made rolling vehicles truly practical. The challenge then became, and has remained, how to carry more with greater comfort, speed, efficiency, and safety.

The concept of a railroad was born in England around 1630 when flanged rails were first used to guide coal wagons. In the early part of the nineteenth century, the revision of this concept into one using flanged wheels on unflanged rails and the concurrent development of the steam locomotive set the stage for the development of modern railroads. That blend of art and science we call mechanical engineering has played a major part in every step of this development.

The Baltimore & Ohio "Old Main Line" and the St. Charles Avenue streetcar line in New Orleans were two early efforts at practical railroads in the United States, the former an intercity route powered first by horses and later by steam and diesel locomotives, and the latter a local carrier that experimented with several power sources before settling on electricity. Both lines remain in service.

The continuing need for power to move heavier trains at faster speeds with greater efficiency has been the genus for several landmark locomotives. These include Texas & Pacific No. 610, an early "Super Power" locomotive that revolutionized modern steam locomotive design, and Southern Pacific No. 4294 and Norfolk & Western No. 611, two later applications of these same concepts to meet two vastly different needs. The New Haven's AC electrification of its New York–

New Haven main line in 1907 pioneered main-line electrification in America. Almost thirty years later, Pennsylvania No. 4800, the prototype for a fleet of 139 electric locomotives that were arguably the best ever built, began operation. The early diesel-electric locomotives are represented here, too. The *Pioneer Zephyr* combined a lightweight diesel engine with a train built with new materials and techniques to usher in the "streamline age." Electro-Motive FT freight diesel No. 103 has aptly been called "the diesel that did it," for this was the locomotive that showed how diesels could outperform steam in freight as well as passenger service.

Where conditions were not suitable for conventional designs, engineers developed other technologies to meet the needs. The rough, often temporary track used by logging railroads needed more flexible engines than the conventional rod-type locomotives. Geared steam locomotives, such as the Shay, Climax, and Heisler designs, provided the answer. The Mt. Washington Cog Railway and the Manitou & Pikes Peak Cog Railway conquered mountain grades too steep for adhesion through the use of rack-and-pinion drive systems. The Monongahela and Duquesne inclines in Pittsburgh combined the concepts of railroad and hoist. In San Francisco, endless cables moving under the hilly streets transmitted power from a central powerhouse to the famous cable cars. When space was not available at street level, the streetcars were taken underground, as illustrated by the New York IRT subway. Decades later, the elevated monorail system at Disneyland demonstrated another technology for the comfortable and efficient transport of large numbers of people.

Throughout railroading history, safety and reliability have always been primary goals of the railway mechanical engineer. While many of the landmarks in this section incorporated improvements in safety over their predecessors, none exemplify this quest more than the Pullman Car *Glengyle*. Its all-steel design still is recognized as one of the most significant advances in car building and passenger safety. Railway engineers realized that proper maintenance was essential for safe, reliable operation, and a vast array of specialized facilities were built to accomplish this, including the Burlington Route's roundhouse and shops in Aurora, Illinois. This facility, with machinery that could produce almost any needed part, once was one of the largest railroad shop complexes in the Midwest, and it is typical of the massive resources needed to keep the trains running. The maintenance procedures for railway locomotives and cars have changed quite a bit from those in use when this complex was built, and the design of modern shops reflects that change, but today's shops clearly have their ancestry in shops like these, and they remain an essential component of safe, reliable operation.

It may be that no activity is more closely associated with mechanical engineering than railroading. No doubt, it is the presence of large machinery, not only in motion, but also moving from one place to another, that inspires this connection. Railroads are a highly visible example of our technological progress and its effect on the nation. This may be adequate to define the link to engineering for the layperson, but it does not explain the attraction these mechanical creations have for so many. Perhaps that has something to do with the scale of railroad locomotives and cars. These are large, powerful machines, but they can be approached closely and their details appreciated. Large as they may be, trains do not obliterate the people who use and control them. It is a very human scale. Distinctive sounds and aromas abound to augment the visual images. Finally, there is the immutable connection between the train and its track. In no other mode of travel are the vehicle and its path so totally defined and linked. Stretching over the horizon or just around a bend, even a vacant track stirs immediate images of the trains it hosts. This may be the stuff of legend and lore, but it is indelibly linked to the progress of railway mechanical engineering.

Some of railroading's glamour may have been superseded in the minds of many by that of jet aircraft or space shuttles, but a certain fascination with railroads and railroad equipment seems to be perennial. It is a fascination that is well deserved. This is the technology that tied a vast continent together into a great nation. Some of the very best of the art and science of mechanical engineering is represented by the Historic Mechanical Engineering Landmarks described in this section. From the "Old Main Line" of the B&O to the Disneyland Monorail System, they constitute a brilliant heritage of mechanical engineering creativity.

Baltimore & Ohio Railroad Old Main Line

Baltimore, Maryland

When the first segment of the Baltimore & Ohio Railroad opened to passenger traffic on May 24, 1830, it marked the first common-carrier railroad service in the United States. Three times a day, horses hauled cars along the one-and-a-half-hour route between Pratt Street in Baltimore and Ellicott's Mills, a distance of 13 miles (21 km). By January 1837, steam locomotives linked Baltimore with Harpers Ferry on the Potomac River, connecting the interior of America with the Eastern seaboard and providing an outlet for the agricultural and mineral products of the Shenandoah and Potomac river valleys. Construction of the road witnessed the birth of countless engineering innovations, winning for the B&O national and even international fame as the "university of railroading."

Following the American Revolution, Great Britain ceded the vast Northwest Territory (comprising what would become the states of Ohio, Indiana, Illinois, Michigan, Wisconsin, and part of Minnesota) to the United States. In 1803 President Thomas Jefferson purchased the Louisiana Territory, giving the United States title to the Mississippi watershed and most of the land east of the Rocky Mountains. The Ohio and Mississippi rivers provided a trade route for produce from the nation's heartland, boosting New Orleans to prominence but threatening the dominance of Eastern cities on the far side of the Appalachian Mountains.

To overcome this mountain barrier and to open the interior to settlement, the National Road—the first interstate highway—was built under the auspices of the federal government between 1808 and 1817. Next, two rival companies began construction of projects intended to replace this crowded and inadequate road. On the Fourth of July, 1828, in Washington, President John Quincy Adams laid the first stone of the Chesapeake & Ohio Canal, while on the same day, in Baltimore, the sole surviving signer of the Declaration of Independence, ninety-year-old Charles Carroll, turned the first spadeful of earth for the Baltimore & Ohio Railroad.

In July 1827, a group of borrowed Army topographical engineers began surveying possible routes between Baltimore and the Ohio River—380 miles (611 km) of rugged terrain. Lieutenant Colonel Stephen H. Long, Captain William Gibbs O'Neill, and Jonathan Knight, a civilian government engineer who became chief engineer of the B&O, formed the railroad's senior engineering management; Caspar W. Wever, a Pennsylvanian who had directed construction of a portion of the National Road, served as superintendent of construction. As horses plodded the route between Pratt Street and Ellicott's Mills, the builders pushed west to Frederick, Maryland, reaching the Potomac River at Point of Rocks in 1832.

Baltimore & Ohio's 1929 reproduction of the 1832 locomotive *Atlantic* pulling two Imlay coaches on the Old Main Line. *Courtesy B&O Museum Archives.*

By 1830, the essentials of "modern" steam locomotives had been developed in England. No existing locomotive, however, could conquer the B&O's impossibly sharp curves. In 1830 Peter Cooper built a demonstrator to be tested on the line; the single-cylinder *Tom Thumb* (the name came later) proved steam's efficacy. Encouraged, B&O directors advertised to find a more efficient locomotive, stipulating a coal- or coke-fired boiler, a maximum steam pressure of 100 psig (689 kPa), a weight limit of 3.5 tons (3.17 t), and the ability to draw a 15-ton (13.6-t) load at 15 miles per hour (24 km/hr). The winner was the *York* (1831), built by Phineas Davis of York, Pennsylvania. From this prototype followed the *Atlantic* (1832), the forerunner of a fleet of geared, four-wheel, vertical-boiler locomotives that became the backbone of B&O service by the mid-1830s.

Since leaving Baltimore, the railroad had undergone constant improvement; every mile brought some innovation dictated by unforeseen conditions. The track, for example, evolved from iron-strap rail to rolled-iron T rail; the ties, from granite to wood. Bridging rivers, the B&O engineers discovered, was best done not by prohibitively expensive stone arches but by timber trusses and, later, the iron trusses of Wendel Bollman and Albert Fink. Meanwhile, to tap the coal fields in the Cumberland area the railroad developed locomotives and cars specially designed for heavy tonnage and steep grades.

Thus, the railroad-engineering concepts that would open the American West and transform the nation owed their origin to the Baltimore & Ohio Railroad. That pioneering project led, as well, to division of the American engineering profession into civil and mechanical branches.

Location/Access

The "Old Main Line" between Baltimore and Harpers Ferry, West Virginia, a distance of about 80 miles (129 km) is littered with what railroad historian Herbert Harwood has called "the richest and most concentrated collection of historic railroad structures anywhere," including the nation's oldest railroad station, at Ellicott City, Maryland (1831); in *Impossible Challenge* (see below), Harwood provides a superb guide to these survivals. In Baltimore, the Mt. Clare Station (1851) at Pratt and Poppleton streets (901 West Pratt Street, Baltimore, MD 21223) forms the entryway for the B & O Railroad Museum, which contains rolling stock and one of the most important historic locomotives in the United States. Phone (410) 752-2490. Hours: Wednesday–Sunday, 10 A.M. to 4 P.M. Admission fee.

FURTHER READING

Herbert H. Harwood, Jr., *Impossible Challenge: The Baltimore and Ohio Railroad in Maryland* (Baltimore: Barnard, Roberts and Company, Inc., 1979).

Edward Hungerford, *The Story of the Baltimore and Ohio Railroad, 1827–1927* (New York: G. P. Putnam's Sons, 1928).

John F. Stover, *History of the Baltimore & Ohio Railroad* (West Lafayette, Ind.: Purdue University Press, 1987).

St. Charles Avenue Streetcar Line

New Orleans, Louisiana

Today, it's a bus named Desire and a streetcar named St. Charles, for Tennessee Williams's legendary streetcar line and all but one other have disappeared from the Crescent City. Happily, the St. Charles streetcar line still operates daily on its six-and-a-half-mile (10-km) route, carrying residents and tourists between the central business district and the city's Carrollton neighborhood. It is the last streetcar operating in New Orleans and the oldest surviving street railway in the United States, having operated continuously since 1834 using horse, steam, and, ultimately, electric power.

Incorporated as the New Orleans & Carrollton Rail Road (NO&C) on February 9, 1833, the line was conceived as part of a sophisticated land-development scheme. Its promoters would use "an English invention, the steam powered Locomotive, rolling on a road of iron rails" to provide "a certain speedy and easy transportation" to developing parts of the city. The first section of the NO&C, a horsecar line operating along St. Charles between Canal and Jackson, opened in 1834. Two steam locomotives ordered from England arrived soon thereafter, while four others were ordered from William Norris of Philadelphia.

One of the Thomas-built streetcars approaches a stop along St. Charles Avenue in New Orleans' Garden District during the 1970s.

Following a broad, crescent-shaped route dictated by the course of the Mississippi River, the street railway played a vital role in the development of the city's Garden District, where wealthy Americans built townhouses and mansions on the old plantation tracts. By 1840, the population of New Orleans was 102,000, having more than doubled in ten years and making it, briefly, the fourth-largest city in the country.

During the Civil War, the U.S. Military Government seized control of the streetcar line, and it was near bankruptcy at war's end. The government leased it to an investors group led by General P. G. T. Beauregard for twenty-five years. The new lessees abandoned the use of locomotives, substituting "bobtail" cars pulled by mules. Beauregard experimented with cable traction (using an overhead-cable-powered car he patented in 1869) and with ammonia-powered locomotives, but these proved impractical and were soon abandoned. Horses and mules continued to provide the motive power until the line's electrification in 1893.

Today the St. Charles line is operated by the Regional Transit Authority of New Orleans. Thirty-five 900-series cars currently in service are the direct descendants of the all-steel (excepting the floor and roof) 400-series cars built in 1915 by the Southern (later Perley A. Thomas) Car Company of High Point, North Carolina. Powered by 600-volt direct current from the Valence Substation (1909), they operate on the original right-of-way with the 5-foot, 2½-inch (1,590-mm) gauge adopted in 1929.

Location/Access

The St. Charles Avenue streetcar operates twenty-four hours a day between Canal Street and Carrollton Avenue. The round-trip ride takes one and one half hours, traveling much of the way along the grassy median of the city's most beautiful boulevard and passing through the Garden District, rich with nineteenth-century architecture and lavish formal gardens. Take a box lunch for a picnic in Audubon Park, directly opposite Tulane and Loyola universities, whose 340 acres (137 ha) stretch from St. Charles Avenue to the Mississippi River.

FURTHER READING

J. L. Guilbeau, *The St. Charles Street Car or The New Orleans & Carrollton Rail Road*, rev. ed. (New Orleans: self-published, 1977).

Chicago, Burlington & Quincy Railroad Roundhouse and Shops

Aurora, Illinois

The Chicago, Burlington & Quincy Railroad roundhouse and back shops are all that remain of what was once one of the Midwest's largest railroad shop facilities. The Aurora shops turned out more locomotives and cars—including the Pullman hotel car *City of New York* of 1866 and the *Delmonico* of 1868, the first full diner—than any other Burlington facility. The Jauriet firebox, which improved the combustion efficiency of high-sulfur Illinois coal, and the Kerr coal chute, which improved locomotive coaling, were both developed here.

The Chicago, Burlington & Quincy traces its origin to the Aurora Branch Railroad, chartered in 1849, which linked the small crossroads of Aurora with Turner Junction 12 miles (19 km) to the north (the present city of West Chicago). There, the railroad connected with the Galena & Chicago Union, a predecessor of the Chicago & North Western. In 1854 the state of Illinois granted a charter allowing the merger of the Aurora & Chicago Railroad (successor to the Aurora Branch Railroad) with the Central Military Tract, Northern Cross, and Peoria & Oquawka railroads, each of which controlled critical segments of trackage to the Mississippi River. The new road, the Chicago, Burlington & Quincy, was the first to link the trade center of Chicago with the Mississippi River.

Following the merger, the railroad's need for new shops became critical. At a cost of $150,000, the Burlington erected a complex of seven new buildings at Aurora, including a roundhouse and machine shop. The roundhouse, of buff-colored limestone quarried in nearby Batavia, was built in three sections: the first section, a half-circle containing 22 stalls, in 1856; a quarter-round, containing 8 stalls, in 1859; and a final 10-stall section in 1866.

By 1857, the Burlington's Aurora shops employed about 350. About half the road's locomotives were based there, the total number growing from 58 in 1858 to 165 in 1872. Between 1871 and 1910, at least 250 locomotives were built

Shop crew and locomotives inside the Chicago, Burlington & Quincy Roundhouse, ca. 1880. *Aurora Historical Society photograph, Library of Congress Collections.*

at the shops, including the American (4-4-0) type and Class "E" (0-4-0). The Burlington's first Mogul (2-6-0) locomotives were designed and constructed here in 1888.

During the twentieth century, the Aurora shops made the transition from the construction and repair of steam locomotives to streamlined passenger cars. In 1970 the Chicago, Burlington & Quincy was merged into the Burlington Northern. Declining railroad service led the company to close the complex in 1974.

Location/Access

The former CB&Q Aurora shops, located at North Broadway between Spring and Pierce streets in Aurora, Illinois, today serve as a regional transit center. The concourse (formerly the back shop) contains displays illustrating the history of the shops and the city of Aurora.

FURTHER READING

Bernard G. Corbin and William F. Kerka, *Steam Locomotives of the Burlington Route* (Red Oak, Iowa: self-published, 1960).

Richard C. Overton, *Burlington Route: A History of the Burlington Lines* (New York: Alfred A. Knopf, 1965).

Mount Washington Cog Railway

Mount Washington, New Hampshire

Rising 6,288 feet (1,917 m) above sea level, Mount Washington is the highest point in New England. The weather at the treeless summit is often called the worst in the world; the strongest winds ever recorded, 231 miles per hour (372 km/hr), swept across Mount Washington in 1934. As early as 1642, the European explorer Darby Field climbed to the highest point. In the nineteenth century, with the construction of railroads and hotels, tourism in New Hampshire's White Mountains was established on a grand scale. The highlight of any visit was a trip to the summit of Mount Washington aboard the Mount Washington Cog Railway.

The Mount Washington Cog Railway is unique on two counts: it is the pioneer cog railway of the world, and it is still in its original condition, never having been modernized by the substitution of electric or diesel power for steam (the case with other cog railways in the world). The sight, sound, and smell of the original steam operation of 1869 are still to be experienced—a rare adventure.

In 1857 inventor Sylvester Marsh (1803–84) climbed the mountain with a friend. Overtaken by nightfall and bad weather, they lost their way, finally stumbling, exhausted, onto the Tiptop House. The experience reputedly convinced Marsh, a New Hampshire-born civil and mechanical engineer who had devoted his career to the meatpacking and grain-handling businesses, of the need for an easier and safer way to ascend the mountain than the carriage road.

Mount Washington Cog Railway.

In 1858 Marsh petitioned the New Hampshire legislature for a charter to build a railroad to the summit of Mount Washington. His proposal was greeted with skepticism, one member reputedly offering an amendment "that the gentleman be further authorized to extend his railroad to the moon." In 1861 Marsh patented an improved locomotive for ascending inclined planes. As conventional adhesion between powered wheels and smooth rails was ineffective on such slopes, Marsh proposed to drive his locomotive by a cogwheel engaging a central, toothed rail and to hold the engine to the track through a system of spring-plates and friction-rollers. A lever-pawl and cam would prevent the engine from running backward. A model engine and car helped enlist investors—largely the railroads operating in or leading to the White Mountains—and in 1865, Marsh organized the Mount Washington Steam Railway Company. Construction began the following year.

All materials had to be brought to the base camp (present-day Marshfield) by ox train from Littleton, 25 miles (40 km) away. Marsh erected a water-powered sawmill on the Ammonoosic River to provide cross ties and wooden members for the extensive trestling required to scale the mountain's rugged terrain. The line was laid out with a gauge of 4 feet, 8 inches (1,423 mm) and built in 12-foot (3,657-mm) sections numbered from 1 at the foot to 1,200 at the summit of the mountain. Located between the two running rails, the rack rail consisted of two strips of angle iron about 6 inches (152 mm) apart, bolted to a center stringer and connected every 4 inches (102 mm) by heavy steel bolts, which engaged the teeth of the driving and braking cogwheels of the engine.

Marsh designed the first locomotive, built by Campbell, Whittier & Company of Roxbury, Massachusetts, and shipped to the site in sections in 1866. Christened "Hero," the odd-looking locomotive was soon renamed "Peppersass" (peppersauce) for its resemblance to a cruet. Two cylinders powered the front axle that carried the cogwheel; the vertical boiler hung on trunnions so that it would remain upright no matter what the grade. In addition to the ratchet safety-pawl to check rollback, the engine could be slowed down on the descent by compressing air in the cylinders. A successful trial run on August 29, 1866, helped stimulate investor interest, and on July 3, 1869, the first train reached the summit. A decade later, the cog railway was handling more than seven thousand passengers a year and had become an essential stop on the grand tour of New England.

The Mount Washington Cog Railway starts partway up the mountain at an elevation of 2,688 feet (819 m) and climbs 3,600 feet (1,097 m) to the summit of 6,288 feet (1,916 m). (This summit elevation is roughly that of Manitou Springs, where the Manitou & Pikes Peak Cog Railway—see p. 252—starts its climb.) The 3.5-mile (5.6-km) railway has an average grade of 25 percent and a maximum grade of 37.4 percent; the latter section, climbing the steepest shoulder of the peak, is named Jacob's Ladder, an allusion to the biblical ladder ascending to heaven.

Except for the war years of 1918 and 1943–45, the Mount Washington Cog Railway has operated continuously since 1869. Jacob's Ladder was substantially reconstructed following a hurricane in 1938, and turnouts were added in the early 1940s to allow the operation of two or more trains at the same time, but otherwise the railway is little changed. The journey to the summit is still slow, cold, cindery, and noisy—just as it was over a century ago. An 0-2-2-0 four-cylinder locomotive (now with horizontal boiler) noses the single passenger car ahead of itself (there is no coupler), with the speed averaging 2 miles an hour (3.2 km/hr) on the 70-minute uphill trek. At the top, on a clear day, travelers are still rewarded with a view that P. T. Barnum called "the second greatest show on earth."

Location/Access

The Mount Washington Cog Railway departs from the Marshfield Base Station, 6 miles (9.7 km) east of U.S. Route 302. Trains operate daily from Memorial Day to Columbus Day. "Old Peppersass" is on display at the base station. Reservations (recommended) and information: phone (603) 846-5404.

FURTHER READING

F. Allen Burt, *The Story of Mount Washington* (Hanover, N.H.: Dartmouth Publications, 1960).

M. F. Sweetser, *Views in the White Mountains* (Portland, Maine: Chisholm Brothers, 1879).

Monongahela and Duquesne Inclines

Pittsburgh, Pennsylvania

The Allegheny and Monongahela rivers carve out precipitous bluffs on the west and south sides of Pittsburgh. In the nineteenth century, commerce and industry claimed the level triangle defined by the two rivers, while home builders looked to the hillsides. Pittsburgh's large German population settled on Coal Hill (now known as Mount Washington), where land was cheap because of its inaccessibility. Land developers, meanwhile, planned inclines, or inclined railways, to overcome the rugged topography.

Based on the *steilbahn* (or funicular) of the old country, the inclined railway consists of a double, sloped track on which a pair of cars are moved by a connecting steel cable wound on a powered drum. In addition to the hauling cable, there is a second, safety cable. One car ascends while the other descends, counterbalancing each other. For passenger transport and light freight, the usual gauge was 5 feet (1,524 mm). The first inclined railway, completed in 1854, handled coal from a mine on Coal Hill (Mount Washington) to a rail line below. Further progress was slow, but inclines flourished after the Civil War, and by the turn of the century,

Pittsburgh had at least fifteen of them, carrying freight, passengers, and teams up and down the hillsides and offering visitors panoramic views of the "Iron City."

The first passenger incline, built to serve Mount Washington, was the Monongahela—or "Old Mon" as it is affectionately known—completed in 1870. It was designed and built by John J. Endres, a German mechanical engineer who had settled in Cincinnati in 1866, and is believed to be the first passenger incline in the United States. The original wood structure was rebuilt by Samuel Diescher, a Hungarian-born engineer, using iron in 1882. Diescher designed most of Pittsburgh's other inclines as well as those in other cities. (Later, Diescher would assist G. W. G. Ferris in the design and construction of the Pittsburgh-built, 1,000-passenger Ferris Wheel, erected in Chicago for the World's Columbian Exposition of 1893.)

The hoisting plant of the Monongahela Incline, as described by *Street Railway Journal* in 1891, consisted of two 12-inch-by-20-inch (304-mm-by-508-mm), Pittsburgh-built Millholland engines. Departing from the standard practice, each car had a separate hoisting rope and drum. The drums were 8 feet, 10 inches (2,692 mm) in diameter, made of cast-iron with wooden lagging on the hoisting surface. A separate freight incline (no longer extant) was built next to the passenger incline in 1883. Electric motors replaced the original steam engines in 1935.

A mile west of Old Mon, the Duquesne Incline serves the hilltop neighborhood of Duquesne Heights. It was built by Samuel Diescher for the Duquesne Incline Plane Company and opened for business in 1877. The original structure, of wood and iron, was rebuilt entirely of iron in 1888. The Duquesne Incline was

The Monongahela Incline, looking toward Mount Washington from West Carson Street, 1969. *Photograph by H. H. Harwood, Jr.*

The Duquesne Incline, from the upper station on Mount Washington, 1975. *Photograph by H. H. Harwood, Jr.*

powered by a double steam engine of 70 horsepower (52 kW) with 14-inch (355-mm) cylinders and 24-inch (609-mm) stroke. The engine operated a shaft carrying a driving pinion 30 inches (762 mm) in diameter, engaging to a main driving gear 12 feet (3,657 mm) in diameter. The cars were connected to a steel wire cable of 1¼-inch (32-mm) diameter, which was wound around a single drum. A Westinghouse electric motor replaced the steam engine in 1932, but both the drum and drive are original.

Pittsburgh's last surviving inclined planes remain in everyday service. Both have been owned by the Port Authority of Allegheny County since 1964. The Duquesne Incline is leased for one dollar a year to the Society for the Preservation of the Duquesne Heights Incline, a group formed in the early 1960s when declin-

Pittsburgh's Inclines

	Monongahela	*Duquesne*
Length, ft (m):	640 (195)	793 (242)
Elevation, ft (m):	375 (114)	400 (122)
Grade, degrees:	28.5	30.5
Gauge, ft (mm):	5 (1,524)	5 (1,524)
Date opened:	May 1870	May 1877
Repowered with electric motors:	1935	1932
Speed, mph (km/hr):	6 (9.6)	4 (6.4)

ing revenues and lack of maintenance threatened to close the incline, then under private ownership. Concerned citizens, most of them residents of Mount Washington, launched a fund-raising drive and established an agreement with the owner to repair and reopen the incline if sufficient money was raised. The society raised the necessary funds, while volunteers from the community cleaned, rehabilitated, and refurbished the incline, which returned to service after a shutdown of only a few months.

Location/Access

Pittsburgh's inclines operate daily; both have observation decks at the top and offer panoramic views of the Golden Triangle, where the Monongahela and Allegheny rivers meet to form the Ohio River. Monongahela Incline: board on West Carson Street near the Station Square Mall; phone (412) 231-5707. Duquesne Incline: board on West Carson Street southwest of Fort Pitt Bridge (lower station) or at 1220 Grandview Avenue (upper station); phone (412) 381-1665.

FURTHER READING

P. G. Eizenhafer, *100-Year History of Pittsburgh Inclines–1863 to 1963* (Pittsburgh: Monongahela Inclined Plane Co., 1963).

"The Inclined Planes," *Street Railway Journal Souvenir* 7 (October 1891): 37–40.

"Modern Hill Climbing," *Scientific American*, September 18, 1880, 1.

Ferries & Cliff House Railway

San Francisco, California

In October 1973, the American Society of Mechanical Engineers inaugurated its program of designating historic mechanical engineering landmarks by recognizing San Francisco's cable railway, the last operating cable railway in the world. City dignitaries, ASME members and friends, and the media gathered in the upper courtyard of the city's cable-car powerhouse at Washington and Mason streets to witness the presentation of a plaque honoring the Ferries & Cliff House Railway (F&CH), one of the city's first cable railways, and its designer, Howard Carleton Holmes.

In 1869 wire-rope manufacturer Andrew Hallidie (1836–1900) reputedly conceived the idea of cable street railways while watching a heavily laden horsecar struggle to make it up a steep hill. Hallidie's idea: to attach a rail car to an endless cable running continuously in a conduit between rails in the street (see sidebar).

In 1872 Hallidie and three partners formed a corporation to build a line up San Francisco's Clay Street hill. With the cable running at 9 miles per hour (14.5

Powerhouse and car barn, Ferries and Cliff House Railway, San Francisco, 1981.
Photograph by Jet Lowe, Library of Congress Collections.

km/h) on Clay Street's 20 percent grade, the trial run on August 2, 1873, represented the first cable line to operate successfully in the United States. The Clay Street Hill Railroad Company began regular operations the following month.

The 1870s saw the transition from animal to mechanical power in many American cities. At the height of their popularity, cable-car systems could be found from New York to Los Angeles. Except for locations with very steep grades, however, cable traction was technologically obsolete by 1888; electric streetcars enjoyed much simpler power distribution, and the overhead trolley wire was much cheaper to erect and maintain than the underground cable and conduit.

In San Francisco, cable-car lines proliferated following Hallidie's convincing demonstration on Clay Street. By 1890, nine competing cable lines were operating some six hundred cars over 110 miles (177 km) of track. The Ferries & Cliff House Railway, organized in the 1880s, represented the amalgamation of two projects: a north-south line across Nob Hill and an east-west line connecting the Ferry Building with the Cliff House resort on the Pacific. The design of the system was entrusted to Howard Carleton Holmes (1854–1921), a civil engineer educated in the San Francisco public schools who had earlier designed the powerhouse for the Oakland Cable Railway. The initial system of the F&CH, put into service in 1888, included today's Powell-Mason line, with a powerhouse at Mason and Washington streets. F&CH promoters also purchased Hallidie's historic Clay Street line, which Holmes rebuilt and expanded before resigning in 1892.

Consolidations had reduced the number of cable railways to four on the eve of the devastating earthquake and fire of 1906. In 1944 the San Francisco Municipal Railway assumed control of virtually all transit operations in the city, including

Ferries & Cliff House Railway: How It Works

An electric motor (formerly steam engines) drives a system of sheaves and pulleys that keeps four endless cables moving at a constant speed of 9½ miles per hour (15.3 km/h) through conduits 18 inches (457 mm) below the street. The cable car is controlled by a "grip"—in essence, a huge pair of pliers that grabs onto the moving cable when the "gripman" hauls on a lever inside the car. When clamped onto the cable, the car moves at the constant speed of the cable whether going down- or uphill.

The cars are separated from the cable only at the moment of a deliberate stop. To stop the car, the gripman releases the cable and applies conventional wheel brakes by stepping on a large foot pedal. These are backed by track brakes, four pine blocks (two per truck) that can be pressed down on the rails. As a last resort, there is an emergency brake that drives a steel wedge into the cable slot to bring the car to a halt.

The cable car has technological cousins. The elevator, the inclined (funicular) railway, and the ski lift operate on similar principles. What distinguishes the cable car is its ability to connect to and disconnect from the constantly moving, endless cable. The 1¼-inch-thick (32-mm) wire-rope cables (steel-wrapped hemp) last between three and ten months, depending on location, season, traffic, and passenger volume.

the remaining cable-car lines, which it targeted for removal. A hard-fought citizens campaign, however, succeeded in preserving three lines: Powell-Mason, which operates between the terminal at the intersection of Market and Powell streets, and Fisherman's Wharf via Mason Street; Powell-Hyde, which operates between the Market and Powell streets terminal and Aquatic Park via Hyde Street; and California Street, which operates between Van Ness Avenue and the financial district via California Street, passing over Nob Hill.

The heart of the present operation is the cable-car barn and powerhouse at Washington and Mason streets. Holmes's original three-story powerhouse of 1887 and much of the equipment were destroyed in the disaster of 1906. The present car barn and powerhouse was erected in 1907. The work of four Corliss steam engines—two 500-horsepower (373-kW) vertical and two 450-horsepower (336-kW) horizontal—today is performed by a single 740-horsepower (552-kW) electric motor. Four cables varying in length from 9,150 feet (2,789 m) to 21,500 feet (6,553 m) keep 23 to 31 cars in service on the system's 10.7 miles (17.2 km) of cable railway. The cars carry an average of 35,000 passengers each day.

San Francisco can boast more than a century of cable-car operation with only two fatalities. Although today's system is greatly reduced in size, in operation it is little changed from the 1880s. Fortunately, the cable cars were not seriously affected by the earthquake of 1989 and remain the city's principal tourist attraction.

Location/Access

The city's three cable-car lines offer service from 6 A.M. to 1 A.M. daily. The cable-car barn and powerhouse, located at Washington and Mason streets, have been refurbished as a cable-car museum (1201 Mason Street, San Francisco, CA 94108). Visitors can observe the cable-winding machinery in operation. Hours: daily, 10 A.M. to 6 P.M.; October–April, daily, 10 A.M. to 5 P.M. Admission free.

FURTHER READING

George W. Hilton, *The Cable Car in America*, rev. ed. (San Diego, Calif.: Howell-North Books, 1982).

"Howard Carleton Holmes" (obituary), *American Society of Civil Engineers Proceedings* 48 (1922): 132–34.

Charles Smallwood, Warren Edward Miller, and Don DeNevi, *The Cable Car Book* (Millbrae, Calif.: Celestial Arts, 1980).

Christopher C. Swan, *Cable Car* (Berkeley, Calif.: Ten Speed Press, 1978).

Manitou & Pikes Peak Cog Railway

Manitou Springs, Colorado

Discovered in 1806 by Lieutenant Zebulon Pike, Pikes Peak is not the highest mountain in Colorado but it is the most widely known because of its commanding location and easy accessibility. The summit may be reached by trail, by an automobile toll road opened in 1916, or by a cog railway in operation since 1891. A latecomer to the field—operation began some twenty-two years after Mount Washington—the cog railway to Pikes Peak reaches the highest elevation, 14,100 feet (4,298 m), of any rack railway, and its 8¾-mile (14.1-km) length is exceeded only by one in Switzerland.

Although the first surveys for the route date from 1870, progress was slow. An early company, organized in 1883, made a good start but collapsed following failure of the New York bank that held its money. In 1888 a new firm bearing the present corporate name was organized. It was entirely financed by mattress king Zalmon G. Simmons of Kenosha, Wisconsin. Simmons and his heirs remained the sole owners of the railway until 1925.

Construction began at the top on September 25, 1889. The narrow roadbed had to be graded by hand using picks, shovels, and wheelbarrows, and the uncomfortably thin air at higher elevations caused an almost constant turnover in the largely immigrant workforce. The line climbs from 6,571 feet (2,003 m) at Manitou Springs (just outside Colorado Springs) to the summit, a rise of 7,539 feet (2,298 m). The average grade is 16.2 percent; the maximum, 25 percent. The standard-gauge road uses the Abt rack system, patented by Roman Abt (1850–1933) of Lucerne, Switzerland, in 1882.

Manitou & Pikes Peak Cog Railway, ca. 1900.

Both Sylvester Marsh, who built the world's first cog railway up Mount Washington in 1869, and Klaus Riggenbach, who built the second in 1871 in Switzerland had used racks in which the "teeth" were pins connecting the upright flanges of angle irons. The Abt design featured either two or three conventional racks—staggered, with open, upstanding teeth—into which meshed two or more driving pinions carried on the locomotive. At Pikes Peak only two racks are used; the pitch of the teeth is 2.35 inches (59 mm). The thickness of the steel—i.e., the width of the rack—varied with the grade, graduating from ⅞ inch (22 mm) for grades up to 12.5 percent to 1¼ inches (32 mm) for the steepest grade.

The original motive power consisted of three 26-ton (23.6-t) Baldwin tank locomotives, with the bearing frames inclined 9 degrees so that the boiler was approximately level on the 16 percent average grade. The bore and stroke of the two-cylinder engine were 17 and 20 inches (430 and 510 mm), with three double pinions or cogwheels for the drive. A fourth Baldwin engine was delivered in 1893. The Vauclain compound with cylinder bores of 9 and 15 inches (230 and 380 mm) and 22-inch (560-mm) stroke was a decided improvement, decreasing both running time and coal consumption. As on other cog lines, the engine nosed the car uphill—there were no couplers. On its descent, the train was controlled by compressing air in the cylinders and a steam brake.

In 1925 the Pikes Peak Auto Highway Company acquired the cog railway from the Simmons family and subsequently modernized its stable of motive power. In 1936 the company built a 24-passenger "streamliner" in its own shops; it was powered by a 175-horsepower (130-kW) gasoline engine (later changed to a Cadillac V-8 engine). The experiment with internal-combustion engines led the company gradually to phase out steam power in favor of diesel-electric power. The first diesel-electric locomotive, built by General Electric and powered by three General Motors model 6-71 engines, went into service in 1939. Four more diesel-electric locomotives followed until 1956. These were equipped with electro-dynamic, air, and hand brakes, although only the first type were required for the trip downhill, when the engines were turned off. Since General Electric was not interested in building one-off units, beginning in 1963 the cog railway obtained new Swiss railcars with Cummins diesel engines; these self-propelled units each carried 76 passengers.

In the 1970s, two twin-unit, diesel-hydraulic trains seating 206 passengers each (the earliest steam trains had carried only 50 passengers) were purchased from the Swiss Locomotive Works of Winterthur. They went into regular service with the dedication of the cog railway as a Historic Mechanical Engineering Landmark in May 1976 and were the first articulated trains on any rack road in the world. Each train, run by a crew of three, has four diesel-hydraulic drive systems of 300 horsepower (224 kW) for a total of 1,200 horsepower (895 kW) per train. In 1975 passing tracks were installed at two points on the route, allowing multiple trains to operate on a continuous schedule.

Steam's last revenue trip on the Manitou & Pikes Peak Railway was made in September 1958. An enlightened management spared four of the oldest steam engines from the wrecker's torch. They are on display at the railway's Manitou Springs station, at the Cheyenne Mountain Zoo in Colorado Springs, and in the Colorado Railway Museum in Golden.

Location/Access

The Manitou & Pikes Peak Railway depot is at 515 Ruxton Avenue in Manitou Springs, 6 miles (10 km) west of Colorado Springs. From May through October, trains depart at 9:20 A.M. and 1:20 P.M.; additional trains run throughout the day from mid-June to mid-August from 8 A.M. to 5:20 P.M. Passengers must purchase tickets thirty minutes before departure; the round-trip ride lasts three hours and ten minutes. Advance reservations are advised. Call the Cog Rail Depot at (719) 685-5401, or write to P.O. Box 1329, Colorado Springs, CO 80901.

FURTHER READING

Morris W. Abbott, *The Pike's Peak Cog Road* (San Marino, Calif.: Golden West Books, 1972).

Claude and Margaret Wiatrowski, *Cog Wheel Route, The Manitou and Pike's Peak Railway* (Colorado Springs, Colo.: MAC Publishing Co., 1982).

Geared Locomotives of the Roaring Camp & Big Trees Narrow Gauge Railroad: Shay *Dixiana*, Climax *Bloomsburg*, and Heisler *Tuolumne*

Felton, California

Beginning in the 1880s and until they were supplanted by motor trucks in the 1960s, geared steam locomotives hauled heavy loads on the rough, temporary tracks of logging and mining companies. In a geared locomotive, power from the cylinders is transmitted to the wheels through a line shaft and reduction gears. By this means, the small, high-speed steam engines produced a steadier resultant pull at low speed than rod engines. The use of flexible, four-wheeled trucks also gave these locomotives the ability to negotiate sharp curves, rough roadbeds, and light, temporary rails with fewer derailments.

Three geared locomotives dominated the field: Shay, Climax, and Heisler. The oldest and most popular was the Shay, designed by Michigan logger Ephraim Shay (1839–1916) and built by the Lima Locomotive Works of Lima, Ohio. Lima turned out more than thirty-three hundred Shays between 1878 and 1945. Varying in size from narrow-gauge "tea kettles" to large machines for Class 1 railroads, the Shay was designed for dozens of different jobs.

The earliest Shays were crude affairs, with wooden frames, vertical boilers, and two cylinders. By the mid-1880s, the Shays were improved with horizontal boilers and steel frames. Larger Shays were equipped with three-cylinder engines that produced more power and resulted in smoother operation. To compensate for the weight and position of the engine (located on the locomotive's right side), the boiler was set off center to the left as a counterbalance, giving the locomotive an odd, lopsided appearance when seen from the front. A Shay thrashing its slow course up a steep grade amid smoke and steam, its cylinders beating rapidly and rhythmically, is an unforgettable sight.

Second of the geared engines to come along and a serious competitor to the Shay was the Climax, patented by George Gilbert and built by the Climax Manufacturing Company of Corry, Pennsylvania, beginning in 1888. The early Class A Climax had a vertical, two-cylinder marine-type engine mounted on the centerline of the frame behind the boiler back head. A line shaft ran through the center of the engine, driving each axle through a pair of bevel gears. A selective transmission gave the engineer a choice of gear ratios.

Most Climax engines were of the two-truck design, although in later years the company produced large three-truck engines that looked more like conventional rod locomotives. With a cylinder mounted on each side of the boiler at a 30-degree angle, the pistons drove a crankshaft running crosswise under the boiler. Power was transmitted through bevel gears to a longitudinal center shaft and again through bevel gearing to the axles. The cylinders were not large, but with the reduction gearing they provided plenty of pulling power and used a minimum of steam.

The Geared Locomotives of Roaring Camp & Big Trees

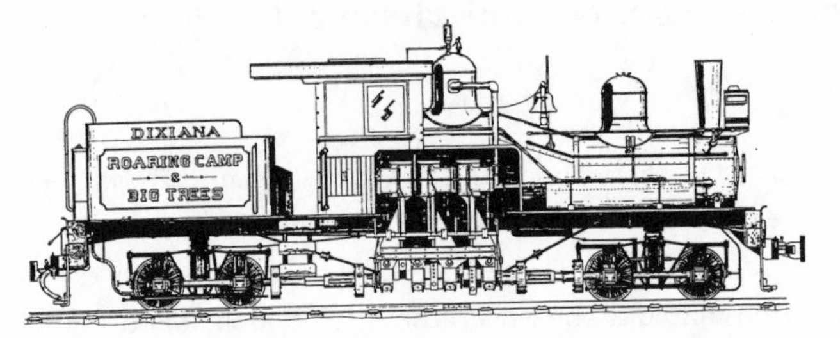

Shay *Dixiana*—The "Dixie" was outshopped by Lima (Ohio) Locomotive Works on October 12, 1912. She served on six different short-line railroads, including the Smokey Mountain Railroad in Tennessee and a narrow-gauge mining road near Dixiana, Virginia (which gave the locomotive its name), before heading west to California. A two-truck engine, with three 10-inch-by-12-inch (254-mm-by-304.8-mm) cylinders, the Dixie weighs 42 tons (38 t). Its 29.5-inch (750-mm) drivers give the locomotive a tractive effort of 17,330 pounds (7,861 kg).

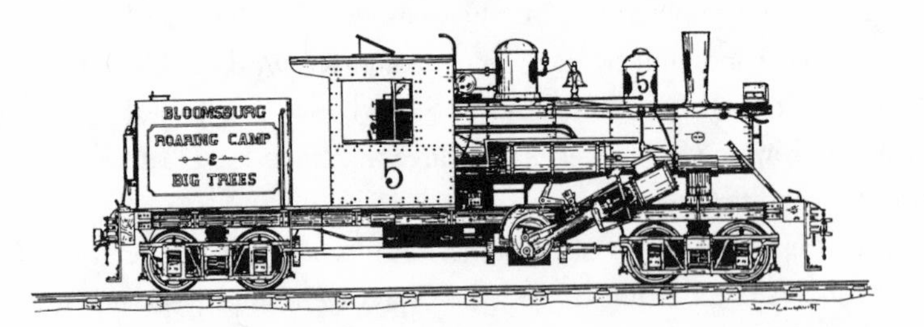

Climax *Bloomsburg*—Built by the Climax Manufacturing Company of Corry (Pennsylvania) for the Elk River Coal & Lumber Company of Swandale, West Virginia, in 1928, the *Bloomsburg* was last operated on the Carroll Park & Western Railroad in Bloomsburg, Pennsylvania, from which it received its present name. Originally standard gauge, the two-truck locomotive was later converted to 42-inch (1,070-mm) gauge. With 33-inch (840-mm) drivers and two 12.25-inch-by-14-inch (311-mm-by-360-mm) cylinders, the locomotive weighs 50 tons (45 t) and has a tractive effort of 22,000 pounds (9,072 kg).

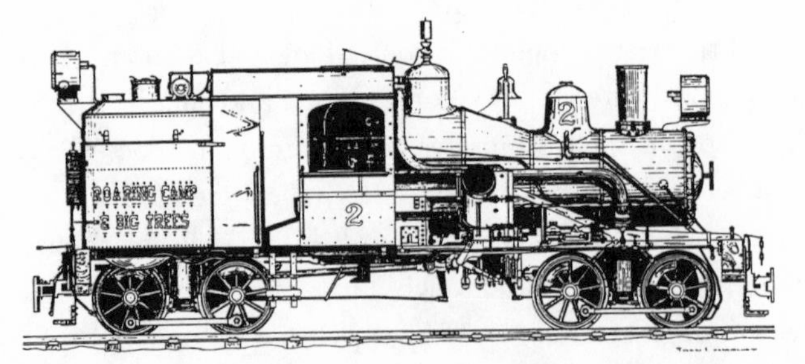

Heisler *Tuolumne*—The oldest operating Heisler engine in the world was built by Stearns Manufacturing Company in Erie, Pennsylvania, in 1899. The *Tuolumne* was ordered by the Hetch Hetchy & Yosemite Valley Railroad for operation at the West Side Flume & Lumber Company sawmill near Tuolumne City, California. The two-truck engine weighs 37 tons (33.6 t) and has a tractive effort of 14,000 pounds (6,350 kg). It has 36-inch (940-mm) drivers and two 10-inch-by-15-inch (250-mm-by-380-mm) cylinders.

Drawings by Joan Lengquist.

Although many companies included a Climax in their rosters for special jobs, few bought them in quantity. Their flying main rods had a tendency to cause vibration in the engine that crews disliked; the engine's detractors liked to say that a Climax would disintegrate itself, the railroad, and the crew with equal impartiality! The Climax works produced just more than a thousand locomotives before closing in 1928.

The third of the geared triumvirate, and the last to make its appearance, was the Heisler, developed by Charles L. Heisler in the early 1890s. The two-cylindered Heisler used a cross-wise V-type engine driving a center-line shaft geared to the inner axle in each truck. Side rods carried power to the outer axles. Early Heislers were of the two-truck type, while later designs were often the larger, three-truck type.

The Heisler was designed to perform much like a rod engine, while retaining the advantages of a geared engine. It was capable of fair speeds on good track. Its faster performance was offset by unusually large cylinders, which sometimes taxed the steaming capacity of the boiler. The Stearns Manufacturing Company of Erie, Pennsylvania, outshopped its last Heisler in 1941, bringing its total production of Heislers to about 625.

The argument over which noisy, gear-driven engine was best—the Shay, the Climax, the Heisler, or their few other competitors—had become moot by the 1930s, when tractors and motor trucks began usurping their duties. Today, the few operable survivors serve the tourist rather than the logging industry. The three most popular types are all represented on the roster of the Roaring Camp & Big Trees Narrow Gauge Railroad, a steam excursion line that began operation in 1963.

Location/Access

Located on Graham Hill Road, a few miles north of Santa Cruz on Route 9, the Roaring Camp & Big Trees Narrow Gauge Railroad offers an hour-long, 6½-mile (10.4-km) excursion through thousand-year-old redwood forests to the top of Bear Mountain. The three landmark locomotives take turns with two others on the railroad's roster. Information and reservations: phone (408) 335-4400.

FURTHER READING

Kramer A. Adams, *Logging Railroads of the West* (Seattle: Superior Publishing Company, 1961).

The Heisler Locomotive (Lancaster, Pa.: Benjamin F. G. Kline, Jr., 1982).

George W. Hilton, *American Narrow Gauge Railroads* (Stanford, Calif.: Stanford University Press, 1990).

Eric Hirsimaki, *Lima: The History* (Edmonds, Wash.: Hundman Publishing, Inc., 1986).

John T. Labbe and Vernon Goe, *Railroads in the Woods* (Berkeley, Calif.: Howell-North Books, 1961).

Thomas T. Taber and Walter Casler, *Climax, An Unusual Steam Locomotive* (Morristown, N.J.: Railroadians of America, 1960).

Interborough Rapid Transit System (Original Line)

New York City, New York

During the late nineteenth and early twentieth century, traffic congestion led many of the world's largest cities to plan new commuter railways, either above or below ground. In Europe, subways opened in London (1863), Glasgow (1886), Budapest (1896), Paris (1900), and Berlin (1902). In New York, the Interborough Rapid Transit (IRT), the city's first subway and the first completely electrically signaled railroad in the United States, opened in 1904.

As early as 1870, Alfred Ely Beach built an experimental one-block tunnel under Broadway from Warren to Murray Street, through which he operated a pneumatically driven railroad car to demonstrate the practicality of subways. Beach's proposal to build a line from lower Broadway to the Bronx won popular approval but met opposition from property owners, and the venture failed. New York's plans for underground rapid transit remained stalled throughout the latter half of the nineteenth century. (North America's first subway, in fact, opened not in New York but in Boston, beneath Tremont Street, in 1897.) Finally, on March 24, 1900, financier August Belmont and Mayor Robert A. Van Wyck broke ground at Borough Hall in Manhattan for New York City's first subway.

The general plan for the IRT called for the subway to tunnel north from City Hall, then up the East Side of Manhattan to Grand Central Terminal. It would then turn west under 42nd Street to Times Square and proceed north beneath Broadway to 145th Street. The length of the initial route, of which about 2 miles (3.2 km) was on viaducts, was 13.5 miles (22 km). Subsequent plans for the subway's extension to Brooklyn and the Bronx necessitated the construction of tunnels under the East and Harlem rivers.

Unusual engineering challenges had to be met, including the support of

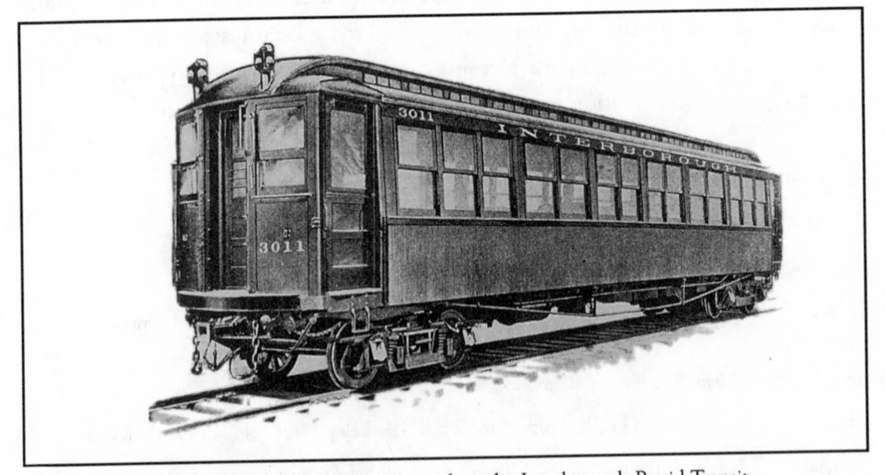

One of the first composite-construction cars used on the Interborough Rapid Transit subway when it began operation in 1904.

towering buildings and heavy street railway and vehicular traffic. Also confounding the engineers was a labyrinth of water, gas, and steam mains, sewers, pneumatic tubes, and electrical conduits beneath the streets. The usual cut-and-cover method of construction—in which workers cut a trench along the route, build the railway, then roof it with steel girders—was supplemented by the use of heavy cast-iron, shield-driven tubes under the rivers.

The IRT powerhouse, a vast French-Renaissance confection designed by Stanford White, occupied the block bounded by West 58th and 59th streets, and Eleventh and Twelfth avenues, adjacent to the North River. It was equipped with nine 8,000- to 11,000-horsepower (5,966- to 8,203-kW) compound steam engines direct-connected to 5,000-kilowatt generators; three steam turbines direct-connected to 1,875-kilowatt lighting generators; and two 400-horsepower (298-kW) engines direct-connected to 250-kilowatt exciter generators to provide field current for the traction and lighting generators. The subway was powered by alternating current converted to direct current at substations for supplying the third rail.

Four car builders filled the subway's initial order for 500 cars. These were of steel-and-wood construction, as an all-steel car had yet to be built (see "Pullman Sleeping Car *Glengyle*," p. 262) The IRT subsequently prepared plans for an all-steel car, awarding an initial contract for 200 cars and eventually replacing its composite fleet with safer steel cars. One of the project's most perplexing problems—how to avoid crashes—was solved by an advanced, practical method of signaling that allowed trains to operate at headways of as little as two minutes.

The first line of the IRT opened to great public enthusiasm on October 27, 1904. Regular service between City Hall and 145th Street and Broadway began the following day. At the end of the year, an average of 300,000 passengers used the new subway daily. Between 1905 and 1908, the Bronx and Brooklyn extensions opened, enlarging the city's first subway to 23.5 miles (38 km) in length. Today New York City has the largest subway system in the United States. About 3.2 million people ride subway trains each day.

Location/Access

Plaques commemorating the opening of the first New York City subway are on permanent display in the Brooklyn Bridge Subway Station, near City Hall. The Transit Museum—at 130 Livingston Street, 9th Floor, Brooklyn, NY 11201; phone (718) 243-5839—is open Tuesday through Friday, 10 A.M. to 4 P.M., and weekends, noon to 5 P.M.

FURTHER READING

Interborough Rapid Transit Company, *The New York Subway: Its Construction and Equipment* (New York: Interborough Rapid Transit Company, 1904).

Benson Bobrick, *Labyrinths of Iron: A History of the World's Subways* (New York: Newsweek Books, 1981).

Alternating-current Electrification of the New York, New Haven & Hartford Railroad

Greenwich, Connecticut

The alternating-current electrification of the New York, New Haven & Hartford Railroad was the first major electrification of a main-line steam railroad. It enabled the New Haven to use electric locomotives over considerable distances for both local and through service, and was a forerunner of today's modern electrified railroads, providing clean, reliable, and efficient mass transit for a densely populated and highly industrialized region.

Early electrification of steam railroads was undertaken largely to solve the problems of smoke and cinders in tunnels, covered terminals, and urban areas. In 1903 the New York State Legislature passed a law prohibiting the use of steam locomotives in New York City south of the Harlem River after July 1, 1908. The New York Central Railroad chose electric operation using a 660-volt, direct-current system with third-rail pickup for its approach to Grand Central Terminal. The New Haven, which also operated into Grand Central, was expected to follow suit. Instead, it announced a contract with Westinghouse for the installation of an 11,000-volt, 25-cycle, single-phase AC electrification of its main line between Woodlawn, New York, and Stamford, Connecticut.

The superiority of alternating current for large-scale electric power transmission was almost universally recognized, but opinion on AC versus DC systems for

Alternating-current electrification of the New York, New Haven & Hartford Railroad.

heavy railroad traction remained divided. Until the electrification of the New Haven, nearly all railroad electric-power experience had been with direct current because the variable-speed performance characteristics of DC motors generally were considered superior for railroad service. But because of the relatively low voltages used, DC systems required heavy currents and, to avoid transmission losses, frequent and expensive substations. The New Haven boldly decided on high-voltage, alternating current for its lines into New York. Because of the ease with which voltage could be stepped up or down by transformers, AC electrification would allow more efficient and economical distribution of power, particularly over long distances.

Construction of the historic system began in 1905. The New Haven built its own generating plant at Cos Cob on the Mianus River in Greenwich, Connecticut. The initial installation consisted of twelve Babcock & Wilcox water-tube boilers supplying steam to three Westinghouse-Parsons steam turbines; the turbines were direct-connected to 3,000-kilowatt Westinghouse generators. A fourth 3,330-kilowatt generator was added shortly after the first three were installed. The station supplied 11,000-volt power directly to the overhead system, a unique triangular—cross-section catenary. Steel bridges, spaced at 300-foot (91-m) intervals and spanning from four to twelve tracks, supported the catenary.

Late in 1905, Westinghouse and Baldwin Locomotive jointly completed the first of 35 Class EP-1, double-truck locomotives. They were 37 feet, 6½ inches (11.44 m) long and employed a box-cab configuration with operating cabs at both ends. The locomotives not only had to overcome the problems of a virtually untested AC system, they had to operate over a 660-volt, DC third rail on the 12 miles (19 km) of New York Central track between Woodlawn and Grand Central. Each locomotive was equipped with two pantographs for overhead AC current collection; a third, smaller DC pantograph for bridging third-rail gaps on the New York Central system; and eight third-rail shoes. Control circuits permitted the transition between AC and DC operation without stopping.

The overhead wire from Cos Cob to New York was energized in April 1907, and the first regular train to run on electric power operated from Grand Central Terminal to New Rochelle, New York, on July 24, 1907. Service was extended to Port Chester, New York, the following month, and to Stamford, Connecticut (33 miles, or 53 km, from Grand Central), by October.

The New Haven's electrification came none too soon. Commuter traffic was increasing rapidly as the suburbs it served burgeoned. Between 1910 and 1920, New Haven commuter traffic into and out of Grand Central more than doubled; by 1924, the New Haven and its Westchester subsidiary were hauling close to 2 million passengers into or out of New York each month.

With the extension of electrification to New Haven in 1914, fourteen Bigelow-Hornsby water-tube boilers and four additional turbo-generators were added to the Cos Cob plant, bringing the station's total generating capacity to

35,400 kilowatts. The original power-supply system, by which 11,000 volts had been fed direct to the distribution system from the generators, was replaced; transformers now stepped the power up to 22,000 volts for distribution to substations, which reduced it to 11,000 volts for the contact wire.

Owing to its worsening financial condition, the New Haven failed to meet its original goal—AC catenary stretching unbroken between New York and Boston. But its pioneering AC electrification was a remarkable engineering accomplishment that set the industry standard for the next half century and led the way for such larger electrification projects as the Pennsylvania Railroad's program of the 1930s.

Location/Access

The Cos Cob Power Station, now abandoned, stands adjacent to the New Haven tracks on the west bank of the Mianus River in the town of Greenwich, about 1 mile (1.6 km) from Long Island Sound. No original equipment remains. The catenary is still in use, though it now carries power at 25 kv, 60-cycle for Amtrak's AEM-7 and E60CP locomotives as well as for self-propelled commuter trains. The 60-cycle power is purchased from the local utility.

FURTHER READING

E. H. McHenry, "Heavy Electric Traction on the New York, New Haven & Hartford Railroad" and "The Overhead Construction of the New Haven Railroad," *Street Railway Journal* 30 (17 August 1907): 242–54.

———, "Electric Locomotives of the New York, New Haven & Hartford Railroad," *Street Railway Journal* 30 (24 August 1907): 278–85.

———, "Cos Cob Power Station, New York, New Haven & Hartford Railroad Company," *Street Railway Journal* 30 (31 August 1907): 308–16.

Arthur J. Manson, *Railroad Electrification and the Electric Locomotive* (New York: Simmons-Boardman Publishing Company, 1923).

William D. Middleton, *When the Steam Railroads Electrified* (Milwaukee: Kalmbach Books, 1974).

Pullman Sleeping Car *Glengyle*

Dallas, Texas

The *Glengyle* is the earliest-known survivor of the fleet of all-steel sleepers built by Pullman beginning in 1907 as a marked improvement over the wooden cars then in use. Some eight thousand steel cars were built for Pullman use during the "heavyweight" era (1907–31). They proved safe, durable, and efficient, even if they failed to offer the privacy and comfort American passengers were beginning to demand.

Completed in January 1911, the Pullman sleeping car *Glengyle* was among the first all-steel sleepers to roll out of the Pullman shops.

For more than a century, "Pullman" was the first word in luxury rail travel. Although George M. Pullman did not invent the first railroad car with sleeping accommodations, by the 1890s the Pullman Palace Car Company, with its extravagantly ornate wooden sleepers, enjoyed a virtual monopoly on the business. But as trains grew longer, faster, and more numerous, it became apparent that a stronger and safer car was needed.

The decision to construct a tunnel under the Hudson River between Hoboken, New Jersey, and Manhattan, together with not-uncommon reports of wrecks and fires involving wooden cars in the subways of Paris and New York, led the Pennsylvania Railroad (PRR) to order new steel coaches in 1907. With encouragement from PRR, Pullman Company directors and engineers adopted all-steel construction from then on, and steel gradually became the standard for all builders.

In 1907 Pullman designed an all-steel, 12-section, 1-drawing-room sleeper. The *Jamestown,* built of steel shapes available commercially, proved much too heavy but led to the erection of a new steel-car plant at Pullman's Chicago works, where sophisticated special shapes could be fabricated to produce lighter cars without sacrificing strength or comfort.

In the *Carnegie,* completed in 1910, Pullman had the practical steel sleeping-car design it was looking for, one that would serve as a model for the sleepers the company would build until the advent of lightweight, streamlined cars in the 1930s. The 12-section, 1-drawing-room car was 82 feet (25 m) long including vestibules; 9 feet, 10⅜ inches (3 m) wide; and 14 feet, ½ inch (4.3 m) tall. It weighed 68½ tons (62 t)—almost 12 tons (11 t) less than the *Jamestown.*

The steel car allowed Pullman to standardize all major parts—including frame, trucks, and roof—as well as interior fixtures and electrical and mechanical components. Using these "building blocks," it was possible for Pullman to construct a wide variety of car types while retaining the manufacturing and maintenance advantages of standardization.

For most sleepers, Pullman retained the open-section plan. Two facing seats by day converted into a lower berth at night, while an upper berth opened from a compartment above the window that provided storage space for linens and the lower-berth mattress during the day. Heavy curtains provided privacy; toilet and washroom facilities for men and for women were located at opposite ends of the car. Despite criticism that it afforded neither privacy nor comfort, the open section remained the most common plan for cars built during the heavyweight era, providing the setting for countless silver-screen comedies and blue jokes.

A group of all-room steel sleeping cars, including the *Glengyle*, rolled out of the Pullman shops during December 1910 and January 1911. The ten cars of Lot Number 3867 each contained seven compartments and two drawing rooms. Their overall dimensions were the same as the *Carnegie*, but they were about 3,000 pounds (1,300 kg) heavier due to different interior arrangements. Pullman built only seventy cars like the *Glengyle*, all between 1911 and 1923, representing less than 1 percent of the total production. But the *Glengyle* and its siblings were among the most luxurious. The cars made their debut on the Pennsylvania's prestigious New York—Florida trains.

In the 1930s, the *Glengyle* was modified with the addition of air conditioning, replacement of exterior body panels, and the upgrading of its trucks and air brakes, but its interior survived remarkably intact. Following the close of a long career in 1957, the sleeping car served as a dormitory for Texas steelworkers. In 1964 the Texas & Northern Railroad donated it to the Age of Steam Railroad Museum.

Location/Access

The *Glengyle*, along with two later Pullman heavyweight sleeping cars, the *McQuaig* and the *Goliad*, are part of a heavyweight-era passenger train displayed at the Age of Steam Railroad Museum, Fairground-State Fair of Texas, Washington and Parry streets. Mailing address: P.O. Box 153259, Dallas, TX 75315-3259; phone (214) 428-0101. Hours: Thursday and Friday, 10 A.M. to 3 P.M.; Saturday and Sunday, 11 A.M. to 5 P.M. Admission fee.

FURTHER READING

Arthur D. Dubin, *Some Classic Trains* (Milwaukee: Kalmbach Publishing Company, 1964).

William W. Kratville, *Steam, Steel & Limiteds* (Omaha: Kratville Publications, 1983).

John H. White, Jr., *The American Railroad Passenger Car* (Baltimore: The Johns Hopkins University Press, 1978).

Texas & Pacific No. 610, Lima "Super-Power" Steam Locomotive

Palestine, Texas

"Super Power" was the Lima Locomotive Works' solution to providing more speed and power for America's railroads. Brainchild of William E. Woodard (1873–1942), Lima vice president of engineering, the A-1 demonstrator of 1925 combined higher boiler pressure and steaming rate, a larger and more efficient firebox, and a higher superheating-to-evaporating surface ratio to develop high horsepower with economical fuel consumption. It was an immediate success, outperforming its contemporaries by a wide margin and serving as the prototype for the modern, high-speed locomotive through the end of the steam age. Texas & Pacific No. 610, on display at the Texas State Railroad Historical Park, is the sole surviving example of the earliest form of superpower steam locomotive.

Following World War I, increasing traffic forced American railroads to reevaluate all facets of their operations, including train size and speed. It was clear that stronger motive power was needed. American steam locomotives were rated according to their drawbar pull, or tractive effort. To many, it seemed obvious that if the pull on the drawbar was what moved a train, the answer to

Texas & Pacific No. 610 (Lima "Super-Power" Steam Locomotive) in excursion service for the Southern Railway at Manassas, Virginia, August 1977. *Photograph by T. N. Colbert, collection of H. H. Harwood, Jr.*

SPECIFICATIONS

Texas & Pacific No. 610

Builder: Lima (Ohio) Locomotive Works, 1927
Type: 2-10-4
Length:
- 60-foot, 4-inch engine (18.4 m)
- 38-foot, 8-inch tender (11.8 m)
- 99-foot engine and tender (30.2 m)

Height: 15 feet, 5¾ inches above top of rail (4.7 m)
Weights, in working order:
- On drivers: 300,000 lbs. (136,077 kg)
- On pilot truck: 41,800 lbs. (18,960 kg)
- On trailing truck: 106,200 lbs. (48,171 kg)
- Total engine: 448,000 lbs. (203,208 kg)
- Tender: 125,500 lbs. (56,925 kg)
- Total engine and tender: 573,500 lbs. (260,133 kg)

Cylinders, diameter and stroke:
- 29 inches by 32 inches (73.7 cm by 81.3 cm)

Drivers: 63 inches in diameter (160 cm)
Boiler pressure: 255 psig (1,758 kPa)
Heating surfaces:
- Firebox: 473 square feet (44 m^2)
- Flues: 4,640 square feet (431 m^2)
- Superheater: 2,100 square feet (195 m^2)
- Total: 7,213 square feet (670 m^2)

Tractive effort:
- Engine: 84,000 lbs. (38,000 kg)
- Booster: 13,000 lbs. (5,900 kg)
- Total: 97,000 lbs. (3,900 kg)

greater power lay in increased tractive effort. Woodard, however, demonstrated that the boiler was really the heart of the steam locomotive and, thus, the starting point for advances in locomotive power and performance. He was the guiding force behind developments that would put Lima squarely in the forefront of steam-locomotive technology.

The Lima-built Michigan Central No. 8000, a heavy 2-8-2 completed in June 1922, provided a test vehicle for Woodard's ideas and confirmed his belief that a high-performance boiler was central to improved locomotive performance. Among other features, No. 8000 incorporated a high-capacity boiler with a larger firebox and increased superheater area. Woodard was pleased with his first effort but not yet satisfied. He went to work on a new "super" locomotive and had it down on paper by 1924.

The A-1 demonstrator of 1925—with its cast-steel cylinders, large boiler,

Elesco feedwater heater, and articulated, four-wheel trailing truck supporting an enormous firebox—was like nothing else on the rails. Would it perform the same way? A test run on April 14 on the Albany Division of the Boston & Albany, one of the B&A's toughest operating districts, proved just how good a locomotive it was.

B&A Mikado No. 190, an H-10 sister of the 8000, left Selkirk Yard at 10:57 A.M. with 46 cars and 1,691 tons (1,534 t) in tow. The A-1 departed at 11:44 A.M. with 54 cars equaling 2,296 tons (2,083 t) and promptly began to narrow the gap, overtaking the laboring "Mike" about 1 P.M. and arriving in North Adams Junction at 2:02 P.M., ten minutes ahead of it. Not only had the A-1 pulled one-quarter more tonnage in fifty-seven minutes less time, it had consumed one-third less coal and water. Subsequent tests on several other roads yielded similar results. No other locomotive could come close to matching the A-1's performance. Lima coined a new name for the remarkable engine that came to be synonymous with its builder: "Super Power."

The first commercial order for superpower locomotives—ten 2-10-4s—was delivered to the Texas & Pacific Railroad late in 1925. Assigned Class I-1 by the railroad (and known as the "Texas" type), the new locomotives burned oil instead of coal and contained a combustion chamber and two Nicholson thermic syphons—appliances missing from the A-1—making its boiler even more efficient.

Following impressive operating savings compared with the railroad's G-1s (2-10-2s), T&P placed a second order for fifteen 2-10-4s (Nos. 610–624, including the present landmark engine), delivered in June 1927. By 1930, the railroad had seventy of these Texas locomotives, now its standard heavy main-line freight power. Other railroads hopped on the superpower bandwagon, among them Boston & Albany, Illinois Central, Southern Pacific, Chesapeake & Ohio, and Norfolk & Western. So did other builders; both Alco and Baldwin built their own versions of "super-power" locomotives based on the principles laid down by Lima's A-1 of 1925.

The Texas & Pacific's I-1s reigned supreme until the arrival of diesels in the late 1940s. Nine were rebuilt into more modern I-2s. No. 610, fortunately, was not among them (though the 610 did receive new Baldwin disc main drivers in 1938, together with new lightweight nickel-steel rods and new crossheads). After more than a million miles (1.6 million km) of service on the T&P, the 610 was put on display in Fort Worth and later restored for service as part of the bicentennial American Freedom Train traveling exhibition. After pulling excursion trains of the Southern Railway, the 610 was donated to the Texas State Railroad Historical Park.

Location/Access

Texas & Pacific No. 610 is on permanent display at the Texas State Railroad Historical Park (on Route 84 between Palestine and Rusk), P.O. Box 39, Rusk, TX 75785; phone (903) 795-3351. Built in 1896 to serve the industries of the state penitentiary at Rusk, the Texas State Railroad today operates restored steam-powered trains on the 50-mile (80-km) round-trip run between Palestine and Rusk. The 610, too large to operate on TSR tracks, is stationary.

FURTHER READING

W. W. Baxter, "New Type Locomotives for Texas & Pacific," *Railway Review* 77 (19 December 1925): 905–12.

"An Epoch Making Advance in Locomotive Design," *Railway Review* 76 (2 May 1925): 799–810.

Eric Hirsimaki, *Lima: The History* (Edmonds, Wash.: Hundman Publishing, Inc., 1986).

Charles M. Mizell, Jr., "T Is for Texas, Texas & Pacific, and TWO-TEN-FOUR," *Trains*, February 1978, 22–32.

Pioneer Zephyr

Chicago, Illinois

Built for the Chicago, Burlington & Quincy Railroad in 1934, the *Pioneer Zephyr* marked the end of one era and the beginning of another. For underneath the *Zephyr's* streamlined, stainless-steel exterior—the result of wind-tunnel tests conducted at the Massachusetts Institute of Technology and the invention of the "shotweld" process—lay a revolutionary diesel-electric power plant that eventually would replace the traditional steam locomotive, changing forever the sight, sound, and smell of railroading. The revolution it sparked can hardly be exaggerated: In 1934, when the *Zephyr* made its first run, some fifty thousand steam locomotives were still at work on America's railroads; by 1961, the industry was completely dieselized.

The diesel engine is named after inventor Rudolf Diesel (1858–1913), who completed his first working prototype in 1897 following years of research. The diesel engine has neither a carburetor nor an electric ignition system. Instead, air is compressed in the cylinder. At the top of the piston's stroke, the air temperature has risen to 1,000°F (537°C), spontaneously igniting the fuel oil injected into the combustion chamber at just that point.

Diesel engines had been used for stationary and marine purposes since the turn of the century, but their low power-to-weight ratio made them impractical for lightweight trains. In 1932 a team at General Motor's Winton Engine Corporation working under the direction of GM vice president Charles F. Kettering perfected a two-stroke, eight-cylinder diesel engine.* One key to success was the development of new alloys having a strength-to-weight ratio that made the engine

* Reciprocating internal combustion engines operate on either the two-stroke or four-stroke cycle: the former having a power stroke every revolution of the crankshaft; the latter, every second revolution.

Chicago, Burlington & Quincy's *Pioneer Zephyr,* here shown new in 1934, was America's first diesel-powered, stainless-steel streamliner.

light enough to mount on a locomotive frame and, at the same time, powerful enough to pull a train. The engine was based on a design that had been developed for navy submarines. Its weight was further reduced by fabricating the engine frame from rolled-steel plates, cut to size by oxyacetylene torch and joined by electric-arc welding. (This was one of the earliest applications of fabricated construction, now widely used in medium- and heavy-machinery construction.)

In 1933, the Burlington ordered a lightweight, stainless-steel train from the Edward G. Budd Manufacturing Company of Philadelphia. Burlington President Ralph Budd (no relation to the railcar manufacturer) decided to power it with a 660-horsepower (492-kW) Winton Model 201A Diesel engine. The 8-cylinder engine, of 8-inch (203-mm) bore and 10-inch (254-mm) stroke, was direct-connected to a 750-volt DC generator, which supplied power to the 300-horsepower (224-kW) traction motors mounted on each axle of the forward truck. The engine output in excess of that needed for propulsion powered an auxiliary 25-kW generator for lighting, air-conditioning, and other electrical functions. "Since the automobile industry, through its brilliant achievements in the design and manufacture of private automobiles, has been responsible for the losses of railway passenger traffic, it seemed logical to try adopting as many of its ideas as practicable," Budd later wrote.

The diesel-powered *Zephyr* rolled out of the Budd shops on April 7, 1934. (Ralph Budd suggested the name, after Zephyrus, god of the west wind in Greek mythology. Burlington's motto was "Everywhere West.") Capable of traveling at speeds up to 120 miles per hour (193 km/hr), the train consisted of three articulated cars (i.e., the ends of adjoining cars were carried on a single truck) measuring 197 feet (60 m) long overall, weighing approximately 208,000 pounds (94,347 kg), and having a total seating capacity of 72. The first car contained the diesel engine, railway post office apartment, and mail storage area; the second, a baggage room, buffet, and 20 coach seats; the third 40 coach seats and 12-seat observation lounge.

Following its christening on April 18, the train barnstormed for several weeks before making its famous nonstop, dawn-to-dusk run between Denver and Chicago on May 26, 1934. On the 13-hour, 1,015-mile (1,633-km) trip, the *Zephyr* attained a maximum speed of 107 miles per hour (172 km/hr) and averaged 77.6 miles per hour (124.9 km/hr). The *Zephyr* arrived in Chicago's Grant Park in time to cap off the "Wings of a Century" transportation pageant at the Century of Progress Exposition, thrilling spectators and providing a publicity bonanza for the Burlington Route.

On November 11, 1934, the *Zephyr* became the first diesel-powered passenger train in America to enter regular service, making a daily round trip of 502 miles (808 km) between Kansas City, Missouri, and Lincoln, Nebraska. Burlington directors, meanwhile, took steps to put the railroad in the forefront of the diesel revolution, ordering three similar trains: two *Twin Zephyrs* for service between Chicago and Minneapolis, and the *Mark Twain Zephyr* for service between St. Louis and Burlington, Iowa. Meanwhile, the original *Zephyr*, renamed the *Pioneer Zephyr* to distinguish it from the growing fleet, completed its first year of operation with flying colors: of 365 days, it had been out of service only 11, representing an astonishing availability record of 97 percent, compared with about 70 percent for the line's steam locomotives. Operating and maintenance costs, meanwhile, averaged 35 cents per train mile, compared with 59 cents for the steam-hauled trains it had replaced. Just as important, the high-speed streamliner, with its clean, temperature-controlled ride and modern decor, had succeeded in luring passengers back to railroad travel, boosting passenger revenues while giving a Depression-weary nation a much-needed psychological lift.

Throughout the 1930s and 1940s, the Burlington continued to substitute diesel-electric locomotives for steam. By 1954, all of the line's passenger trains in regular service and about 95 percent of its freight trains were diesel-powered. The *Pioneer Zephyr* continued to serve the West and Midwest, logging some 3.2 million miles (5.1 million km) before it was retired in 1959.

Location/Access

The *Pioneer Zephyr* is on permanent display at the Museum of Science and Industry, 57th Street and Lake Shore Drive, Chicago, IL 60637; phone (312) 684-1414. Hours: June–August: daily, 9:30 A.M. to 4 P.M.; September–May: Monday–Friday, 9:30 A.M. to 4 P.M., and Saturday–Sunday, 9:30 A.M. to 5:30 P.M. Admission fee.

FURTHER READING

A. N. Addie, "The History of the Diesel-Electric Locomotive in the United States," in ASME Rail Transportation Division, *Railway Mechanical Engineering: A Century of Progress* (New York: American Society of Mechanical Engineers, 1979).

E. C. Anderson, "The Burlington Zephyr," Transactions of the American Society of Mechanical Engineers (Railroads) 56 (1934): 659–66.

Ralph Budd, "The Burlington Zephyr," *Civil Engineering* 4 (August 1934): 383–87.

John F. Kirkland, Dawn of the Diesel Age: The History of the Diesel Locomotive in America (Glendale, Calif.: Interurban Press, 1983).

Richard C. Overton, *Burlington Route: A History of the Burlington Lines* (New York: Alfred A. Knopf, 1965).

Pennsylvania Railroad GG1 Electric Locomotive No. 4800

Strasburg, Pennsylvania

The Pennsylvania Railroad's GG1 locomotive No. 4800, built by Baldwin and General Electric, was the prototype for a 139-unit fleet built between 1934 and 1943 to serve the Pennsylvania's electrified lines. The GG1 served longer frontline duty than any other class of locomotive in history—steam, electric, or diesel. It owed its success in large part to a flexible suspension system that provided equal traction for all driving wheels regardless of track condition.

To relieve congestion, in 1913 the Pennsylvania Railroad (PRR) decided to electrify its two most heavily traveled suburban routes operating out of Philadelphia's Broad Street Station. By this time, the superior characteristics of single-phase AC electrification for heavy main-line service had been well established by the New Haven (see "Alternating-current Electrification of the New York, New Haven & Hartford Railroad," p. 260), and the Pennsylvania chose the AC system for its Philadelphia suburban lines.

By the mid-1920s, the Pennsylvania's eastern main lines carried some of the densest freight and passenger rail traffic in the United States. To increase capacity, the railroad announced a massive electrification program in 1928. Altogether, some 325 miles (523 km) of railroad—between New York and Wilmington, Delaware, and between Trenton, New Jersey, and Columbia, Pennsylvania—would be electrified. The program gradually was expanded, and by 1935 electrified track stretched all the way to Washington, D.C., and west to Harrisburg, giving the Pennsylvania more track miles under catenary than any other railroad in the country.

With expansion of its electrified track came the need for new electric motive power. An early fleet of P5a locomotives by Baldwin-Westinghouse showed serious shortcomings, especially tracking problems at high speeds. In 1934 the PRR ordered two new prototypes from Baldwin. The first, Class R1 locomotive No. 4800 with a 2-D-2 wheel arrangement, followed the railroad's traditional practice of obtaining maximum horsepower from as few axles as possible on a rigid frame. The second prototype, Class GG1 No. 4899, employed a 2-C+C-2 wheel arrangement on an articulated frame—i.e., a pair of frames, each with a four-wheel pilot truck

Pennsylvania Railroad GG1 electric locomotive No. 4800 with passenger train in the 1930s. *Courtesy National Museum of American History.*

and three driving axles—an arrangement first employed by General Electric in 1929 for locomotives built for the Cleveland Union Terminal and later for the New Haven's EP-3a's.

GG1 No. 4899, delivered in September 1934, was an impressive unit. Supported by two cast-steel frames hinged at the center, the GG1 measured 79 feet, 6 inches (24.2 m) between coupler faces and weighed more than 230 tons (209 t). Twin AC traction motors drove each pair of 57-inch (145-cm) driving wheels through a geared quill drive, providing a total of 4,620 horsepower (3,445-kW) at a maximum speed of 100 miles per hour (161 km/hr). The driving axles were fitted into roller-bearing boxes that could move vertically in pedestal jaws in the frame; thus, while the motors were rigidly fixed to the frame, the quill drive allowed the wheels and driving axles to move freely. Although little about the GG1 was new, they represented the refinement of proven technology and its incorporation into a well-balanced design. The result was one of the smoothest-riding locomotives ever built, with firmness and stability at high speed and superior tracking ability.

Following ten weeks' competition on the Pennsy's test track near Claymont, Delaware, the GG1 emerged victorious over the R1. The two prototype units exchanged numbers. Industrial designer Raymond Loewy created the distinctive "cat's whiskers" gold striping and persuaded the railroad to weld all subsequent GG1 bodies (the prototype had been riveted), giving the locomotive a classic, streamlined appearance. On January 28, 1935, No. 4800 pulled the first electrically powered train between Washington and Philadelphia, covering the 134 miles (216 km) in 110 minutes at an average speed of 73 miles per hour (117 km/hr).

The GG1's enormous power enabled the Pennsylvania to increase train speeds, reduce running times, and keep passenger trains of more than twenty cars on time, even under the most demanding conditions. No. 4800 routinely operated at speeds of 100 miles per hour (161 km/hr), logging nearly 5 million miles (8 million km) during its forty-five-year career. It remained in service on the PRR—later Penn Central and Conrail—until its retirement in 1979. In 1980 the Friends of GG1 4800 rescued the famous locomotive by buying it from Conrail at its scrap price of $30,000. It was restored to the original Loewy livery and presented to the Railroad Museum of Pennsylvania.

Location/Access

PRR locomotive No. 4800 is on display at the Railroad Museum of Pennsylvania, Rte. 741E (near Gap Pike), P.O. Box 125, Strasburg, PA 17579; phone (717) 687-8628. Hours: May–October: Monday–Saturday, 9 A.M. to 5 P.M., and Sunday, noon to 5 P.M.; November–April: Tuesday–Saturday, 9 A.M. to 5 P.M., and Sunday, noon to 5 P.M. Admission fee. The museum contains an extensive collection of Northeastern (especially Pennsylvania) railroad equipment, including another, welded GG1.

FURTHER READING

William D. Middleton, *When the Steam Railroads Electrified* (Milwaukee, Wis.: Kalmbach Books, 1974).

Alvin F. Staufer, *Pennsy Power: Steam and Electric Locomotives of the Pennsylvania Railroad, 1900–1957* (Carollton, Ohio: Standard Printing & Publishing Co., 1962).

Karl R. Zimmermann, *The Remarkable GG1* (New York: Quadrant Press, 1977).

Electro-Motive FT Freight-service Diesel-Electric Locomotive

St. Louis, Missouri

The bulldog-nosed, dark green machine—a broad yellow stripe running down its side, "GM" on its snout—rolled out of General Motors' sprawling Electro-Motive Division (EMD) plant at La Grange, Illinois, in November 1939. During the next eleven months, this four-unit EMD-103 demonstrator locomotive would sail through an 83,000-mile (133,500-km) road test to become the prototype for the world's first mass-produced diesel freight locomotives. In short order, diesel would triumph over steam to usher in a new era for America's freight-hauling railroads.

No. 103 (its shop serial number) was the brainchild of Electro-Motive chief engineer Richard M. Dilworth. It consisted of four 1,350-horsepower (1,007-kW) units for a total horsepower of 5,400 (4,027 kW). Each unit was carried on two

The bulldog-nosed Electro-Motive FT diesel-electric locomotive.

4-wheel trucks, with a traction motor for each axle. This design differed significantly from that of passenger diesels (two 6-wheel trucks, each with two driving axles and a center, idler axle), and provided more tractive power. And, while EMD passenger diesels were geared to run a maximum of 117 miles per hour (188 km/hr), the FT was held to 75 miles per hour (121 km/hr), resulting in greater tractive effort at low speeds.

EMD invited railroads to try it out on their lines. From November 1939 to October 1940, the experimental unit ran freight hauls in thirty-five states over 83,000 miles (133,500 km) of track on twenty Class 1 railroads. One road after another challenged the 103 with steep grades, tight curves, extreme temperatures, and capacity tonnage.

The barnstorming diesel consistently beat steam schedules for comparable tonnage hauls, proving Dilworth's claim that the No. 103 was superior to any steam engine in the country and putting to rest questions about diesel's ability to do hard work. The diesel was also a steady worker, never once missing its assigned run and stopping for fuel only once every 500 miles (805 km). Besides savings in fuel (diesel fuel costs were about half those of steam), railroads stood to realize additional savings from the elimination of water tanks, ash pits, coaling towers, roundhouses, and other costly steam-support equipment.

No. 103 and its 1,350-horsepower (1,007-kW) FT siblings were subsequently replaced by other, more powerful diesel locomotives. But the almost 1,100-member fleet had a long and useful life. The 103 units became Southern Railway Nos. 6100, 6150, 6151, and 6104, serving on Southern's Cincinnati, New Orleans, & Texas Pacific line until retirement in 1960. Southern donated the No. 6100 to the National Museum of Transport near St. Louis in 1961.

Location/Access

EMD 103 (previously Southern Railway No. 6100) is on permanent display at the National Museum of Transport, 3015 Barrett Station Road, St. Louis, MO 63122; phone (314) 965-7998. The museum contains a large assortment of locomotives and railroad rolling stock from the pre–Civil War era to the 1950s as well as exhibits of other transportation modes. Hours: daily, 9 A.M. to 5 P.M. Admission fee.

FURTHER READING

David P. Morgan, "The Diesel that Did It," *Trains*, February 1960, 18–25.

Franklin M. Reck, The Dilworth Story: The Biography of Richard Dilworth, Pioneer Developer of the Diesel Locomotive (New York: McGraw-Hill Book Company, Inc., 1954).

Franklin M. Reck, On Time: The History of Electro-Motive Division of General Motors Corporation (La Grange, Ill.: Electro-Motive Division of General Motors Corp., 1948).

Norfolk & Western No. 611, Class J Steam Locomotive

Roanoke, Virginia

In 1944 *Railway Age* described the characteristics of the modern steam passenger locomotive. It would have the ability to haul heavy trains—on steep grades if necessary—at speeds of from 45 to 70 miles per hour (72 to 113 km/hr), a design suitable for top speeds of 90 to 100 miles per hour (145 to 161 km/hr), and the ability to run over long distances, day in and day out, with only minor terminal servicing. It would be of the 4-8-4 type, of 70,000 pounds (31,750 kg) tractive force, with a total weight (engine and tender) of more than 800,000 pounds (362,800 kg). And it would have mechanical lubrication for all major moving parts, so that it could operate more than 1,000 miles (1,609 km) without replenishment of oil reservoirs.

Railway Age might well have been describing Norfolk & Western's Class J locomotive, among the most advanced and powerful of any 4-8-4 passenger locomotive, designed by the railroad's Roanoke (Virginia) Shops and built there between 1941 and 1950. Their superb performance and reliability allowed them to operate 15,000 miles (24,000 km) per month, even on the N&W's relatively short, mountainous routes, delaying the day when diesel-electrics would prevail.

The J was designed for a maximum tractive effort of 80,000 pounds (36,287 kg), the greatest of any 4-8-4. Under controlled tests, the J realized an average drawbar horsepower of 5,028 (3,749 kW) at a speed of 41 miles per hour (66 km/hr)—a performance unsurpassed even today by modern single-unit diesel locomotives—and pulled a 15-car, 1,015-ton (921 t) passenger train at speeds up to 110 miles per hour (177 km/hr) on level, tangent track.

Norfolk & Western No. 611 being fired up for an excursion run at Lynchburg, Virginia, in 1983.

In addition to producing a high tractive effort, the J's comparatively small, 70-inch (1,780-mm) drivers allowed the boiler to be unusually large in diameter without exceeding the N&W's clearance limits. The J also had the largest firebox and longest combustion chamber of any 4-8-4 burning eastern bituminous coal. A unique side-rod-and-driver counterbalancing design, together with stiffened centering of the leading and trailing trucks, permitted speeds in excess of 100 miles per hour (161 km/hr), while a mechanical, pressurized lubrication system, feeding oil to 220 points, allowed the engines to operate 1,300 miles (2,092 km) between refills.

For up to eighteen years, the Js pulled the *Powhatan Arrow, Pocahontas,* and *Cavalier* on their daily runs between Norfolk and Cincinnati. They also ran on the

SPECIFICATIONS

Norfolk & Western No. 611, Class J Steam Locomotive

Builder: Norfolk & Western Roanoke (Virginia) Shops (1950)
Type: 4-8-4
Cylinders: 27 inches by 32 inches (686 mm by 813 mm)
Drivers: 70 inches (1,778 mm)
Boiler pressure: 300 psig (2,068 kPa)
Engine weight: 494,000 lbs. (224,073 kg)
Engine and tender weight: 889,260 lbs. (403,359 kg)
Tractive effort: 80,000 lbs. (36,287 kg)

N&W portion of the joint N&W–Southern Railway routes, pulling the *Pelican*, *Birmingham Special*, and *Tennessean* between Washington, D.C., and southern cities.

The Norfolk & Western was the last major railroad to convert to diesel power, finally discontinuing steam operation in 1960. The N&W Roanoke Shops built the last standard-gauge steam locomotive in the United States—0-8-0 switcher No. 244—in December 1953. In all, the craftsmen of the Roanoke Shops built 447 steam locomotives over sixty-nine years.

In 1959 Norfolk & Western donated No. 611, the sole survivor of fourteen Class J locomotives, to the city of Roanoke. From 1981 to 1982, for the occasion of the city's one hundreth birthday, the railroad (now Norfolk Southern Corporation) restored the 611 to operating condition. It pulled excursion trains until 1994, when Norfolk Southern ceased this type of operation. No. 611's last run was to the Virginia Museum of Transportation in Roanoke.

Location/Access

No. 611 was donated to the Virginia Museum of Transportation, 303 Norfolk Avenue, S.W., Roanoke, VA 24016; phone (540) 342-5670. Hours: Monday–Saturday, 10 A.M. to 5 P.M., and Sunday, noon to 5 P.M.

FURTHER READING

Arthur M. Bixby, Jr., "The Norfolk & Western's Roanoke Shops and Its Locomotives," *Railroad History* 137 (Autumn 1977): 20–37.

Lewis Ingles Jeffries, *N&W: Giant of Steam* (Boulder, Colo.: Pruett Publishing Co., 1980).

O. Winston Link and Tim Hensley, *Steam, Steel & Stars* (New York: Harry N. Abrams, Inc., 1987).

Richard E. Prince, *Norfolk & Western Railway, Pocahontas Coal Carrier* (Millard, Neb.: R. E. Prince, 1980).

Ron Rosenberg with Eric H. Arthur, *Norfolk & Western Steam (The Last 25 Years)* (New York: Quadrant Press, 1973).

Southern Pacific No. 4294 "Cab-in-Front" Articulated Steam Locomotive

Sacramento, California

Southern Pacific No. 4294, a 4-8-8-2 "cab-in-front" articulated locomotive built in 1944 by the Baldwin Locomotive Works of Eddystone, Pennsylvania, is the sole survivor of its type. It represents the culmination of a series of cab-in-front steam locomotives that grew out of the ever-increasing demands for greater tractive power and the need to reduce heat and smoke in the cab, especially in the 38-mile (61-km) series of snowsheds between Truckee and Blue Canyon in Cali-

Southern Pacific No. 4294 "Cab-in-Front" articulated steam locomotive.

fornia, on the Sacramento-Reno line, where drifts up to 200 feet (61 m) pile up in the passes.

In 1909, faced with moving ever-increasing tonnage over the Sierras, the Southern Pacific ordered two Mallet compound 2-8-8-2 locomotives from Baldwin, Class MC-1, Nos. 4000 and 4001. These were really two engines in one; one boiler served two pairs of cylinders driving independent groups of wheels. Functionally, the new locomotives lived up to expectations. Problems arose, however, when they were operated through snowsheds and tunnels. The hot gases of the exhaust almost asphyxiated the engine crew.

SPECIFICATIONS

Southern Pacific No. 4294, Class AC-12 "Cab-in-Front" Locomotive

Builder: Baldwin Locomotive Works (1944), Eddystone, Pennsylvania
Type: 4-8-8-2
Engine and tender length: 125.5 feet (38.2 m)
Drivers: 63 inches (1,600 mm)
Cylinders (4): 24 by 32 inches (610 by 810 mm)
Engine weight: 657,900 pounds (298,417 kg)
Engine and tender weight: 1,046,900 pounds (474,861 kg)
Boiler pressure: 250 psig (1,723.7 kPa)
Tractive effort: 124,300 pounds (56,381 kg)

Charles Browning, Jr., engineer of tests and chief chemist in the SP's Sacramento Shops, made a daring suggestion: turn the locomotive around and run with the cab in front, with the crew ahead of the heat and smoke. This was possible because the SP burned oil instead of coal. With oil, it would not matter whether the tender was adjacent to the cab and firebox. Howard Stillman, SP chief mechanical engineer, directed the drafting of new plans with cabs in front.

Baldwin delivered 15 Class MC-2 locomotives of the new design, Nos. 4002–4016, late in 1909. These followed the basic specifications of the earlier locomotives but were slightly heavier. They allowed the engineer and fireman to see farther down the track and eliminated the problem of heat and smoke in the cab. (Not all crew members approved; some believed they would be more vulnerable in the event of a collision. One engineman reportedly told his road foreman, "I don't want a caboose in my lap some day," to which the road foreman replied, "Mister, if you do your job right, that won't happen.") All but twelve of SP's subsequent articulated locomotives were built with cabs in front. They came to serve as a kind of trademark of the Southern Pacific, the only major railroad in the United States to use the type.

In 1927 SP rebuilt No. 4041 into a simple articulated locomotive, with all four cylinders using high-pressure steam. It was so successful that plans were made to convert the other Mallets to simple engines. Instead, as state-of-the-art details and appliances were added to the plans, the 4-8-8-2 Class AC-4 emerged. This design was used for 195 locomotives in eight classes, AC-4 through AC-8, and AC-10 through AC-12. These were built between 1928 and 1944, and carried Nos. 4100–4294. Long and heavy, the 4-8-8-2s were the largest, most powerful locomotives on the Southern Pacific. They were used to haul heavy freight and passenger trains over the steep grades of the Sierra and Cascade mountains. In all, Baldwin built 256 cab-in-front locomotives for the Southern Pacific.

The last such locomotive to operate was No. 4274, on a farewell excursion November 30–December 1, 1956. On October 19, 1958, AC-12 No. 4294, the last built (1944), was placed in the station plaza at Sacramento, next to the *C. P. Huntington*, as a reminder of the days when they ruled the summits. The two locomotives illustrated the beginning and the end of steam on the Southern Pacific. Both have since been moved to the California State Railroad Museum.

Location/Access

Southern Pacific No. 4294 is now on permanent display at the California State Railroad Museum, which occupies a block-long site along the Sacramento River in Old Sacramento. Exhibits in the History Building (111 I Street, Sacramento, CA 95814) include twenty-one restored locomotives and railroad cars as well as interpretive exhibits on the history of railroad travel in California and the West. Hours: daily, 10 A.M. to 5 P.M. Admission fee. Phone (916) 448-4466.

FURTHER READING

G. M. Best, "The Southern Pacific Company," *The Railway and Locomotive Historical Society Bulletin* 94 (March 1956): 7–154.

John B. Hungerford, Cab-in-Front: The Half-Century Story of an Unconventional Locomotive (Reseda, Calif.: Hungerford Press, 1959).

Disneyland Monorail System

Anaheim, California

During a visit to Europe shortly after the opening of Disneyland in 1955, Disney engineers saw an experimental monorail near Cologne, Germany. Developed by Swedish industrialist Dr. Axel L. Wenner Gren and built by Alweg Corporation in 1957, the 1.25-mile (2-km) line featured lightweight cars on rubber-tired wheels traveling on a concrete beamway. Disney envisioned a similar, all-electric "highway in the sky" that would weave together the most ambitious expansion program in the young park's history. In 1958 Disney and Alweg joined efforts to develop a working prototype for Disneyland.

The idea of a railway with only one rail instead of two was not new. English engineer Henry Robinson Palmer patented a monorail in 1821; the horse-drawn cars hung down like saddle bags on both sides of a framework carrying an elevated rail. Frenchman C. F. M. T. Lartigue put a single rail on top of a series of triangular trestles and gave his cars vertical and lateral support; an electric Lartigue line opened in central France in 1894. Adapting the Lartigue design, F. B. Behr built a 3-mile (4.8-km) electric line near Brussels in 1897. The world's oldest operating monorail, the 8.25-mile (13-km) Schwebebahn in Wuppertal, Germany, opened in 1901; it uses cars hung from an overhead rail.

The ⅔-scale circular line at Disneyland, opened in 1959, consisted of eight-tenths of a mile (1.3 km) of track and two trains. The cars were equipped with 600-volt, direct-current, 100-horsepower (75-kW) traction motors. The cars straddled a single, I-shaped, reinforced-concrete beamway. Each truck had six rubber-tired wheels—two driving and braking wheels on top of the beamway, and two guiding and stabilizing wheels on each side.

The Disney monorail wound around Tomorrowland, over the Submarine Lagoon, and past a replica of the Matterhorn. Unlike the Alweg monorail, whose beamway followed a long curve without grades, the Disney-Alweg system was purposely designed to negotiate sharp curves and climb grades of up to 7 percent, to give visitors an interesting ride. Top speed was about 20 miles per hour (32 km/hr).

In 1961 the monorail was extended to a length of 2.5 miles (4 km), linking the park with the nearby Disneyland Hotel. For its many long, straight sections,

The Disneyland monorail winds past the Matterhorn at the Anaheim, California, theme park.

the new line used prestressed-concrete girders to support the beamway. Top speed increased to 45 miles an hour (72 km/hr).

The newest Disney trains, operating at Disneyland and other Disney parks, feature fiberglass bodies, making them lighter and more energy-efficient than their predecessors, while an onboard computer system controls train functions and keeps daily maintenance records. Designed by Walt Disney Imagineering, the new trains were manufactured by Messerschmitt, Bolkow & Blohm of Germany using the same basic running gear as the original trains.

Location/Access

Disneyland is located at 1313 Harbor Boulevard, Anaheim, CA 92803; phone (714) 999-4565. Take the Anaheim exit of the Santa Ana Freeway (I-5). Hours: year-round: Monday–Friday, 10 A.M. to 7 P.M., and Saturday and Sunday, 10 A.M. to 6 P.M.; extended summer hours: Monday–Friday, 9 A.M. to 10 P.M., Saturday until midnight, and Sunday till 10 P.M. Admission fee.

FURTHER READING

"Disneyland: Building for Fun Is Serious Work," *Engineering News-Record* 162 (7 May 1959): 34–36.

"Fun Monorail Grows Up to Rapid Transit," *Engineering News-Record* 166 (8 June 1961): 38–39.

Road and Off-Road Transportation

INTRODUCTION by R. Michael Hunt

Three of the landmarks in this section illustrate the invention and application of the "crawler" track. To anyone who has watched a bulldozer at work or seen military tanks racing across a rugged landscape, the crawler track may seem a rather obvious device. But it was not invented until the beginning of the twentieth century and then appeared almost simultaneously in Maine and California.

Consider the process of practical invention. First, there has to be a problem that needs to be solved or a significant benefit to be gained. Then, there needs to be someone with a possible solution, who has sufficient persistence and resources to bring the solution to at least the point where it can be demonstrated. And the technology base must be there with the materials, processes, and tools required. It is no accident that the period from about 1875 through 1925 was the golden age of mechanical invention. By then, the modern materials (e.g., steel and aluminum) were becoming readily available, analytical methods had progressed so that the strength and performance of machines could be assessed in advance without the cut-and-try approach of earlier years, machine tools and factories could produce parts in great quantity with great precision, the barons of business were ready to invest heavily when faced with the prospect of large profits, and the telegraph and telephone kept people informed of the latest developments wherever they happened. Given the need for the right infrastructure, it is not surprising then that similar inventions often appeared in different places at about the same time. In addition, there was no alternative to mechanical technology in that era; if a machine had to perform complex operations, then implementation had to be by mechanical means.

To set the stage, imagine that it is now the beginning of the twentieth century: Gottlieb Daimler has demonstrated that an internal-combustion engine may be used to move a vehicle, and Henry Ford has demonstrated his "Quadracycle." But these vehicles are puny and primitive, little more than gasoline-engined, four-wheeled bicycles. For real hauling, traction engines—steerable steam locomotives running on a road—are king. But they are big and very heavy.

Wood is in great demand. The rising expectations and incomes of the working and middle classes create an enormous demand for lumber for construction as well as wood for furniture and other products. In the north woods of Maine, the traction engines hauling the lumber out bog down as the snow falls, the ground thaws, or the rains come. To spread the load on the rear driving wheels over a larger area so that they will be less prone to sink into the trails, Alvin Lombard in 1903 introduces an engine with long flexible traction belts replacing the rear wheels. The front wheels, which steer the vehicle, are replaced with a sled. By 1915 more than two hundred have been sold for logging. But Lombard does not exploit the crawler track, and the hauler is eventually replaced by the truck. Ironically, it is the bulldozer that constructs good logging roads, which makes hauling by truck practical.

In California's San Joaquin valley, there is much rich land that is too soft to be farmed by heavy steam traction engines, even when equipped with large, wide wheels. In 1904 Benjamin Holt devises tracks to replace the driving wheels, retaining the front wheels for steering. In time, the steam engine is replaced by the gasoline engine, then the diesel, and "caterpillar" tractors are steered by speeding up or slowing down the track on one side of the vehicle relative to the other, so eliminating the front wheels. By 1920 the essential configuration evolved to the design familiar today, and the company continues as Caterpillar, Inc.

Returning to the present day, the transporters of Cape Kennedy are a unique application of the crawler track. The launch rockets are assembled vertically and launched vertically. The question is how to move rockets from assembly to launch while keeping them vertical. The answer is to build them on a large platform that is carefully lifted and maintained level as it is slowly transported to the launch site. Here the crawler treads are one possible solution to supporting the weight and providing the traction. Their advantage is that they spread the load over a large area so that the cost of the roadway is kept low, and they average out the effect of any slight bumps.

The fourth landmark is less visible than the crawler track. Lift your foot from the accelerator pedal in a gasoline-engined car and you feel the engine helping to slow to you. This effect is much reduced in a diesel engine, with its unrestricted, unthrottled air intake, placing more reliance on the brakes of a diesel-powered vehicle such as a large truck. Brakes convert the kinetic (moving) energy of a vehicle into heat and become less effective when used for long periods because they become hot. This characteristic is unimportant in most driving situations because the brakes cool down between uses, but there can be a significant loss of braking power ("fade") on long downgrades. The Jacobs brake addressed this need by altering the valve timing in a diesel engine so that the engine compression provides braking. Interestingly, Clessie Cummins's invention was recognized by a company outside the engine industry and not by the big players.

Lombard Steam Log Hauler

Patten, Maine

The Lombard steam log hauler emancipated horses from the killing work of hauling sleds of logs over rough roads in the north woods and greatly expanded the area from which timber could be profitably harvested. Designed and patented by Alvin O. Lombard (1856–1937), this steam-powered hauler was the first practical vehicle employing the lag, or crawler, tread that would become standard for engine-driven agricultural and construction equipment, and military tanks.

Lumbering in Maine was limited to timber growing near enough to water to permit its transportation by horse or oxen to streams or rivers on which it could then be floated to sawmills. Only those varieties that would float—spruce, fir, cedar, and pine—were harvested, leaving uncounted acres of maple, birch, beech, and ash, which, with the equipment then available, could not be brought to market. With the development of the self-propelled vehicle in the 1890s, the time was ripe for a mechanical means of hauling logs. But a log hauler must run on snow, ice, and muddy roads, and the wheels of the horseless carriage would not provide the necessary traction or bearing.

Johnson Woodbury, a farmer and millwright living in Patten, Maine, attempted to solve the problem. He was inspired by a treadmill used to provide power for a threshing machine, the essential feature of which was an endless belt of wooden lags running over two pairs of trucks to which the lags were geared. But

The steam-powered Lombard log hauler, developed by Maine mechanic Alvin O. Lombard in 1900, greatly expanded the area from which timber could be harvested profitably.

without the funds or facilities to develop a tractor that would run on snow, Woodbury's plan languished. He reputedly related his idea to Alvin O. Lombard, a Waterville mechanic who already had several inventions to his credit. Lombard went to work with such diligence that, legend has it, he emerged from his shop two days later with a drawing and a working model of a traction engine for hauling logs. He arranged for the Waterville Iron Works to build a prototype, and on Thanksgiving Day 1900, the first machine made its trial run.

Lombard's contribution was not the engine but the means of locomotion. Patented in 1901 (No. 674,737), his invention consisted of three parts: a carriage and a traction member on each side of the carriage. Each traction member consisted of two cogged driving wheels; a flexible traction belt, which fit over and meshed with the driving wheels; and two antifriction roller belts. The traction belts consisted of a series of cogged sections with ribbed faces. This lag tractor tread kept the machine from slipping and sliding, and assured solid footing for the hauler.

The steam-powered Lombard hauler could make its way through the woods with 300 tons (272 t) of logs at a speed of 4 to 5 miles per hour (6 to 8 km/hr). It hauled logs on trains of bobsleds, averaging 8 sleds to a train, although on good roads it could haul trains of 10 or 12. Coal was the usual fuel, but on smaller jobs many haulers burned wood. A Lombard crew consisted of an "engineer," a fireman, a conductor, and a steerer.

Following sale of the first log hauler in 1903 to a logger in Waterville, Maine, Lombards rapidly came into general use in the logging regions of the northern United States. In 1904 Lombard granted a license to the Phoenix Manufacturing Company of Eau Claire, Wisconsin, to manufacture the log haulers in exchange for a royalty fee of $1,000 per engine. By 1915, more than two hundred Lombard log haulers had been sold. The first such machines had an upright boiler and two upright engines; these were soon replaced by a horizontal boiler and engine. Later, gasoline engines replaced steam. In 1934 Lombard built a diesel-powered log hauler, but by then heavy trucks had made the log hauler obsolete. The first diesel log hauler was also the last.

A few survivors still stand in the Maine woods, abandoned and rusting. A few others are preserved in museums.

Location/Access

A Lombard steam log hauler, restored to running condition, is on display at the Patten Lumberman's Museum, Shin Pond Road, Patten, Maine 04765; phone (207) 528-2650. Hours: Memorial Day–Labor Day: Tuesday–Sunday, 10 A.M. to 4 P.M. Admission fee. (A gasoline Lombard is in the Maine State Museum in Augusta.)

FURTHER READING

Walter M. MacDougall, "Lombard's Iron Monster," *Yankee*, March 1965, 72.

Holt Caterpillar Tractor

Stockton, California

Although Alvin O. Lombard of Maine was the first to manufacture steam engines with crawler tracks (see "Lombard Steam Log Hauler," p. 284), his invention was limited to logging. On the West Coast, Benjamin Holt (1849–1920), president of the Holt Manufacturing Company of Stockton, California, developed the modern, all-purpose, track-type tractor to meet the demands for power created by large-scale farming. The Caterpillar tractor would revolutionize agriculture, the construction industry, even military warfare.

During the 1890s, Holt manufactured mammoth steam-powered traction engines to replace horses in pulling farm machinery. But the heavy machines mired down in the soft soil of California's San Joaquin Delta; conventional wheels simply had too little ground-bearing surface. To solve this problem, in 1904 Holt had his mechanics remove the rear drive wheels on one of his engines and replace them with a pair of tracks 9 feet long and 24 inches wide (274 by 610 mm). The track shoes consisted of 2- by 4-inch (50- by 100-mm) wooden slats bolted to an endless chain driven by sprockets. The result was the world's first practical track-type tractor. Watching the tractor return from its first field test, company photographer Charles Clements observed, "It crawls just like a caterpillar." "Caterpillar it is," Holt replied. "That's the name for it."

The track-type tractor's unique ability to traverse difficult terrain made it ideal for farming, logging, road building, and moving dirt on canal and irrigation

Holt Caterpillar Tractor.

projects. *Farm Implement News* (1905) described Holt's achievement, observing that, on land where people could not walk without sinking to their knees, "the new traction engine was operated without a perceptible impression in the ground. . . . This tract of land had been useless for crop raising for several years because no way was found to plow it, but the platform wheel engine has brought the land into use again. . . ."

After testing six different models of track-type steam engines, Holt sold its first machines in 1906 to a Louisiana developer for use in clearing land in the Mississippi Delta. Two years later, Los Angeles city engineer William E. Mulholland used Holt tractors to help build the 233-mile (375-km) Los Angeles Aqueduct across the Mojave Desert, firmly establishing the Caterpillar's reputation. During World War I, the American, British, French, and Russian governments purchased thousands of Holt crawler tractors for hauling artillery and supplies. British Army General Ernest D. Swinton credited the Caterpillar track-type tractor for giving him the idea for a new weapon, the armored tank.

The Holt "75" Caterpillar tractor, built in 1920, is representative of the earliest gasoline-powered, track-type tractors. Of all the models of track-type tractors Holt built between 1904 and 1925, model 75, designed for "larger farms or bigger power jobs," according to an early Holt advertisement, was the most widely used. (The number 75 indicated the unit's horsepower.) The tractor, with its four-cylinder engine, was used on a farm in San Joaquin County. In the mid-1950s, Holt Bros., a Caterpillar dealership in Stockton, purchased it for display in their showroom. It remained there until 1974, when it was presented to the Pioneer Museum (now the Haggin Museum) in Stockton, California.

The Caterpillar Tractor Company, organized in 1925, consolidated the Holt and Best tractor companies, longtime rivals. Today Caterpillar Inc., with headquarters in Peoria, Illinois, is a multinational company. Caterpillar's sixty thousand employees worldwide design, manufacture, and market earthmoving, construction, and materials-handling machinery, and engines for a wide variety of applications.

Location/Access

The Holt Caterpillar is on display in the Holt Memorial Hall at the Haggin Museum, 1201 North Pershing Avenue, Stockton, CA 95203; phone (209) 462-4116 or 462-1566. Holt Memorial Hall illustrates the contribution of Stockton industry to agricultural and industrial technology. Also on display are a 1904 Haines-Houser combine and Benjamin Holt's experimental shop. Hours: Tuesday–Sunday, 1:30 P.M. to 5 P.M.; closed Mondays.

FURTHER READING

Reynold M. Wik, *Benjamin Holt and Caterpillar: Tracks and Combines* (St. Joseph, Mo.: American Society of Agricultural Engineers, 1984).

Jacobs Engine Brake Retarder

Bloomfield, Connecticut

Long, steep downhill grades provide anxious moments for truck drivers and add extra hours behind the wheel. The advent of the Jacobs engine brake retarder—better known as the Jake Brake—put an end to the anxiety and, at the same time, safely raised average speeds by converting the power-producing diesel engine into a power-absorbing air compressor.

The need for an engine brake retarder was dramatically—and terrifyingly—demonstrated to its inventor, Clessie L. Cummins, in 1931. Cummins and two associates were trying to prove the worth of the automotive diesel engine by driving a Cummins diesel-powered truck from New York to Los Angeles. Descending Cajon Pass, a 35-mile (56-km) stretch of mountainous downgrade on old U.S. 66 leading into San Bernardino, California, the brakes would not hold, and Cummins could not get the truck into a lower gear. Cummins narrowly missed colliding with a freight train crossing the road ahead of him. The truck set a new coast-to-coast speed record. Its driver, meanwhile, vowed someday to make its diesel engine work going downhill just as well as going uphill.

It is easy to use the gasoline engine as a brake by simply closing the throttle and allowing the wheels to drive the engine; the energy that the wheels have to provide to draw the air past the closed throttle is sufficient to slow the vehicle down without the driver touching the brakes. A diesel engine, however, has no throttle. Theoretically, a diesel engine could be used as a brake if the exhaust valve were kept closed during what normally would be the exhaust stroke, until the piston reached the end of the stroke. At that point, the exhaust valve would be

The Jacobs engine brake retarder, better known as the Jake Brake.

The Engine without the Jake Brake

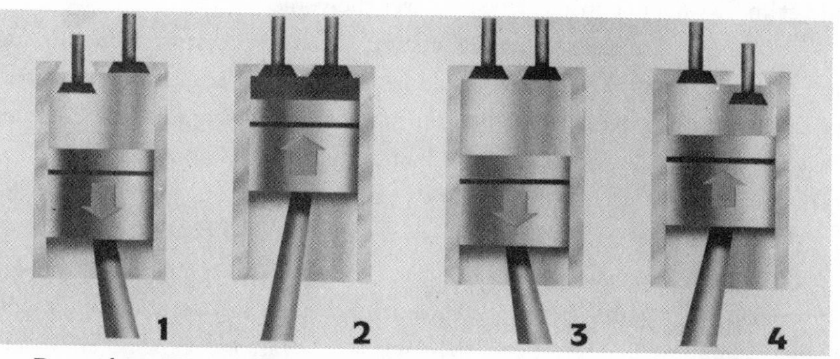

During the piston's normal compression stroke (2), air compressed in the cylinders raises internal temperature to almost 1,000°F (537°C). At this point, fuel is injected and combustion occurs, raising the temperature even further. Pressure from this expanded gas forces the piston down (3). The engine has thus produced the positive power needed to turn the crankshaft.

The Engine with the Jake Brake

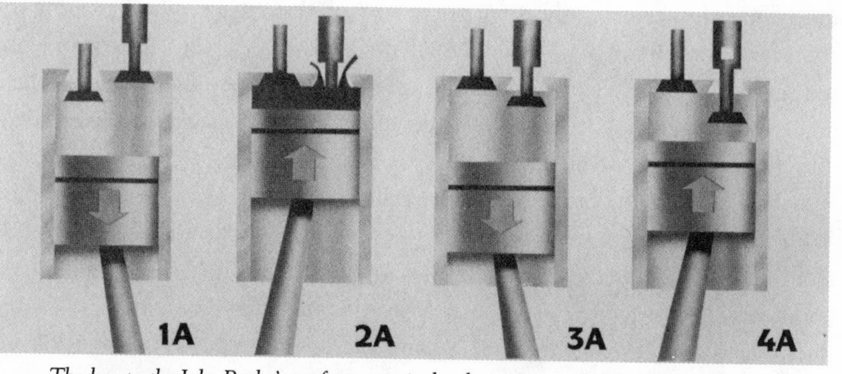

The key to the Jake Brake's performance is the slave piston. Hydraulically actuated, the slave piston opens the engine's exhaust valve (2A) near the end of the compression stroke. The compressed air, which normally forces the piston down even though no fuel is added, is vented through the exhaust system (2A). By releasing this energy, the Jake Brake prevents any positive power from being exerted on the piston. The vehicle's forward momentum provides the energy needed to return the piston to its bottom position (3A), thus completing the process of generating maximum retarding power with a Jake Brake.

Source: *Jacobs® Engine Brake Retarder, Bloomfield, Connecticut, October 17, 1985,* commemorative brochure (New York: The American Society of Mechanical Engineers, n.d.).

opened, releasing the air compressed in the cylinder. Then, when the piston descended on what normally would be the following power stroke, the pressure in the cylinder would be insufficient to power the vehicle wheels.

To accomplish this, the natural motion of the rocker arm, which is arranged to keep the exhaust valve closed, would have to be overcome. Earlier attempts by other inventors to produce a diesel engine brake involved complicated arrangements that altered the valve controlling mechanism. Cummins wanted to avoid this by taking advantage of the existing, and perfectly timed, motion of the diesel engine.

Beginning in 1955, Cummins, now retired, studied concepts based on hydraulically transferring a timing signal from an engine-driven pump. But the signal provided by this method was not sufficiently accurate. In 1957, during a sleepless night, Cummins seized on an idea: Why not take advantage of the accurate timing signal provided by the cam that drives the fuel injectors? At the moment that it would be necessary to open the exhaust valve in the braking mode, the injector cam for that cylinder would be poised to operate the injector (but not to inject fuel). Cummins first proposed to transfer this motion mechanically, but hydraulic transfer proved more practical.

How does it work? A switch on the dashboard activates the brake, opening the exhaust valve near the end of the upward piston stroke. The stored energy in the cylinder is released to the atmosphere through the exhaust pipe—creating the characteristic popping noise when a truck descends a steep hill—instead of being transmitted to the wheels of the vehicle (see sidebar). When the driver no longer requires braking action, the switch is operated again, and the pressure is removed from the hydraulic system.

The first engine brake retarder of the prototype design was installed on a Sheldon Oil Company truck in Suisun, California, in 1959. When delivering oil to an asphalt plant at the foot of the Sierras near Lake Tahoe, the veteran driver normally was forced to pass the turn-off to the plant and come back when the truck had slowed enough to turn around! With the engine brake, the driver needed to use his wheel brakes briefly only two times on the descent and made the turn-off easily.

Cummins Engine Company (to whom Clessie Cummins was contractually bound to show his ideas) and other manufacturers of "three cam" diesel engines turned down the opportunity to buy or license the idea. But Jacobs Manufacturing Company, a leading maker of drill chucks looking to diversify, purchased an option on the invention and, following further testing, established a new division to manufacture the brake in 1960. The first production units for the Cummins NH series engines left the factory in 1961.

Today, Jake brake retarders are available for all major U.S. makes of heavy-duty diesel truck engines. They are also standard on several U.S. military vehicles. Cummins's idea for making a diesel engine work downhill as well as uphill has contributed to greater control and safer operation of heavy trucks worldwide.

Location/Access

The Jacobs engine brake retarder is displayed in the lobby of the Jacobs Manufacturing Company, 22 East Dudley Town Road, Bloomfield, CT 06002; phone (203) 243-1441.

FURTHER READING

C. Lyle Cummins, Jr., "A History of the Jacobs Engine Brake," *Diesel Car Digest* 6 (Winter 1981): 18–24.

C. Lyle Cummins, Jr., and G. S. Haviland, "The Jacobs Engine Brake—A New Concept in Vehicle Retarders," paper #387A, presented at the Society of Automotive Engineers national meeting, Portland, Oregon, August 1961 (available SAE International, 400 Commonwealth Drive, Warrendale, PA 15096).

Crawler Transporters of Launch Complex 39

John F. Kennedy Space Center near Titusville, Florida

On May 25, 1961, President John F. Kennedy told Congress: "I believe that this nation should commit itself to achieving the goal, before this decade is out, of landing a man on the moon and returning him safely to earth." Critical to America's successful landing on the moon in 1969 was the development, in the mid-1960s, of two crawler transporters. These, the world's largest land vehicles, solved one of the National Aeronautical and Space Administration's most difficult engineering problems: how to move the thirty-six-story Apollo-Saturn space vehicle from the Vehicle Assembly Building to the launchpads of the Kennedy Space Center's Launch Complex 39.

The task of selecting and developing one of three transport methods—a barge and canal system, a rail system, or a land crawler—fell to Donald Buchanan, the mechanical engineer who was the launcher systems and umbilical tower design section chief. Bucyrus-Erie Company was asked to study the crawler concept after launch operations officials visited Paradise, Kentucky, to watch a Bucyrus-Erie crawler shovel, used for surface-mining coal, in action. In May 1962, Bucyrus-Erie made its final presentation to the National Aeronautics and Space Administration (NASA). The proposed crawler would travel 1 mile per hour (1.6 km/hr) under load, while a hydraulic leveling system would keep the platform within 10 inches (254 mm) of the horizontal when moving on a 5-percent grade. Following extensive studies of all three concepts, Buchanan decided on the crawler transporter because of its advantages of cost and flexibility over the barge and rail systems.

With a bid of $8 million, Marion Power Shovel Company won a contract to develop and build two crawler transporters. (The award was not without con-

Larger than a baseball infield, the crawler transporter (here, without the thirty-six-story Apollo-Saturn space vehicle it was designed to transport) makes its way along the special roadbed of Launch Complex 39 at the Kennedy Space Center. *Courtesy NASA.*

troversy; Bucyrus-Erie bid $11 million, closer to the eventual $13.6 million price tag.) Ohio-based Marion had experience building large, self-propelled strip-mining shovels, but the design and construction of a land vehicle expected to carry a load of 8,800 tons (8,000 t) was without precedent. Even the largest mining shovel had a chassis only 48 feet (14.6 m) square; by comparison, the crawler transporters would be 131 feet long and 114 feet wide (39.9 by 34.7 m). The trickiest feat was balancing the thirty-six-story Apollo-Saturn vehicle. The motion of the transporter, the height of the load, variations in the level of the roadway, wind—all would combine to throw the cargo off balance. Marion had to adjust its initial designs, adding a separate power system, distinct from the diesel engines that powered the treads, for load-leveling, jacking, steering, and ventilating.

Marion built the two crawlers in Ohio, then took them apart for shipment to Kennedy Space Center. Labor disputes delayed the first test of the vehicle from November 1964 to January 1965. On January 23, the crawler transporter moved under its own power for the first time. The gargantuan tractor resembled "a steel sandwich held up at the corners by World War I tanks," according to one observer. It was larger than a baseball infield and weighed close to 3,000 tons (2,720 t). Two 2,750-horsepower (2,051-kW) diesel engines powered 16 traction motors, which moved the 4 double-tracked treads. Each tread had 57 "shoes"; each shoe measured 1 foot by 7.5 feet (304 mm by 2,286 mm) and weighed 1 ton (0.9 t). Following delays to solve a problem with bearing friction, the first crawler transporter was ready for service in 1966.

In the space-shuttle era, the crawler transporters have become the workhorses of Launch Complex 39. A team of eleven engineers and technicians stationed in the crawler cab, on the ground, and in a central control room operates the vehicle. Working together, they position the crawler under the mobile launch platform in the Vehicle Assembly Building. Sixteen hydraulic jacks (four at each corner) raise the crawler up for contact with the mobile launcher. En route to the launchpad, a trip of seven hours, the crawler transporter detects any unevenness in the specially designed "crawlerway," or roadbed; the leveling system automatically maintains the mobile launch platform and launch vehicle within plus-or-minus 10 minutes of arc—i.e., in the case of the Apollo-Saturn, the top of the 363-foot (110-m) space vehicle did not deviate from the vertical by more than a foot (305 mm). On the launchpad ramp, the same system compensates for the 5-percent grade.

Once the launcher and space vehicle are set down and bolted to the pad pedestals, the crawler transporter withdraws to a safe distance. Following launch, the operation is reversed, and the crawler transporter returns the mobile launch platform to the Vehicle Assembly Building. The crawler transporters of Launch Complex 39 have carried all of the Apollo, Skylab, Apollo-Soyuz Test Project, and space-shuttle vehicles from the Vehicle Assembly Building to the launchpad.

Location/Access

The crawler transporters may be seen as part of a two-hour bus tour of the Kennedy Space Center that departs from KSC's Visitor Center every twenty minutes. Information and reservations: Visitor Center, Kennedy Space Center, Florida 32899; phone (407) 452-2121.

FURTHER READING

Charles D. Benson and William Barnaby Faherty, *Moonport: A History of Apollo Launch Facilities and Operations*, The NASA History Series (Washington, D.C.: National Aeronautics and Space Administration, 1978).

Air and Space Transportation

INTRODUCTION by J. Lawrence Lee

When one thinks of airplanes and spacecraft, it probably is the aeronautical or, more recently, aerospace engineer who first comes to mind as the designer of such machines. A closer look often reveals that engineers practicing in other disciplines, particularly mechanical engineers, have made extensive contributions to successful aircraft and spacecraft. In several instances, it has been the mechanical engineer's contribution that made a particular design practical, or even possible.

In its purest sense, aeronautical engineering is concerned with the flow of a compressible fluid—air—around a solid object. It seeks to produce a vehicle of useful size and proportion, maximizing its lift and minimizing the drag upon it. In a similar manner, aerospace engineering in its purest form addresses the additional rigors of space travel. To build a practical vehicle, however, problems must be solved in the areas of structural design, materials selection, avionics, and a host of mechanical devices and systems, such as engines, landing gear, and control systems. The Historic Mechanical Engineering Landmarks in this section celebrate some of the flying machines that were made possible largely through breakthroughs in mechanical design. This is by no means intended to be a comprehensive list of landmark aircraft and spacecraft but rather is meant to focus on a group of particular mechanical achievements that have contributed greatly to our conquest of air and space.

Probably the most distinctly mechanical of the flying machines are helicopters. Although they are now common, these versatile machines were virtually unknown prior to World War II. Conceptual designs for such machines date back to the Renaissance, but it was Igor Sikorsky's VS-300 that pioneered the basic design concepts for a practical helicopter. The secret of its success was in its flexible power-transmission and rotor-control machinery. Similar designs can be seen in helicopters being produced today.

During the 1960s, the world was amazed at pictures sent from space by *Surveyor*, *Viking*, and *Mariner* spacecraft, but few knew that these voyages were made possible by a remarkable rocket engine called the RL-10. The RL-10 powered a second-stage booster that was launched atop an Atlas or Titan first-stage rocket; thus, its operation was generally out of camera range and usually unsung. Nevertheless, it was a major advance in rocket technology, and it paved the way for larger hydrogen-fueled engines, including the space shuttle main engines.

Some of the most powerful machines ever built have been the rockets used to boost spacecraft above the limits of Earth's atmosphere into orbit and beyond. Rocket boosters are, in reality, little more than powerful engines with large fuel tanks, but the operating environments and exotic fuels involved make tremendous demands on every component. This was still a young technology in the 1950s and 1960s, and the engineers who designed the Atlas and Saturn V rockets were called upon to advance that technology in very large steps. There were some dramatic early failures, but the successes were extraordinary moments in engineering as well as human history. No one who witnessed an Atlas launching John Glenn into orbit aboard *Friendship 7* or the awesome spectacle of a Saturn V lifting *Apollo 11* and its crew toward the moon could possibly forget it.

The advances in air and space travel since the Wright brothers' flight of 1903 have been nothing short of spectacular. Exploration of the heavens will undoubtedly continue to stir our curiosity and renew the excitement of discovery for decades, perhaps even centuries, to come. It is truly the adventure of our age. As with adventures of the past, new technologies will be required to make these journeys possible, and engineers will be called upon to provide them. If the landmarks described here are any indication of how well mechanical engineers will continue to meet these challenges, the future of this adventure looks as bright as a rocket's flame.

Sikorsky VS-300 Helicopter

Dearborn, Michigan

The Focke-Achgelis Fa-61, first flown in Germany in 1936, is widely considered the world's first practical helicopter. With its two side-by-side rotors, it accomplished maneuvers—vertical takeoff and landing; hovering; and forward, backward, and sideways flight—now taken for granted. But Igor Sikorsky's VS-300, which made its maiden flight from the grounds of Vought-Sikorsky Aircraft in Stratford, Connecticut, on September 14, 1939, opened the modern phase in the history of vertical flight and paved the way for production helicopters that could perform useful work. Sikorsky's arrangement of one main rotor and one tail rotor quickly became the predominant type, putting the United States in the lead in the field of vertical flight.

Igor Sikorsky (1889–1972) made his first attempt to build a helicopter as a young aviation designer in Tsarist Russia. He built two prototypes in 1909 and 1910, but for lack of powerful engines they proved impractical and Sikorsky turned his talents to fixed-wing aircraft. Among his triumphs was the famous *Il'ya Muromets*, the world's first four-engine bomber, in 1914.

Following Lenin's rise to power in 1917, Sikorsky left Russia, settling in the United States in 1919. After struggling to find steady employment, he built a successful aircraft company from humble beginnings on a Long Island chicken

Tethered to the ground, Igor Sikorsky tests his VS-300 helicopter. *Courtesy Smithsonian Institution.*

SPECIFICATIONS

Sikorsky VS-300 (Final Variant)

Manufacturer:
Vought-Sikorsky Aircraft, Stratford, Connecticut
Engine: 1 Franklin 4AC-199
Brake horsepower: 100
Fuselage length: 27 feet, 10 inches (8.5 m)
Fuselage width: 4 feet (1.2 m)
Overall height: 8 feet (2.4 m)
Main rotor diameter: 28 feet (8.5 m)
Tail rotor diameter: 7 feet, 8 inches (2.3 m)
Landing gear tread: 10 feet (3 m)
Weight, empty: 1,043.5 pounds (473.3 kg)
Weight, useful load: 247.5 pounds (112.3 kg)
Gross weight: 1,325 pounds (601 kg)
Maximum speed: 59.8 miles per hour (96.2 km/hr)
Cruise speed: 49.5 miles per hour (79.7 km/hr)
No. built: 1 (6 variants)

farm. With the advent of lighter materials and improved engines, Sikorsky, now engineering manager of Vought-Sikorsky Aircraft, a division of United Aircraft Corporation, again turned his attention to helicopters.

In the spring and summer of 1939, Sikorsky designed and built a research helicopter to test his theories. On September 14, 1939, he made the first test flight of the VS-300—ten seconds long at an altitude of only a few inches and tethered to the ground by ropes lest the awkward bird leap too high. The crude helicopter had a three-blade main rotor and a single tail rotor (to offset torque), an open frame of welded steel tube, a 75-horsepower (56-kW) Lycoming engine, and four landing wheels. Though the helicopter flew, it had control problems and shook violently. There followed two years of trial-and-error experimentation and four distinct configurations. As the design evolved, Sikorsky worked to establish helicopter flight endurance records, attracting spectators and the press. On May 6, 1941, Sikorsky set a world helicopter endurance record by staying aloft for 1 hour, 32 minutes, and 26 seconds.

In its final form, the VS-300 had a single main rotor with full cyclic pitch control—i.e., the pitch of each blade could be changed during rotation, so that by increasing or decreasing the pitch of the blades, the helicopter could travel in any direction—a single tail rotor to control torque, a 100-horsepower (75-kW) Franklin engine, and a tricycle undercarriage. In this configuration, the VS-300 made its first flight on December 8, 1941, the day after the attack on Pearl Harbor. In the meantime, the U.S. Army awarded Vought-Sikorsky a contract to build an experimental two-place, closed-cabin helicopter with a larger and more powerful engine. Production of the XR-4, delivered in 1942, marked the beginning of a new industry.

With their special ability to hover, helicopters quickly proved their usefulness, serving as air taxis, performing rescue missions (especially at sea), and assisting in the erection and operation of offshore oil and gas rigs. In wartime, they offered tactical mobility, carrying troops, weapons, and materiel into the battle zone and ferrying out casualties.

On October 7, 1943, Igor Sikorsky presented the VS-300—the American prototype of these all-but-indispensable machines—to Henry Ford's Edison Institute in Dearborn, Michigan.

Location/Access

The Sikorsky VS-300 is on display at the Henry Ford Museum, 20900 Oakwood Boulevard, Dearborn, MI 48214; phone (313) 271-1620. Hours: daily, 9 A.M. to 5 P.M. Admission fee.

FURTHER READING

Dorothy Cochrane, Von Hardesty, and Russell Lee, *The Aviation Careers of Igor Sikorsky* (Seattle: University of Washington Press for the National Air and Space Museum, 1989).

Atlas Launch Vehicle

San Diego, California

Originally designed to be a weapon, the Atlas launch vehicle achieved greater distinction when the military veteran was reconditioned to take a new, more peaceful path as a vital part of the U.S. space program.

In 1946 the U.S. Air Force awarded a contract to the Consolidated Vultee Aircraft Corporation (later the Convair Division of General Dynamics Corporation), San Diego, to develop a long-range missile. By mid-year, a team of Convair engineers headed by Karel J. "Charlie" Bossart had completed a design for the MX-774 research rocket, modeled after the German V-2 of World War II. The V-2 was a one-stage, liquid-fueled rocket with a range of 200 miles (322 km), an altitude of 50 miles (80 km), and speeds of 3,500 miles per hour (5,632 km/hr). The word "accuracy" hardly applied; the V-2 often missed its target by as much as 10 miles (16 km). Convair had to do much better.

The biggest problem was weight. There can be no unnecessary pounds on a ballistic missile aimed at a target thousands of miles away. Bossart decided on a separable nose cone (warhead) and a pressure-stabilized airframe that needed no internal bracing. To achieve reliable ignition, he decided on a unique one-and-a-

An Atlas launch vehicle lifts off the launch pad at Cape Canaveral, Florida, with a manned *Mercury* spacecraft. All four Mercury orbital missions during 1962 and 1963 utilized Atlas rockets.

half-stage system consisting of three main rocket engines and two verniers, or small rockets mounted near the rear of the propellant tank to fine-tune the missile's direction and velocity.

To improve flight control, engineers decided to gimbal, or swivel, the engines. To track and control the missile's course during powered flight, a Convair electronics team led by James W. Crooks came up with the idea of a radio-inertial system with ground-based receiving stations. Three partially successful test launches of MX-774s were made at White Sands, New Mexico, in 1948 before the project was canceled due to federal belt-tightening.

In 1951, following the outbreak of war in Korea, Convair was awarded a new contract for the secret MX-1593. Bossart dubbed it "Project Atlas"—after the Titan of Greek mythology compelled to support the heavens on his shoulders—and reassembled his team. Their challenge: to build a rocket capable of delivering a 3,500-pound (1,588-kg) warhead 6,325 miles (10,179 km) with an accuracy of 2 to 3 miles (3 to 5 km). The task grew more urgent with the news, in August 1953, that the Soviet Union had tested a hydrogen bomb.

In 1954 came a thermonuclear breakthrough: nuclear warheads could be made small, light, and powerful enough to ride atop long-range ballistic missiles.

The air force began a crash program to build both Atlas missiles and a network of eleven Strategic Air Command bases from Maine to California.

By the end of the year, Convair had redesigned the Atlas to its present configuration. It called for a missile 75 feet (23 m) long and 10 feet (3 m) in diameter. It had three liquid oxygen-kerosene rocket engines plus the two verniers. To reduce weight, it had a helium-pressurized tank made of thin-gauge stainless steel. (The Atlas tank, with walls thinner than a dime, actually weighed less than 2 percent of the propellant it carried.) "There has never been a missile like Atlas," *Aviation Week* reported early in 1955. "Big vehicles have been built before . . . but most big structures have been based on progressive experience with smaller ones. Atlas is the first and has very little to draw on but engineering courage."

The first static tests were conducted in 1956. Following two disappointing failures, the Atlas made its first successful flight test on December 17, 1957, soaring from Cape Canaveral, Florida, out over the Atlantic some 600 miles (966 km). The flight boosted American confidence: just weeks earlier, the Soviet Union had launched *Sputnik I* and *Sputnik II*, the first satellites to orbit Earth, proving it already had developed a successful intercontinental ballistic missile (ICBM).

On December 18, 1958, an Atlas missile again lifted off from Cape Canaveral, its true mission a closely guarded secret. This time, it nosed itself into a trajectory parallel with Earth and pushed into orbit. The weapon of war carried a message of peace—the voice of President Dwight D. Eisenhower, beamed by radio signal to Earth: " . . . I convey to you and to all mankind America's wish for peace on earth and goodwill toward men everywhere."

As early as 1955, Convair engineer Krafft Ehricke realized that the Atlas was fully capable of casting small payloads into orbit around the earth; it could even project itself into a satellite orbit. But space planners realized that a more powerful upper stage would be needed to supplement the payload-carrying capacity of the Atlas ICBM. Ehricke proposed a new upper-stage rocket, called Centaur, powered by liquid hydrogen (see "RL-10 Rocket Engine," p. 301).

Alone or in combination with Centaur, the Atlas proceeded to earn honors, boosting the nation's first communications satellites into orbit, then operating successfully through NASA's Mercury program. On February 20, 1962, it boosted the first American, astronaut John H. Glenn, into orbit around Earth.

Location/Access

An Atlas launch vehicle is displayed at Gillespie Fields Airport, 1960 Joe Crosson Drive, El Cajon, CA 92020; phone (619) 596-3900.

FURTHER READING

John L. Chapman, *Atlas: The Story of a Missile* (New York: Harper & Brothers, 1960).

RL-10 Rocket Engine

West Palm Beach, Florida

The RL-10 rocket engine, built by Pratt & Whitney, was the first in the world to use high-energy liquid hydrogen as a fuel. On November 27, 1963, a pair of RL-10s successfully boosted a Centaur launch vehicle into orbit around the earth in the first flight demonstration of this space power plant. The success of the RL-10 rocket engine led to the development of larger hydrogen-fueled engines that made possible America's greatest space engineering achievement, the lunar landing of July 1969.

Liquid hydrogen is an odorless, colorless, frigid (-423°F, -252°C), flammable liquid first produced in 1898 by James Dewar. Because of its high heat of reaction with all oxidizers and the low molecular weight of its combustion products, it provides more energy than any other fuel. This energy, or "specific impulse," is measured in pounds of thrust per pound of propellant per second of burning. Since the thrust developed by a propulsion system is essentially proportional to the exhaust gas velocity—which, in turn, depends on the average molecular weight—hydrogen yields higher gas velocities for a given temperature and provides greater thrust than any other known liquid propellant. But this advantage initially was outweighed by hydrogen's low density—one-fourteenth that of water, which required bulky fuel tanks—and by the difficulty and danger of storing and handling it. Many questioned the advisability of using the highly hazardous fuel associated with the *Hindenburg* disaster of 1937.

RL-10 Rocket Engine.

Pratt & Whitney's development of hydrogen-fueled rocket engines began in the late 1950s as part of the U.S. Air Force's secret Suntan project to determine the feasibility of a hydrogen-fueled jet engine. The hydrogen jet did not materialize, but the liquid-hydrogen research conducted as part of that project formed a foundation for the nation's emerging space program.

The Soviet Union's launch of the first satellite in 1957 had made the scientific investigation of space a high national priority. In the fall of 1958, the air force awarded a contract to the Pratt & Whitney division of United Aircraft Company (later United Technologies, Inc.) to develop an upper-stage rocket engine using liquid hydrogen fuel. Such a fuel was found to be able to boost twice the payload of previous space propulsion systems. Two years later, the National Aeronautics and Space Administration (NASA), to which management control of the project had been transferred, awarded Pratt & Whitney a contract for development of the RL-10. The first engine model was successfully tested in November 1961.

Outwardly, the RL-10 resembles other liquid-fuel rocket engines. Inside, however, it embodies numerous advanced design features. The liquid hydrogen performs two important functions even before it is mixed with liquid oxygen and burned to produce thrust. First, it acts as a coolant, passing through a series of tubes forming the thrust chamber, where the combustion temperature is over 5,000°F (2,760°C) during engine operation. Second, the hydrogen, which has been heated during its passage through the thrust chamber, is passed through a turbine to provide power to pump hydrogen and oxygen (used as an oxidizer) into the system and to drive other engine-mounted accessories.

Following its first successful launch in 1963, the RL-10 rocket engine launched an array of the nation's most sophisticated unmanned spacecraft, compiling a perfect record: more than 150 of these 15,000-pound- (66.72-kN-) thrust engines were used in space launches—including the *Viking*, *Mariner*, and *Pioneer* spacecraft—without a single engine failure. Powered by RL-10 engines, Atlas-Centaur rockets launched seven Surveyor spacecraft to land on, photograph, and analyze the moon's surface in preparation for the Apollo landings. By 1959, NASA was thinking ahead to even heftier propulsion systems (see "Saturn V Rocket," p. 303).

Location/Access

The Pratt & Whitney office that houses the RL-10 is a U.S. government-secured facility and is not open to the public at this time.

FURTHER READING

John L. Sloop, *Liquid Hydrogen as a Propulsion Fuel, 1945–1959*, The NASA History Series (Washington, D.C.: National Aeronautics and Space Administration, 1978).

George P. Sutton, *Rocket Propulsion Elements: An Introduction to the Engineering of Rockets*, 5th ed. (New York: John Wiley & Sons, 1986).

Saturn V Rocket

John F. Kennedy Space Center, near Titusville, Florida
Lyndon B. Johnson Space Center, near Houston, Texas
Marshall Space Flight Center, near Huntsville, Alabama

Soviet cosmonaut Yuri A. Gagarin's historic orbit of the earth on April 12, 1961, accelerated the race for space. Under the leadership of President John F. Kennedy, the United States that year resolved to put a man on the moon by the end of the decade, but no rocket in the nation was equal to the task. In January 1962, the National Aeronautics and Space Administration announced plans to develop a new launch vehicle—the largest and most powerful ever made—to be called Saturn V. An engineering masterpiece, the Saturn V rocket successfully launched seven manned and unmanned spacecraft before launching the Apollo spacecraft that would touch down on the Sea of Tranquility on July 20, 1969, marking humanity's historic first landing on the moon.

The development of the Saturn V rocket, a joint effort of government and industry, was based on the F-1 and J-2 rocket engines developed by the Rocketdyne Division of North American Aviation, Inc., beginning in 1959. The vehicle consisted of three stages and an instrument unit, atop which sat the 45-ton (40-t) Apollo spacecraft or (as for its last launch in 1973) the 120-ton (108-t) Skylab workshop:

The first stage of Saturn V, the S-IC Stage, 138 feet (42.1 m) high and 33 feet (10.1 m) in diameter, provided the boost to an altitude of about 200,000 feet (60,960 m), or 38 miles (61 km), accelerating the vehicle to 7,700 feet (2,347 m) per second, or 5,250 miles per hour (8,447 km/hr). It employed a cluster of five Rocketdyne F-1 engines generating 7.5 million pounds of thrust (33,360 kN). Each F-1 consumed about 2.5 tons (2.27 t) of fuel and oxidizer (kerosene and liquid oxygen) per second; thus, the booster burned almost 2,000 tons (1,814 t) of propellant in 2.5 minutes, its total burn time.

The S-II Stage, 81.5 feet (24.8 m) high and 33 feet (10.1 m) in diameter, provided the second-stage boost. It was powered by five Rocketdyne J-2 engines burning liquid hydrogen and using liquid oxygen as the oxidizer. During its 6-minute flight, the second stage developed nearly 1 million pounds of thrust (4,448 kN), propelling the Saturn V to an altitude of 114 miles (183 km) and accelerating the vehicle to 15,000 miles per hour (21,100 km/hr).

The S-IVB Stage, 58.6 feet (17.9 m) high and 21.6 feet (6.6 m) in diameter, used a single Rocketdyne J-2 engine developing 200,000 pounds of thrust (890 kN). This was capable of re-igniting, so that the thrust could be applied to the payload in two burns. A first burn of 2.75 minutes hurtled the spacecraft into a 118-mile- (190-km-) high orbit of Earth at a speed of 17,500 miles per hour (28,100 km/hr). A second burn of 5.2 minutes accelerated the spacecraft out of

A Saturn V rocket lifts *Apollo 11* and its three astronauts toward the first moon landing from the Kennedy Space Center's Launch Complex 39 on July 16, 1969.

Earth orbit at 25,000 miles per hour (40,200 km/hr) and into a trajectory that would take it to the moon.

The instrument unit (IU), 3 feet (0.9 m) high and 21.6 feet (6.6 m) in diameter, sat atop the third stage. Shaped like a ring or collar and placed around the upper end of the propellant tank for the S-IVB Stage, the IU contained the guidance and control instrumentation that directed the boost phase of the flight.

Atop the IU sat the *Apollo* spacecraft, consisting of the Lunar Module, Service Module, Command Module, and Escape Tower. The rocket and spacecraft, 363 feet (110.6 m) high overall, was assembled inside the Kennedy Space Center's Vehicle Assembly Building, the world's largest building. Following extensive testing, it was carried erect on a mobile launcher by the huge crawler transporter (see p. 291) to the launchpad 1.5 miles (2.4 km) away.

A team of engineers led by Wernher von Braun (1912–77), director of the Marshall Space Flight Center, Huntsville, Alabama, coordinated development of the Saturn V for NASA. The Boeing Company, North American Aviation, Inc., and the Douglas Aircraft Company built the rocket's first, second, and third stages, respectively. IBM assembled the instrument unit.

The first launch of the Saturn V rocket (*Apollo 4*) was made on November 9, 1967, the first manned launch (*Apollo 8*), on December 21, 1968. Following the lunar landing (*Apollo 11*) of July 1969, additional Saturn V rockets powered six more Apollo missions through December 1972. In May 1973, the last Saturn V rocket to fly launched Skylab, America's first space station, into orbit around Earth.

The Saturn V may well be the first and last of its kind, for the powerful rocket was expendable. Thus, each launch represented the loss of hardware estimated to cost $81 million. NASA's space shuttle program now employs a reusable orbiter and solid rocket boosters.

Location/Access

Saturn V rockets are on display 47 miles (76 km) east of Orlando, Florida, at the John F. Kennedy Space Center, Route 405, NASA Causeway, Orlando, FL 32899, phone (407) 452-2121; the U.S. Space and Rocket Center, the world's largest space museum, adjacent to the Marshall Space Flight Center, Route 565 (Governor's Drive), in Huntsville, Alabama, phone (205) 837-3400; and the Lyndon B. Johnson Space Center, 3 miles (5 km) east of Interstate 45, at 1601 NASA Road, Clear Lake City (near Houston), TX 77058, phone (713) 483-4321. All are open daily.

FURTHER READING

Charles D. Benson and William Barnaby Faherty, *Moonport: A History of Apollo Launch Facilities and Operations*, The NASA History Series (Washington, D.C.: National Aeronautics and Space Administration, 1978).

Roger E. Bilstein, *Stages to Saturn: A Technological History of the Apollo/Saturn Launch Vehicles*, The NASA History Series (Washington, D.C.: National Aeronautics and Space Administration, 1980).

Research and Development

INTRODUCTION by Euan F. C. Somerscales

Since engineering is an art with a scientific foundation, it is not possible to design a perfect product on the drawing board. There is, at least, requirement to test it with the expectation that some adjustment will have to be made. Besides the need to test completed systems, it is necessary to provide data on material properties for use in design. These activities can be subsumed under the heading "Research and Development," or, as it is frequently called, R & D.

Engineering research and development uses many of the measurement devices and shares many of the philosophical concepts of the scientist conducting experimental research, such as a concern with the precision and accuracy of measurement. However, engineering experimentation differs in one important, and not very widely recognized, respect from scientific research. In testing an engineering concept, there is a need to simulate the reality the device will experience in use. Of course, tests carried out under the actual conditions of service would be ideal, but this can involve unacceptable expense and even danger to the test personnel. Consequently, engineers make considerable use of what they call "models." Probably the most familiar examples of engineering models are the small reproductions of actual aircraft that are tested in wind tunnels. Other examples include the small-scale representation of the Mississippi River maintained by the U.S. Army Corps of Engineers at its Experiment Station in Vicksburg, Mississippi, small-scale ships that are used to estimate the engine requirements of a new ship design, and transparent plastic representations of the human heart that allow the blood flow in that organ to be studied. Compared to the system they are simulating, these models share a number of common features: their size may be larger or smaller, the duration of events can be shorter or longer, and the materials used in their construction can be different. These factors, together with the control over experimental conditions that working in the laboratory provides, greatly increases the convenience and reduces the cost and danger associated with testing the actual device. In fact, Osborne Reynolds (1842–1912), who laid some of the foundations of model testing, in 1887—after completing model tests on the silting of the estuary of the River Mersey at Liverpool—wrote, "this method of experimenting seems to afford ready means of investigating and determining beforehand the effects of any proposed estuary or harbor works; a means which, after what I have seen, I should feel it madness to neglect before entering in any costly understanding."

The idea, then, in model testing is to construct a device, together with a testing situation, that simulates reality. The performance of the model is then studied and the performance of the real device, called the prototype, is predicted from the behavior of the model. For example, the force required to tow a ship model through a tank of water at a certain speed is used to predict the power of the engines required to move the prototype ship.

As with much in engineering, the use of models must surely go back to the ancients, but the earliest applications probably provided only qualitative results. It seems likely that the first models to provide quantitative information were the model waterwheels and model windmill vanes tested by John Smeaton (1724–1792) in 1752 and 1753.

As soon as quantitative information must be supplied by the model, it becomes essential to know how to construct and operate it in such a way that its observed performance can be related to the expected performance of the prototype. Or, as Smeaton said in 1759, "it is very necessary to distinguish the circumstances in which a model differs from a machine in large; otherwise a model is more apt to lead us from the truth than toward it."

The requirements for assuring that the model results could be meaningfully related to the performance of the prototype were probably first laid down by William Froude (1810–79) in about 1869. However, others, notably Osborne Reynolds, Lord Rayleigh (1892–1919), and Edgar Buckingham (1867–1940) have contributed to the methodology of model testing. Today, the principles of testing engineering models are well understood, but their successful application to particular situations requires considerable insight into the physical processes that are involved.

Model testing is usually important in the early stages of developing an engineering device. Ultimately, the performance of a full-scale prototype must be evaluated. Sometimes, as in the case of ships, this step is omitted because of expense, but where safety is more important than cost, prototype testing must precede the production of the finished item. The landmarks program has recognized the Vallecitos Boiling Water Reactor, and the Experimental Breeder Reactor I in Idaho as examples of the prototype testing stage of the research and development process.

At the design stage, information on material and component performance and life is required. The Cooperative Fuel Research (CFR) engine was developed for such a task. One of the routine tests to which automobile gasoline is subjected is a "knock" test. This evaluates the tendency of the fuel to burn noisily in an engine cylinder with a consequent loss of power and an increase in driver annoyance. While the phenomenon of knock was identified in the early years of this century, if not somewhat earlier, it took thirty years before engineers could agree on an acceptable method of measuring it under standard conditions.

Mechanical engineering, as the preceding illustrates, has its special research needs, but it is frequently—some might say always—a contributor to scientific experiments. Successful apparatus for scientific research requires the application of the principles of engineering design. The landmarks described in this chapter include two examples of this sort, namely, the Stanford Linear Accelerator and the 100-inch Hooker Telescope at the Mount Wilson Observatory. In both examples, components of substantial size had to be made and assembled to exceptionally close tolerances. In the case of the telescope, the production of the mirror and the cutting of the teeth for the driving gear are outstanding examples of the engineer's art. The Stanford Linear Accelerator presented important engineering problems in the design of systems to handle low-temperature liquid helium and to produce reliable vacuums in large spaces, among other challenges.

In mechanical engineering, it is probably true to say that it is in research and development that the human side of the profession is clearest. The physical labor, the anxiety, the need to placate important persons and irritable employers—all are persistent and inescapable parts of research and development. From the outside, engineering, no doubt, looks like a rational manner of mathematics and science, but in truth it is a tale of blood, sweat, and tears.

Alden Research Laboratory Rotating Boom

Holden, Massachusetts

Rotating booms, or "whirling arms," had been around for more than 160 years by the time the one at the Alden Laboratory was built. The first boom is believed to have been devised during the 1740s by Benjamin Robins in England for studying the air resistance of projectiles. In the early 1750s, John Smeaton, the eminent English mechanical and civil engineer, used one to test windmills; and in the 1760s, in France, Jean Charles Borda used a whirling arm to test the movement of various shapes in water.

Professor Charles Metcalf Allen (1871–1950), director of the Alden Hydraulic Laboratory from 1896 to 1950, needed a moving test stand for hydraulic experiments and for rating current meters used for measuring flows in rivers and streams. Assisted by two students at the Worcester Polytechnic Institute (WPI), Allen designed a boom in 1908.

The original boom at the Alden Laboratory was built of wood on a submerged rock foundation in a pond next to the laboratory. It consisted of a 42-foot (13-m) testing arm balanced by a 21-foot (6-m) arm loaded with counterweights. A 24-inch (609-mm) Hercules hydraulic turbine on shore supplied power, which was transmitted to the boom by rope drive, producing tip speeds up to 10 feet (3 m) per second.

In 1910 the original boom was replaced by an 84-foot (26-m), equal-arm boom made of steel. Following the boom's reconstruction, Professor David Gallup

The Alden Research Laboratory rotating boom is still used for hydromechanical testing. *Photograph by Jet Lowe, Library of Congress Collections.*

of WPI tested aircraft propellers by mounting a 75-horsepower (56-kW) electric motor at the center of the boom, which transmitted power through a long drive-shaft and angle drive to a propeller mounted at the end of the boom. The propeller's thrust caused the boom assembly to rotate, while a drag device in the water calculated the power dissipated; the experiment was important as one of the first nonstationary tests to measure propeller efficiency. During World War I, Major Victor E. Edwards used the rotating boom to conduct drag tests on artillery shells; the tests proved valuable for his subsequent studies of artillery shell ballistics at the Aberdeen Proving Ground (see p. 315).

In 1936 the turbine drive was replaced by an electric motor at the center of the boom, doubling the maximum tip speed to 20 feet (6 m) per second. Since then, the boom has been used periodically—for testing current meters, ships' logs (for measuring oceanic travel), pitot tubes (for measuring the flow of fluids), minesweeper paravanes (for cutting cables on mines), and Darrieus rotors (vertical-axis wind turbines)—without any further major changes.

Location/Access

The rotating test boom is located outside and across the street from the Alden Research Laboratory, Inc., 30 Shrewsbury Street, Holden, MA 01604.

FURTHER READING

Charles M. Allen, "Circular Current Meter Rating Station," *Worcester Polytechnic Institute Journal* (May 1909): 221–28.

C. W. Hubbard, "Investigation of Errors of Pitot Tubes," *Transactions of the American Society of Mechanical Engineers* 61 (1939): 477–506.

100-inch Hooker Telescope, Mount Wilson Observatory

near Pasadena, California

On a clear night, the unaided eye can see about 5,000 stars. With a new 100-inch (2,540-mm) telescope, George Ellery Hale (1868–1938), founder and first director of the Mount Wilson Observatory near Pasadena, California, hoped to see 100 million more. In fact, the light-grasp of the 100-inch (2,540-mm) Hooker telescope—almost twice as great as the observatory's 60-inch (1,520-mm) reflector and 2,500 times greater than the "optick tube" with which Galileo began the modern era of astronomy in 1609—completely revised our ideas about the universe. From 1918 until 1948, when it was eclipsed by the Palomar 200-inch (5,080-mm) reflector, the 100-inch (2,540-mm) Hooker telescope dominated dis-

Since 1918, the 100-inch Hooker Telescope at Pasadena's Mount Wilson Observatory has helped unlock the secrets of the expanding universe. *Courtesy Mount Wilson Observatory.*

coveries in astronomy worldwide, revealing for the first time a dynamic and evolving universe and proving conclusively that our own Milky Way is but one of countless other galaxies.

In 1902 Hale persuaded the Carnegie Institution of Washington, a private foundation for scientific research, to build an observatory on the 5,900-foot (1,800-m) summit of Mount Wilson. The site offered long periods of clear, calm weather and the possibility of establishing shops, laboratories, and offices in the city of Pasadena, within easy reach of foundries, supply houses, and sources of electric power. Hale assembled a team of astronomers, engineers, and technicians to design and construct the telescopes for the Mount Wilson Observatory to new standards of performance. The first was set up in 1904.

In 1906 John D. Hooker of Los Angeles made a gift to the observatory to underwrite a telescope mirror of 100 inches (2,540 mm) aperture. The St. Gobain Plate Glass Company of Paris made the 100-inch (2,540 mm) glass disk, which is 13 inches (330 mm) thick and weighs almost 5 tons (4.5 t); George Willis Richey ground the mirror in the observatory's optical shop in Pasadena.

To support the huge mirror without flexure, instrument designer Francis G. Pease devised a telescope mounting of exceptional size. The Fore River Ship Yard in Quincy, Massachusetts, fabricated the largest pieces of the mounting; smaller ones were made in the observatory's own machine shop. D. H. Burnham &

Company of Chicago designed the rotating dome protecting the telescope; the 500-ton (454-t) dome, 100 feet (30 m) in diameter, is mounted on wheels that run on circular tracks.

To rotate the telescope on sidereal (solar) time, a worm gear 18 feet (5.5 m) in diameter is mounted on the south end of the polar axle. The gear, of cast iron with hollow spokes, was fabricated in two halves and bolted together along the diameter. It meshes with a tool-steel worm driven by a governor-controlled, weight-driven mechanical clock. Since tracking errors greater than one-tenth arc second cannot be tolerated, cutting the teeth of the worm gear required great precision. Clement Jacomini, chief instrument maker, cut the teeth with the gear in place on the telescope. Using a microscope and a diamond scriber, he first divided the gear into 1,440 equal segments, then gashed the teeth one by one.

Equipped with the largest and best reflector in the world, the Mount Wilson Observatory quickly became the world center for the study of astrophysics. One of the most important discoveries was that the intrinsic luminosity, or total light output, of a star could be found by inspecting the record made when starlight is dispersed into a spectrum by a prism or diffraction grating. The discovery opened the way to understanding the evolution of the stars and their ages. In 1929, with its aid, the great Edwin Powell Hubble (1889–1953) made perhaps the most important scientific discovery of the twentieth century: that we live in an expanding universe, that the observed velocities of galaxies increase progressively with ever-increasing distances. Hubble described his groundbreaking work, made possible by the Hooker telescope, in *The Realm of the Nebulae* (1936).

The Hooker telescope was closed in July 1985 because of light pollution, but it reopened in recent years, under the management of the Mount Wilson Institute, with a new adaptive optics system.

Location/Access

The Visitors' Gallery is open to the public on weekends year-round. From April to October, tours are given at 1 P.M. on Saturdays and Sundays. Group tours can be arranged upon application to the Mount Wilson Institute, 740 Holladay Road, Pasadena, CA 91106; phone (818) 793-3100.

FURTHER READING

Isaac Asimov, *Eyes on the Universe: A History of the Telescope* (Boston: Houghton Mifflin Company, 1975).

George Ellery Hale, *Ten Years' Work of a Mountain Observatory* (Washington, D.C.: Carnegie Institution of Washington, 1915).

———, *Signals from the Stars* (New York: Charles Scribner's Sons, 1931).

Dennis Overbye, *Lonely Hearts of the Cosmos: The Scientific Quest for the Secret of the Universe* (New York: HarperCollins, 1991).

Cooperative Fuel Research Engine

Waukesha, Wisconsin

The octane rating of gasoline is familiar to most motorists. But few know how it is measured, and fewer still know how it came to be. As the automobile grew in popularity after 1920, motorists were frequently bedeviled by "knock," a thumping or rattling noise in the engine caused by premature combustion, which was not only annoying but also harmful to the engine.

In the early 1920s, the American Petroleum Institute and the Society of Automotive Engineers formed the Cooperative Fuel Research (CFR) Committee to study the characteristics of combustion in gasoline-powered engines. After several years' study, the committee, comprised of representatives of the automotive and petroleum industries as well as of the academic community and the Bureau of Standards, concluded that a standard single-cylinder test engine was needed as a first step to developing a uniform method of rating fuels for knock.

In December 1928, the CFR Committee authorized the Waukesha Motor Company of Waukesha, Wisconsin, to build a prototype test engine. Forty-five days later, on January 14, 1929, the first Cooperative Fuel Research test engine was delivered to Detroit for display at the annual meeting of the Society of Automotive Engineers.

The CFR Committee stipulated that the test engine meet three qualifications: that it be universal, rugged, and low cost. Engineers at Waukesha Motor met the first demand by designing the engine to permit conversion from a fixed-compression L-head cylinder to a variable-compression overhead-valve cylinder without disturbing any part of the crankcase or camshaft drive. The engine manufacturer met the second qualification by designing a heavy, counterweighted, large-bearing crankshaft; the engine's bearing dimensions are ample enough for a heavy-duty engine 60 percent larger operating at three times the speed. Finally, the engine's universality and the engineers' choice of a cylinder dimension of common size—3¼ by 4½ inches (82 by 114 mm), approximating that of the average passenger car—helped keep down the cost of the engine.

The CFR engine was quickly accepted as the standard test engine of the automotive industry. It provided a recognized standard for defining fuel quality and led to the rapid evolution and improvement of both fuels and engines. By November 1931, one hundred CFR test engines had been sold to petroleum and automotive companies for use in research and for routine production control.

Following introduction of the first CFR engines, on the recommendation of committee members, Waukesha engineers redesigned the variable-compression cylinder and modified the cooling system, valve gear, ignition system, and accessibility of certain parts. Such was the quality of the original design, however, that the engine was easily modified.

Cooperative Fuel Research Engine.

The CFR engine measures the detonation, or knock limit, of a given fuel, and thereby determines its octane rating. It is designed to subject a given fuel-air mixture to a wide range of compression ratios, provide precisely timed ignition, and allow accurate determination of engine performance as a function of fuel type, compression ratio, spark timing, and mixture concentration or leanness. The degree of intensity of knock is registered by a bouncing-pin indicator connected to a knockmeter on the engine's control panel.* Along with the test engine, the CFR Committee adopted a common reference fuel and a uniform procedure for laboratory testing.

Many outstanding engineers contributed to the creation of this unique engine. Among them were Harry L. Horning, Waukesha Motor president, and Arthur W. Pope, Jr., Waukesha chief research engineer, both of whom served on the CFR Committee. Howard M. Wiles, a young researcher whose thesis project at Iowa State University had concerned the design and construction of a variable-

* The bouncing pin is a slender steel rod whose lower end rests on the piston of a cylinder pressure sensor. When detonation occurs, the pin is thrown free of the piston, its upper end striking and closing electrical contacts. An electrical circuit converts the duration and frequency of contact closure into a reading on the dial of the knockmeter.

compression single-cylinder engine, was responsible for machining and assembling the first CFR engine, making many of the parts on a small bench lathe in the Waukesha experimental lab. It was Wiles who understood the significance of the first CFR engine and had the vision to preserve it.

During the 1930s, Waukesha Engine produced a cetane unit for testing the detonation characteristics of diesel fuel and, during World War II, a supercharged version for rating aviation fuel. Waukesha, now a division of Dresser Industries, has produced and sold more than five thousand fuel research engines worldwide. The CFR engine of 1929 is still widely used for rating fuels as well as for basic research on exhaust emissions and alternate fuels.

Location/Access

The first CFR engine is on display at the Product Training Center, which is across the street from the headquarters of Waukesha Engine Division, 1000 West St. Paul Street, Waukesha, WI 53186.

FURTHER READING

H. L. Horning, "The Cooperative Fuel-Research Committee Engine," *Society of Automotive Engineers Journal* 28 (June 1931): 637–41.

Aerodynamics Range, Aberdeen Proving Ground

Aberdeen, Maryland

This was the world's first large-scale ballistic range for producing data on the aerodynamic characteristics of missiles in free flight. The Aerodynamics Range made it possible, for the first time, to accurately record not only projectile motion but also the detailed, transient flow structure about the round. Using innovative launch techniques and high-speed photography, ballistic engineers were able to develop important data on projectile, missile, and aircraft aerodynamics in all flight regimes, from subsonic through hypersonic velocities. Ballistic studies conducted here in the 1940s still form a baseline for modern projectile design.

Robert H. Kent and Alexander C. Charters pioneered advanced ballistic measurement techniques at the Aberdeen Proving Ground beginning in the 1930s. World War II provided an incentive for the U.S. Army to incorporate that research into the design and construction of a new aerodynamics range at the army's Ballistic Research Laboratory (BRL). Completed in 1943, the Aerodynamics Range made it possible to study, in detail, the aerodynamics of bodies in supersonic free flight. During World War II, engineers at the BRL used the Aerodynamics Range to test new designs of shells, rockets, guided missiles, and bombs.

Free-flight aerodynamics range, Aberdeen Proving Ground, Aberdeen, Maryland. Studies conducted here in the 1940s still serve as a baseline for modern projectile design.

The Aerodynamics Range is an enclosed facility instrumented to launch a missile in free flight and record its motion over 285 feet (87 m) of the trajectory. It consists of the firing room containing the launcher; the blast chamber, isolating the instrument area from the muzzle blast; the range gallery, containing apparatus for recording the flight of the missile; and the control room, from which operations and data recording are conducted.

Obtaining information about aerodynamic forces demands unusual precision in the measurement of time, distance, and angle. This was achieved by spark photography—freezing motion on film by using a short, intense flash of light from an open spark. Spark photography gives distance precision to 0.001 foot (0.3048 mm), angular precision to 2 minutes of arc, and time-interval precision to 0.1 microsecond.

A missile is launched from a gun mounted in the firing room with its muzzle in the blast chamber. The gun is positioned so that the missile trajectory traverses the field of the spark photography stations; the stations provide position data on the missile at forty-five different points along the trajectory. When the missile is fired, the enclosed range gallery is dark. As the missile approaches each station, it breaks a light beam, triggering sparks of light one-millionth of a second in duration and of sufficient intensity to expose a photographic plate. The light silhouettes the image of the missile, its shock waves, and the fiducial marks of the station on the photographic plates.

The resulting photographs, called shadowgraphs, reveal the missile's trajectory in minute detail. To analyze the range data, it is necessary to know the velocity of the sound and the density of the air at the time the round was fired. The range is air-conditioned to keep these factors reasonably constant. Temperature, humidity, and air pressure readings are taken immediately after each round is fired.

The size of the Aerodynamics Range limits its use to testing missiles 40 mm and less in diameter, traveling at a minimum velocity of 600 feet (183 m) per second. These limitations led to development of the BRL Transonic Range in 1954. The Aerodynamics Range continues in service, however, producing high-quality data on ballistics. It has served as a prototype for similar facilities elsewhere, including those at the Naval Ordnance Laboratory in White Oak, Maryland, and Eglin Air Force Base, Florida.

Location/Access

The Aberdeen Proving Ground is located at the U.S. Army Test & Evaluation Command, Ryan Building, Aberdeen, MD 21005-5055; phone (410) 278-4173. Open daily for group tours. Phone (410) 278-1151.

FURTHER READING

A. C. Charters, "Free Flight Range Methods," in *High Speed Problems of Aircraft and Experimental Methods*, edited by A. F. Donovan, H. R. Lawrence, F. E. Goddard, and R. Gilruth, *High Speed Aerodynamics and Jet Propulsion*, vol. 8 (Princeton, N.J.: Princeton University Press, 1961.

Icing Research Tunnel, NASA Lewis Research Center

Cleveland, Ohio

Since the first flight in 1903, ice formation on aircraft has presented a serious hazard. But with the growth of commercial and military aviation, the problem of icing became critical. Ice builds up quickly on aircraft surfaces, adding weight and impairing aerodynamic efficiency, even leading, in severe conditions, to a crash.

Early in World War II, the National Advisory Committee for Aeronautics (NACA) directed that an icing research tunnel be added to plans for an altitude wind tunnel being built as part of a new $18-million aircraft engine research center in Cleveland. (NACA's three research centers—at Langley, Virginia; Sunnyvale, California; and Cleveland—later became the nucleus of the National Aeronautics and Space Administration. The Cleveland laboratory was renamed the Lewis Research Center, in honor of George W. Lewis, NACA director of research from 1924 to 1947.)

Developed in 1950, the NACA (now NASA) icing research tunnel spray system was capable of simulating a natural icing cloud.

The NACA icing research tunnel (IRT), designed by Alfred Young and Charles Zelanko, and completed in 1944 at a cost of $670,000, followed on the heels of almost two decades of research into the problem of aircraft icing. It is a single-return, closed-throat tunnel with a test section 6 feet (1,820 mm) high, 9 feet (2,740 mm) wide, and 20 feet (6,096 mm) long. It is similar to other subsonic wind tunnels; aircraft components placed in the test section of the tunnel can be subjected to winds up to 300 miles per hour (483 km/hr) to simulate flight conditions. But the IRT has several unique features.

Air temperature in the test section can be varied from 30°F to -45°F (-1.1°C to -42.7°C). A heat exchanger maintains a uniform air speed and uniform air temperature (±1°F, ±0.56°C) for the duration of the test, while a refrigeration plant, together with spray nozzles, duplicate the icing conditions—liquid content, droplet size, and air temperature—that aircraft might encounter. The liquid water content can be varied from about 0.2 to 3.0 g/m³; droplet size, from 5 to about 40 microns. (A micron equals one thousandth of a millimeter.) The 2,100-ton (1,905-t) refrigeration plant, designed by Carrier Corporation to serve both the icing tunnel and the adjacent altitude wind tunnel, was the largest in the world. (HVAC engineers still point to it as among the most difficult and exacting refrigeration systems ever designed.)

Simulating the atmospheric conditions of an icing cloud presented a far more difficult engineering problem. In 1943 no one knew the size of natural cloud droplets or their liquid-water content. In a cooperative flight research program, NACA, the U.S. Air Force, and commercial airlines gradually collected the necessary data by using pressure-type, icing-rate meters and deliberately flying into severe conditions.

But duplicating a natural icing cloud was another matter. The first icing test was performed on June 9, 1944, but not until 1950, after five years of painstaking research and trial-and-error testing, was the IRT spray system able to produce droplets small enough to reproduce realistic icing patterns on aircraft components.

Vernon Gray directed the efforts of NACA fluid systems engineer Halbert Whitaker to redesign the tunnel's spray system. Through trial and error, the two perfected a design consisting of a battery of some 80 spray nozzles mounted on 6 horizontal bars. The system produced an icing cloud about 4 by 4 feet (1,220 by 1,220 mm) in size.

With the ability to simulate natural icing conditions, the IRT was now a viable research tool. Engines, propellers, induction systems, and other components were tested under a variety of controlled icing conditions, leading to the development of thermal anti-icing and thermal cyclic de-icing systems that virtually eliminated icing as a major menace to air transportation.

Ironically, the evolution of jet aircraft, whose greater power allowed them to fly above icing clouds rather than through them, led Lewis Laboratory to phase out its icing research program in 1957. In response to industry demand, however, NASA reinstituted its tunnel testing program in 1978. The icing tunnel was renovated and its instrumentation updated. Today, private industry contracts with NASA to run tests on hardware under development.

Location/Access

The NASA Lewis Research Center, 21000 Brookpark Road, Cleveland, OH 44135, is adjacent to Cleveland Hopkins International Airport, about 20 miles (32 km) southwest of downtown Cleveland. Tours of its power-systems and propulsion research facilities are given on Wednesdays from 2 to 3 P.M.; advanced reservations are requested, and visitors must be at least sixteen years old. The NASA Lewis Visitor Center houses eight exhibit galleries with displays pertaining to the nation's space program. Phone (216) 433-2001. Hours: Monday–Friday, 9 A.M. to 4 P.M.; Saturday, 10 A.M. to 3 P.M.; Sunday, 1 to 5 P.M. No charge.

FURTHER READING

Virginia P. Dawson, *Engines and Innovation: Lewis Laboratory and American Propulsion Technology*, The NASA History Series (Washington, D.C.: National Aeronautics and Space Administration, 1991).

Rotating-arm Model-test Facility

Hoboken, New Jersey

In the early 1930s, Dr. Kenneth S. M. Davidson (1898–1958) could frequently be seen towing model yachts through the swimming pool at Stevens Institute of Technology. Davidson, a professor of mechanical engineering and an avid yacht racer, was trying to find out why some boats were faster than others.

In 1935, with grants from the Research Corporation of America and interested sportsmen, Davidson built a steel towing tank on the Stevens campus. The tank's dimensions—100 feet (30 m) long, 9 feet (2.74 m) wide, and 4½ feet (1.37 m) deep—made it essential that the test models he used be correspondingly small. (Since the days of William Froude, who in the 1870s pioneered the measurement of frictional resistance of model ships as they were towed through water in a tank, only large and costly model basins had been employed to predict ship performance.) Using models about 5 feet (1,524 mm) in length, Davidson developed the special techniques that are required to obtain reliable data using small models.

By 1939, studies of power-driven vessels had taken the lead over sailing yachts at the Experimental Towing Tank, requiring the development of new test methods and apparatus. Contracts with the U.S. Navy and the National Advisory Committee for Aeronautics gave new impetus to the tank, which advanced from the study of only steady, straight-line motion to a broader concept: steering, turning, directional stability, pitching, and rolling characteristics of various hull types. Between 1942 and 1945, the tank engineers put aside its work with merchant vessels and sailing yachts to focus entirely on war problems, including the development of various types of military landing craft.

In 1942 the Experimental Towing Tank was augmented with the installation of a new tank for the investigation of steering and turning. Tank No. 2 was

Recorded data is studied while a technician makes adjustments to a cargo ship model (rear) undergoing tests in the Rotating-arm Model-test Facility.

financed by the Office of Scientific Research and Development. Built of concrete, it is 75 feet (7 m) square and 4½ feet (1.4 m) deep, with a dock for starting models in one corner. Initially, models tested in this tank were self-propelled or manually operated. In 1945, with financial assistance from the U.S. Navy, laboratory staff members Walter Fried and Alfred Muley designed a unique rotating-test arm for the new tank.

The test arm is a hollow aluminum tube, braced and cantilevered from a central vertical supporting column. The arm is 34 feet (10.36 m) long with tracks running its length for moving the towing carriage to the desired radius. The arm, driven by an electric motor at the top of the column, is capable of rotating up to six revolutions per minute, corresponding to a tangential speed of 22 feet (6.7 m) per second at the end of the arm. Measuring devices on the carriage indicate the mechanical forces acting on the model in motion and transmit the data electrically to a central control station; the data help predict dynamic stability, maneuverability, and control of surface ships, submersibles, and airships. A third tank, sponsored by the Bureau of Aeronautics, was completed in 1944. By this date the staff of the Experimental Towing Tank had grown to sixty-three, almost one-third of whom were women.

Pioneering work conducted here has added immensely to our understanding of the hydrodynamics of ship control. The maneuverability of vessels can now be quantified, and hulls designed and modified to achieve the desired performance. The rotating arm has been applied to hundreds of military and commercial projects, including conception and development of the revolutionary 200-foot (61-m) *Albacore* submarine and development of computer-driven simulators to predict ship trajectories. Tests of a family of ship hulls, meanwhile, have helped define the influence of rudder size, block coefficient, draft, breadth, and section shape upon maneuvering qualities.

Although the model-test facility has been upgraded over the years, its basic principles remain unchanged. Following Davidson's death in 1958, the Experimental Towing Tank was renamed in his honor.

Location/Access

Davidson Laboratory, Stevens Institute of Technology, Castle Point Station, 711 Hudson Street, Hoboken, NJ 07030; phone (201) 216-5300. Hours: weekdays, 9 A.M. to 5 P.M.

FURTHER READING

Kenneth S. M. Davidson, "The Growing Importance of Small Models for Studies in Naval Architecture," *Transactions–The Society of Naval Architects and Marine Engineers* 49 (1941): 91–121.

———, "Turning and Course-Keeping Qualities," *Transactions–The Society of Naval Architects and Marine Engineers* 54 (1946): 152–200.

The Experimental Towing Tank Ten-Year Report (Hoboken, N.J.: Stevens Institute of Technology, 1946).

McKinley Climatic Laboratory

Eglin Air Force Base, Florida

The McKinley Climatic Laboratory has an unequaled ability to simulate a wide range of climatic conditions, producing Arctic cold, desert heat, and tropical humidity to order. Since 1947, hundreds of different aircraft and thousands of items of military equipment have been tested in this "weather wonderland"—that is how Brigadier General C. A. Brandt, first director of the laboratory, described it—providing information vital to their safe and reliable performance under extreme conditions.

The Army Air Corps established a cold-weather testing station at Ladd Field in Fairbanks, Alaska, in 1940, but testing was difficult because of unpredictable weather, and the results often were questionable. Colonel Ashley C. McKinley proposed construction of an all-weather testing facility that would permit testing under controlled conditions year-round, and in 1943 Congress assigned development of a cold-weather testing program to the United States Air Proving Ground Command at Eglin Field, Florida. The Climatic Hangar, as it was first called, was completed in 1947. It was renamed in honor of McKinley in 1971.

The McKinley Climatic Laboratory consists of a main hangar and several smaller test rooms located in separate buildings nearby. The hangar, with an unobstructed floor space of 201 by 252 feet (61 by 77 m) and a ceiling 70 feet (21 m) high at the center, encloses 2.9 million cubic feet (82,070 m^3) of space. It was capable of accommodating five B-29s at the same time. Only a small addition, completed in 1969, was necessary to test the C-5A. The hangar doors, each weighing 200 tons (181 t), move on standard-gauge railroad tracks; electric surface heaters maintain a thermostatically controlled temperature above freezing to prevent the doors from freezing in the closed position.

The weather laboratory included an engine and equipment test room, 30 by 133 feet (9.1 by 40.5 m) and 25 feet (7.6 m) high, able to simulate sand and dust storms; several smaller test rooms, including tropic-marine, desert, jungle, cold, and hot test rooms; an all-weather room; and an altitude chamber. The hangar and test rooms can produce temperatures ranging from -70°F (-57°C) to 165°F (74°C), as well as humidity (from 10 to 95 percent), rain (up to 12 inches, or 305 mm, per hour), freezing rain, fog, snow, dust, sand, wind (up to 100 miles an hour, or 161 km/hr), and artificial sunlight.

The entire facility is electrically powered. Refrigeration is produced by three centrifugal chiller systems using R-12 refrigerant. Each system consists of one 1,000-horsepower (746-kW) low-stage compressor, one 1,250-horsepower (932-kW) high-stage compressor, a condenser, three bi-phase tanks, and liquid-filled cooling coils. The high-stage compressors, each with four centrifugal wheels, are normally used for test conditions requiring temperatures down to 0°F (-17.78°C). For lower

The McKinley Climatic Laboratory, located at Eglin Air Force Base, Florida, is a "weather wonderland," producing Arctic cold, desert heat, and tropical humidity to order. *Courtesy Eglin Air Force Base.*

temperatures, the low-stage, three-wheel units are compounded with the high-stage machines. Two ventilation systems deliver conditioned outside air (heated or cooled) to the hangar to replace air exhausted during engine operations.

The size of the hangar allowed engineers to test aircraft in most operating situations, except flying itself. Flaps, landing gear, bomb doors, bomb releases, gun turrets, cameras, and other mechanical equipment all could be operated and assessed for field-readiness. The first large-scale cold testing in the new laboratory, conducted in May 1947, subjected a Fairchild C-82, Boeing B-29, Lockheed P-80, North American P-51, Republic P-47, Sikorsky R-5D helicopter, trucks, tanks, clothing, and other items to a temperature of -70°F (-57°C).

The McKinley Climatic Laboratory put cold-weather testing on a scientific basis, advancing the sciences of climatic simulation and testing, refrigeration, and insulation. Today, in addition to military use, the laboratory conducts a wide variety of nonmilitary testing vital to the national interest. These have included: for the National Aeronautical and Space Administration, tests of insulation on the space shuttle fuel tank; for the Federal Aviation Administration, tests of deicing fluids for aircraft wings and visibility instruments in fog; for the Tennessee Valley Authority, tests of the effect of freezing rain on electrical contacts; for the National Science Foundation, tests of balloon-supported instrumentation systems for cloud studies; and for the U.S. Coast Guard, developmental testing of the 65A Dolphin helicopter and Arctic survival training for personnel. This is still the only facility of its kind. Even snow tires for automobiles have been tested here.

Location/Access

The main gate of Eglin Air Force Base is on U.S. 98 northeast of Fort Walton Beach. Tours of the laboratory may be scheduled by calling the Public Affairs Office at (904) 882-3931, or write to 101 West D Street, Suite 110, Eglin AFB, FL 32542.

FURTHER READING

C. W. Kniffen, "The AAF's Climatic Hangar," *Refrigerating Engineering* 54 (August 1947): 128–30.

"Hangar for Aeroplane Tests in Extreme Climatic Conditions," *Engineering* (London) 166 (9 July 1948): 25–27.

Experimental Breeder Reactor I

Arco, Idaho

Imagine a furnace that can change burning fuel into more fuel, enough to replace the fuel it burns plus a little bit more. This describes a process called fuel breeding, first demonstrated to be technically feasible by the Experimental Breeder Reactor I (EBR-I) in 1953. Developed and operated by scientists of the Argonne National Laboratory between 1947 to 1963, EBR-I was also the first nuclear reactor to produce usable amounts of electricity, demonstrating to the world that the atomic nucleus could serve as a source of power.

During World War II scientists and engineers, led by Enrico Fermi, had worked feverishly to achieve a controlled nuclear chain reaction as a step toward developing America's first nuclear weapons. They became convinced that breeding more fuel than is consumed in a nuclear reactor was a theoretical possibility, but the urgency of war dictated that full attention be given to the weapons program, so interest in the breeder reactor had to be put aside temporarily. But breeder research flourished after the war as the United States Atomic Energy Commission (AEC), established by Congress in 1946, explored the possibilities of civilian uses of atomic power.

In 1947 the AEC approved the construction of a prototype fast-neutron breeder reactor that would demonstrate power generation and attempt to prove the theory of fuel breeding. The task seemed critical, given the scarcity of fissionable material (known uranium reserves then were much lower) and increased Defense Department demands. The experimental reactor, designated EBR-I,

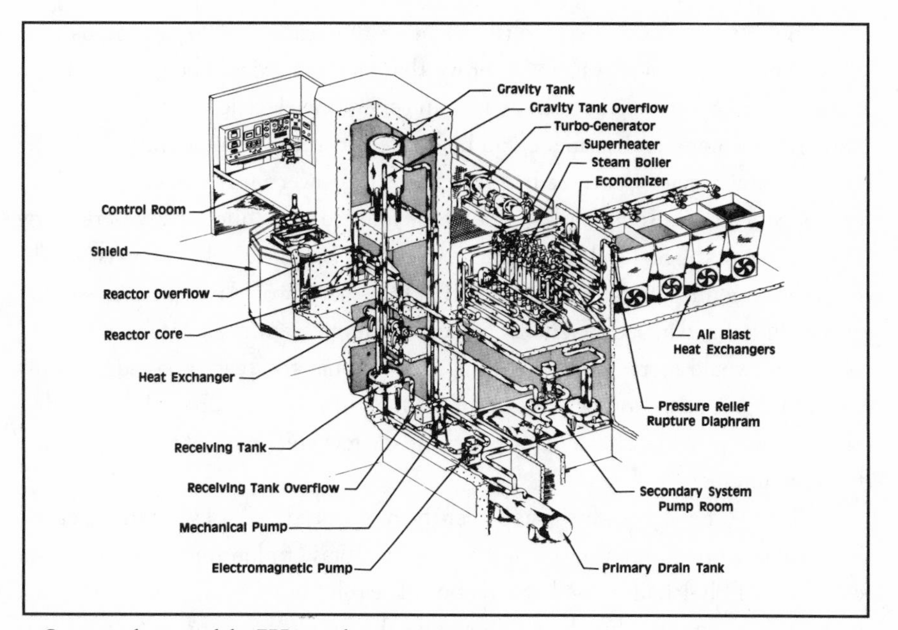

Cut-away drawing of the EBR-1 and supporting systems.

would use fissionable uranium-235 as fuel to convert relatively stable uranium-238 into fissionable plutonium-239. It would be cooled by liquid sodium-potassium. (Thermal reactors use light or heavy water as a coolant, but water moderates the neutrons, thereby decreasing the breeding ratio, a trait undesirable in breeder reactors.) A team led by physicist Walter H. Zinn, a student of Fermi's and first director of the Argonne National Laboratory, planned the fast reactor, one of several projects to be built simultaneously at a new testing station in a remote desert region of eastern Idaho.

EBR-I, built by the Bechtel Corporation, was completed in 1951. Late in May, Zinn arrived for the first attempt to reach criticality—a painstaking procedure because Zinn could only guess at the number of pencil-like rods of uranium-235 that would be needed. Not until August 24, with a little more than 114 pounds (52 kg) of fuel, did Zinn bring the reactor to the point of criticality. Following several months of low-power runs and tests of the control system, on December 20 Zinn gathered his team for a historic experiment: for the first time, they would attempt to produce useful amounts of electricity from nuclear power. Gradually, Zinn increased the power—to 250, to 340, to 410 kilowatts. The chain reaction was producing heat in the "blanket" of natural uranium surrounding the core. At 1:23 P.M., Zinn noted in the log book: "Load dissipater connected to generator. Electricity flows from atomic energy." Security concerns dictated a more cryptic message be sent to the AEC's director of reactor development: "Our boy started his journey today," Zinn wrote in a telegram. "All is well. He was able to undertake the trip without assistance. Merry Xmas."

EBR-I operated at a power level of about 1.2 megawatts thermal, producing

250 kilowatts of electricity, enough to meet the reactor building's needs. But EBR-I's primary mission was not to prove that electricity could be generated by a nuclear reactor—there was nothing new in generating electric power from heat—but whether more nuclear fuel could be created than the reactor consumed. The experimental reactor continued to run at design power until a leak in the heat exchanger caused a temporary shutdown in June 1952. While repairs were being made, some of the fuel rods were removed for analysis. In October, the first results from the Argonne laboratory suggested that the reactor would prove the possibility of breeding. Encouraged, scientists turned to the problem of designing improved cores that would increase the breeding ratio, so that the reactor could not only sustain its own operation but produce more fuel. In June 1953, EBR-I proved that breeding was possible when chemists separated a few milligrams of plutonium from the uranium fuel.

Three other cores were tested over the next decade. The last of these, called Mark IV, used plutonium to produce 1.27 new atoms of plutonium for each atom consumed. EBR-I had proved the technical feasibility of nuclear fuel breeding, promising to extend a hundredfold or more the nation's uranium resources. Following completion of the Mark IV tests in 1963, EBR-I was shut down and decommissioned. In 1966 it was designated a National Historic Landmark by the U.S. Department of the Interior.

Location/Access

EBR-I is located at the Idaho National Engineering Laboratory.

FURTHER READING

Richard G. Hewlett and Francis Duncan, *A History of the United States Atomic Energy Commission*, vol. 2, *Atomic Shield* (University Park, Pa.: The Pennsylvania State University Press, 1969).

William Lanouette, "Dream Machine," *The Atlantic Monthly*, April 1983, 35-52.

Association of American Railroads' Railroad-wheel Dynamometer

Pueblo, Colorado

The Association of American Railroads' railroad-wheel dynamometer tests railroad wheels under controlled conditions exceeding those of normal service. It can apply a maximum brake shoe force of 40,000 pounds (18,150 kg) and can test wheels at speeds up to 178 miles per hour (286 km/hr) and under loads of up to 127,000 pounds (57,606 kg), the largest equivalent inertia wheel load in railroad practice. Its ability simultaneously to apply lateral and vertical loads of up

AAR Railroad-wheel Dynamometer from motor end.

to 15,000 pounds and 60,000 pounds (6,800 kg and 27,215 kg), respectively, makes this machine unique among railroad-wheel dynamometers.

The railroad wheel supports the locomotive or car as it rolls along the rail, guiding it through curves and switches with its flange. The wheel also acts as a brake drum; when the brakes are applied, the brake shoes press against the wheel tread and, through their rubbing friction, slow or stop the train. This friction also causes the rim of the wheel to heat and expand, generating additional stresses in the rim and plate. As running speeds and weights on wheels doubled and doubled again in the late nineteenth century, mechanical and thermal stresses often caused cracks to develop in the cast-iron wheels and, sometimes, disastrous wheel failures. The problem led railway mechanical engineers and metallurgists to begin formal studies of wheel failures, and wheel and brake-shoe design, materials, and manufacture.

To test full-size wheels and brake shoes under controlled conditions, Southern Pacific Railroad designed and built what is believed to be the first full-scale wheel and brake dynamometer in 1891. In 1893 the Master Car Builders' Association (forerunner of the mechanical division of the Association of American Railroads) commissioned a full-scale brake-shoe testing machine in order to establish an industrywide standard for friction and wear for the many makes of brake shoes then in use. As cars became heavier and operating speeds higher, which increased braking heat and put even greater demands on the integrity of the wheels, other dynamometers were built to study braking and its effect on shoes and wheels.

In the early 1950s, to evaluate the performance of its wrought-steel wheels and axles, and cast-iron brake shoes, U.S. Steel Corporation (USS) created preliminary design specifications for a railroad-wheel dynamometer with capabilities far beyond those of existing braking dynamometers. The steelmaker contracted with Adamson United Company of Akron, Ohio, to build the special dynamometer for use at the USS-operated Applied Research Laboratory in Monroeville, Pennsylvania. Joseph M. Wandrisco, USS chief research engineer for railroad products, was responsible for design specifications. Rex C. Seanor, chief engineer of Adamson United, oversaw its construction.

In 1983 the Association of American Railroads purchased the dynamometer from U.S. Steel and moved it to its Chicago technical center. AAR wanted to evaluate the effects of grade or drag braking and rolling loads on internal stresses in various freight-car wheel designs. It also needed a dynamometer to test composition and metal brake shoes for certification and quality control. It was moved to Pueblo, Colorado, in 1996.

The dynamometer can test wheels under a variety of braking and rolling-contact loading conditions. It can conduct grade, stop, and static braking tests with a high degree of precision, and can test axles as rotating cantilever beams. The flanged stub axle, on which the test wheel is mounted, is equipped with strain gauges to measure torque during braking.

The dynamometer's replaceable, friction-driven rail is a continuous ring of heat-treated rail steel bolted to a fabricated wheel. The rail wheel can be oscillated laterally to simulate curving and lateral instability. Ten thick and four thin 64-inch- (1,620-mm) diameter steel flywheel disks can be bolted to the flanged rotor shaft to produce the inertia effect of dynamic wheel loads up to 27,000 pounds (12,250 kg), greater than that of any other railroad dynamometer.

The machine is powered by a 200-horsepower (149-kW) direct-current mill motor to speeds ranging from 0 to 1,500 rpm (0 to 178 mph, or 286 km/hr, for a 40-inch, or 1,020-mm, wheel). A 250-kW generator, driven by a 400-horsepower (298-kW) synchronous motor, supplies power to the motor. In 1987 the dynamometer was computerized to allow automatic control and the digital display of test input and results.

Location/Access

The dynamometer may be viewed upon application to the Association of American Railroads, Transportation Technical Center, Test Center Road, P.O. Box 11130, Pueblo, CO 81001-0130; phone (719) 584-0541.

FURTHER READING

G. F. Carpenter and T. E. Johnson, "AAR Dynamometer," *Proceedings of the 1988 Joint ASME-IEEE Railroad Conference, Pittsburgh, Pa., April 13–14, 1988*, 103–9.

Vallecitos Boiling Water Reactor

near Pleasanton, California

The Vallecitos Boiling Water Reactor (VBWR) was the first privately built nuclear power plant to supply power in megawatt amounts to an electric-utility grid. Although this developmental plant's capacity was small (5 megawatts) and its life brief (only six years), tests conducted here helped pave the way for the larger nuclear plants that followed.

The Vallecitos Boiling Water Reactor was built by General Electric and Bechtel Corporation in 1956 and 1957 for joint operation by GE and the Pacific Gas & Electric Company. It went critical—that is, it achieved a controlled, self-sustaining nuclear reaction—in August 1957 and was connected with the PG&E grid the following October. Until it was shut down in 1963, the VBWR helped develop and test boiling-water reactor fuel, core components, controls, and systems, and provided a valuable training ground for engineers, physicists, and operators.

Following the discovery of nuclear fission in 1939, scientists knew that fission (the splitting of atoms) released far more energy than common chemical reactions, such as burning coal. But to get useful heat from fission, engineers had to find ways to make the uranium fuel stable at high temperatures and to transfer the heat produced. The idea of extracting heat by boiling water, an obvious choice, led to construction of an experimental boiling-water reactor at the Argonne National Laboratory, which proved the soundness of the idea.

In the early 1950s, Commonwealth Edison Company, in cooperation with Nuclear Power Group, Inc., signed a contract with General Electric and Bechtel to design and build the 180-megawatt Dresden Nuclear Power Station in Illinois. The Vallecitos reactor was designed to serve as a pilot plant for the Dresden project. There, GE would test nuclear stability, alternative control systems, instrumentation, heat transfer, and other aspects of boiling-water-reactor operation.

The core of the VBWR, containing the fuel elements, was enclosed in a vessel 7 feet (2,133 mm) in diameter and 20 feet (6,096 mm) high, lined with stainless steel. The initial fuel elements consisted of a mixture of fully enriched uranium oxide and stainless-steel powder clad in stainless steel; this was later replaced with a prototype element for the Dresden plant consisting of enriched uranium-oxide pellets clad in zirconium. In addition to the reactor, the pressure vessel contained the coolant (water) and control rods (rods of neutron-absorbing material inserted into the reactor to control the rate of the reaction).

The reactor was housed in a vapor-tight steel cylinder 48 feet (14.63 m) in diameter and 100 feet (30.48 m) high, with hemispherical ends. A 4-foot- (1,219-mm-) thick concrete floor at grade level separated the reactor vessel, steam generator, pumps, and piping (which were all below ground) from the upper service area. The turbine-generator and auxiliaries were housed in a two-story building

Vallecitos Boiling Water Reactor, June 1957. *Courtesy Pacific Gas & Electric Company.*

adjacent to the reactor. Interestingly, the turbine, a standard marine unit modified to accept saturated steam, started life aboard an American-built tanker.

Vallecitos proved an ideal test facility. Its radiation environment—water at 550°F (287°C) and 1,000 psig (6,895 kPa)—was identical to that in larger boiling-water reactors, while its operating schedule could be readily adjusted to the needs of engineers testing alternative operating cycles, something that could not have been done at a plant dedicated entirely to producing electricity.

When the Vallecitos reactor was shut down in 1963, it had generated 391,000 megawatt-hours of thermal energy. Its generator had been connected to the PG&E grid for 16,614 hours and had delivered 40,400 net megawatt-hours of electricity—all without a single lost-time accident. More importantly, the plant had provided reliable performance data on boiling-water reactors and demonstrated the safety of nuclear power.

Location/Access

The Vallecitos reactor is located at GE's Vallecitos Nuclear Center near Pleasanton, California, approximately 40 miles (64 km) southeast of San Francisco and is not open to the public. The reactor, pressure vessel, and laboratories were still there in 1987, but the fuel, turbine-generator, and instrumentation had been removed.

FURTHER READING

L. Kornblith, Jr., and W. A. Raymond, "Operating Experience with the Vallecitos Boiling Water Reactor," *Electrical Engineering* 78 (April 1959): 334–38.

Stanford Linear Accelerator Center

Stanford, California

High-energy particle physics lies on the borderline between physics and engineering, and the development of particle accelerators—the "atom smashers" of the popular press—to higher and higher energies has depended on the production of reliable and accurate machines, each larger and more powerful than its predecessor.

Particle accelerators are machines used to give energy to beams of electrically charged subatomic particles. The most powerful accelerators are used in research to discover the fundamental components of matter and to study their behavior. In a linear accelerator, or "linac," nuclear particles are injected into one end and accelerated to very high energies by an oscillating electric field.

Modern linacs were pioneered by two groups: one at Stanford, working with electrons; and another at the University of California-Berkeley, working with protons. At Stanford, W. W. Hansen's interest in X-ray problems led him to look for new ways of obtaining high-voltage electrons beginning in the 1930s. Encouraged by the success of two 12-foot (3.7-m) accelerators, dubbed the Mark I and Mark II, Hansen and his colleagues proposed construction of a billion-volt (or 1 GeV, for giga electron volts) accelerator. Funded by the Office of Naval Research and completed in 1952, this 300-foot- (91.4-m-) long machine, designated the Mark III, proved a powerful tool for nuclear research and served as a model for electron accelerators worldwide.

The Stanford Linear Accelerator Center (SLAC), operated under a contract with the U.S. Department of Energy, was established in 1961 as a national laboratory for research in particle physics and the development of new techniques in high-energy accelerators and elementary particle detectors. The basic research tools at SLAC have required the application of technologies involving high-vacuum systems, low temperatures, high-speed electronics, precision small-scale mechanical fabrication, special magnets, and unconventional materials.

The 2-mile- (3.2-km-) long linear accelerator, which began operations for physics research in 1966, can provide electron and positron beams at energies up to 50 GeV, making this the most powerful linac in the world. Construction of traveling-wave accelerators like this one requires machine work of the highest precision. The accelerator is aligned to be straight within 0.020 inch (0.5 mm). If any section of the accelerator is out of alignment, the electromagnetic fields produced in the walls by the passing beam can severely limit the intensity.

The accelerating waveguide, a long conducting tube about four inches (10.2 cm) in diameter, was assembled from cylinders and disks forming individual microwave cavities. These cavities, made of high-purity copper, were machined to a precision within 0.20 inch (5.08 mm), then, to hold high vacuum, brazed in a

One end of a 40-foot (12-m) section of the accelerator, showing a retractable Fresnel alignment target. Electrons are injected into one end and accelerated to high energies by pulsed microwaves. The rectangular waveguide near the top conveys microwave power to the accelerator.

hydrogen furnace into 10-foot (3-m) sections. Each cavity was then "tuned" by slightly deforming the outside using hydraulic rams. There are nearly 100,000 of these cavities in the accelerator.

The problem of "beam break-up" was solved by mounting four 10-foot (3-m) accelerator sections on a large aluminum pipe and aligning them with an optical transit in the laboratory. These 40-foot- (12.1-m-) long sections were then transported to the underground tunnel and connected together. Fresnel targets, built in the end of each section, intercepted a laser beam traveling down the center of the pipe. Each section was then aligned separately using its own jacking screws.

A "switchyard" of magnetic elements at the end of the accelerator can direct the beams to any of several experimental areas, including three magnetic spectrometers capable of analyzing momenta up to 1.6, 8, and 20 GeV. Other SLAC facilities include SPEAR, an electron-positron storage ring facility engaged in research with colliding beams, each of energy up to 3.7 GeV (SPEAR was used in the discovery of the psi particle, for which researchers received the 1976 Nobel

Prize); PEP, a large colliding-beam storage ring completed in 1980; and the SLAC Linear Collider (SLC), which began operating in 1988. The latter machine reaches center-of-mass energies up to 100 GeV, where the recently discovered Z°, the neutral mediator of the weak interaction, can be produced.

In recent years, SLAC has shifted its principal focus from fixed-target experiments to the study of electron-positron annihilation to form hadrons, leptons, and photons. Researchers at SLAC continue to uncover new modes of disintegration and new complexities in the properties of matter.

Location/Access

The Stanford Linear Accelerator Center is located in the scenic eastern foothills of the Santa Cruz Mountains, west of the main campus and about 30 miles (48 km) south of San Francisco (2575 San Hills Road, Menlo Park, CA 94025). It is open to the public for tours. Contact the Public Information Office: phone (415) 926-2204.

FURTHER READING

M. Chodorow, et al., "Stanford High-Energy Electron Accelerator (Mark III)," *The Review of Scientific Instruments* 26 (February 1955): 134–204.

Leon Lederman with Dick Teresi, *The God Particle* (Boston and New York: Houghton Mifflin Company, 1993).

M. Stanley Livingston and John P. Blewett, *Particle Accelerators* (New York: McGraw-Hill Book Company, Inc., 1962).

Communications and Data Processing

INTRODUCTION by William J. Warren

Our ideas live through communication. Our access to the accumulated knowledge of the world depends on our ability to store and retrieve it. Gutenberg's printing press made possible multiple reproduction of works previously copied only laboriously by hand. Thus, the thoughts of scholars were opened to anyone with the skills to read them.

Commerce has also been heavily dependent on communications. Both the typewriter and the telephone revolutionized the business office. Rapid, precise communication is an expected norm rather than a luxury.

Demand for quick access to the news of the day led to the development of many devices. The Paige Compositor was designed to automate the laborious process of hand-setting type. While it worked, the rapid development of other devices quickly supplanted it. Nevertheless it was an important step in the process.

Edison's phonograph gave us the opportunity to enjoy great musical artists in our own homes. It also led down the path of data storage by other than written means.

As the volume of written communication increased, so did the needs of the delivery system. Licking and hand-canceling stamps for thousands of pieces of mail became impractical for both business users and the U.S. Post Office. The Pitney-Bowes postage meter was an early precursor of today's automated mail-delivery service.

One of the greatest impacts on communications in the last fifty years was the invention of the photocopying machine. Quick multiple images allowed distribution of an avalanche of information to all remotely interested parties. Office files bulged with paper. More reasonable storage and retrieval systems were needed. From that necessity came the IBM RAMAC disk file. Storing millions of bits of information on multiple disks with lightning-fast retrieval led to the useful computer. Programs and data equivalent to rooms of files could be stored within a desktop box and accessed in the blink of an eye.

Fast and inexpensive communications have allowed the development of today's world. Mechanical engineers are still in the forefront of making communications easier and more precise.

Edison Phonograph

West Orange, New Jersey

Thomas Alva Edison (1847–1931) stumbled across the phonograph in 1877 while engaged in research on the telegraph at his laboratory in Menlo Park, New Jersey. The young inventor's "automatic telegraph" recorded incoming Morse messages by embossing the dots and dashes on revolving discs of paper. Edison was intrigued to discover that a needle attached to a vibrating diaphragm would indent patterns onto a moving surface and that at high speeds "a light musical, rhythmical sound, resembling human talk heard indistinctly" sometimes emanated from the instrument.

On November 29 that year, Edison produced a sketch of a machine to record and reproduce the human voice. "Instead of using a disc," he later recalled, "I designed a little machine using a cylinder provided with grooves around the surface. Over this was to be placed tinfoil, which easily received and recorded the movements of the diaphragm. A sketch was made and the piecework price $18 was marked on the sketch."

Edison gave the sketch to John Kruesi, one of his top machinists, who built the machine despite his skepticism. The first phonograph was a solid brass and cast-iron instrument, with a 3½-inch (90-mm) grooved cylinder on a 1-foot-(305-mm-) long shaft and a hand crank to turn it. At each end of the cylinder was a diaphragm with a stylus, mounted in an adjustable tube.

On December 6, 1877, with his assistants at his side, Edison put tinfoil around the cylinder and turned the handle of the shaft while shouting into one of the diaphragms:

Mary had a little lamb,
Its fleece was white as snow,
And everywhere that Mary went
The lamb was sure to go.

He turned the shaft back to the starting point, drew away the first diaphragm, adjusted the other to reproduce sound, and again turned the handle. All were amazed to hear Edison recite the nursery rhyme "almost perfectly." Edison later declared, "I was never so taken aback in all my life. Everybody was astonished. I was always afraid of things that worked the first time." Edison filed a patent for the phonograph on December 15, 1877.

The new talking machine received worldwide notoriety and won fame for Edison as the greatest inventor of the age. But after his initial success, Edison put it aside for ten years to work on the first practical incandescent light and electrical distribution system. When he moved to a new laboratory in West Orange in 1887, Edison resumed sustained experimental work on his "favorite invention" and established two factories to meet growing popular demand for the device; by 1914,

Thomas Edison and his new talking machine, ca. 1877.

Edison's phonograph business exceeded $7 million in volume annually. Edison continued to work on the technical improvement of the phonograph for the rest of his life, amassing 195 patents.

Edison's simple and unprecedented instrument allowed, for the first time, the permanent recording and reproduction of sound and brought music, heretofore the luxury of the privileged class, into the everyday life of the common person. From this machine evolved great industries whose products have increased the enjoyment of life for people worldwide.

Location/Access

Edison's experimental phonograph is on display in Thomas Edison's laboratory, part of the Edison National Historic Site administered by the National Park Service, Main Street at Lakeside Avenue, West Orange, New Jersey 07052; phone (201) 736-5050. Laboratory tours are given Wednesday through Sunday, 9:30 A.M. to 4:30 P.M. Admission fee (except for senior citizens and children under sixteen).

FURTHER READING

Matthew Josephson, *Edison: A Biography* (New York: McGraw-Hill Book Company, Inc., 1959).

Andre Millard, *Edison and the Business of Innovation*, Johns Hopkins Studies in the History of Technology (Baltimore: The Johns Hopkins University Press, 1990).

Paige Compositor

Hartford, Connecticut

The Paige compositor, or typesetting machine, named after inventor James W. Paige (1841–1917) of Rochester, New York, was the first machine able to set, justify, and distribute type simultaneously and automatically from a common case. Its history is complicated, and it is hard to assign a definite date to its invention. Paige's original idea for the typesetting machine, developed in 1873, made no provision for justification, which was developed by Charles R. North, a skilled mechanic employed by Paige. In 1877 Paige took his invention to the Farnham Typesetter Company in Hartford, which gradually developed the gigantic compositor.

The Paige compositor has been called a gentle giant. It is 9 feet (2,743 mm) long and weighs 5,000 pounds (2,268 kg). With some 18,000 separate parts, 800 shaft bearings, and an almost uncountable number of springs, it is the most elaborate machine ever built to set type for the printer. For all its bulk and complexity, however, the colossus could set only agate (5½-point) type.

The keyboard of 109 characters was arranged so that whole words could be assembled with a single stroke. A skilled operator, using all fingers of both hands, could set and justify 9,000 to 12,000 ems (a unit of type) per hour. As the operator brought down the words, the machine automatically measured the space occupied by each word; divided the space left in the line by the number of spaces required; forwarded the type to the justifying mechanism, where spaces of the proper thickness were automatically inserted; and pushed the correctly justified line onto the receiving galley or raceway.

The Paige compositor could set, justify, and distribute type automatically, but it was almost immediately eclipsed by the Linotype machine.

Distribution was accomplished independently. Three columns of dead type could be placed in the distributor galley, located beneath the machine, at a single time. There, the machine ejected broken, inverted, or otherwise defective characters before advancing them to the composing section.

The first completed Paige compositor, built at the Pratt & Whitney works in Hartford, was unveiled in 1887. After further experimentation and protracted delays owing to insufficient capital, a second machine was announced in 1894. But by then, Ottmar Mergenthaler (1854–99) had perfected his Linotype machine, which, instead of reusing foundry-cast type, cast new type for each application. The Paige machine's backers, among them Samuel Clemens (Mark Twain), reputedly lost more than $2 million in the ill-fated venture. Its inventor, meanwhile, died penniless.

Only two Paige compositors were built. The first, of 1887, is the only survivor. Mergenthaler Linotype Company gave the other, of 1894, to Cornell University, which donated it for scrap during World War II. The Paige made only a brief commercial appearance, when it was tested in the offices of the *Chicago Herald* in 1894.

Location/Access

The Paige compositor of 1887 is displayed at the Mark Twain House, 351 Farmington Avenue, Hartford, CT 06105; phone (203) 493-6411. Hours: June-October: daily, 9:30 A.M. to 4:30 P.M.; November–May: Tuesday–Saturday, 9:30 A.M. to 4 P.M., and Sunday 1 to 4 P.M. Admission fee.

FURTHER READING

Lucien Alphonse Legros and John Cameron Grant, *Typographical Printing-Surfaces: The Technology and Mechanism of Their Production* (London: Longmans, Green, and Co., 1916).

Pitney-Bowes Model "M" Postage Meter

Stamford, Connecticut

The introspective inventor envisioned a machine that could print stamps on envelopes, doing away with the costly and time-consuming process of buying, licking, and sticking, and protecting businesses from stamp theft. Arthur Hill Pitney (1871–1933), a Chicago wallpaper-store clerk, evidently knew nothing of a half dozen earlier attempts to build a postal meter when he devised his "postage-stamp device" in 1901. Postal officials greeted his invention with skepticism, but Pitney obtained a patent (No. 710,997) the following year and shortly thereafter formed the Pitney Postal Machine Company to develop the idea.

Pitney's invention consisted of a printing mechanism mounted in a frame,

Pitney-Bowes postage meter, ca. 1920. *Courtesy Pitney-Bowes.*

with a series of "numbering-wheels" for consecutively numbering each printed impression. The device featured a hand crank and an automatic lock to prevent tampering. The lock would provide security for postal revenues. The meter was to be set by the postmaster and sealed; once a predetermined number of envelopes had been posted, the machine was to be returned to the postmaster, who would break the seal and reset the machine for further operation.

Pitney was invited to Washington to demonstrate his device to a special committee of the Post Office Department. The U.S. Post Office tested Pitney's improved machine —it was now electrically operated, automatically sealing and stacking envelopes as well as printing the indicia —from November 1903 to January 1904. But despite the test's success, postal officials did nothing about Pitney's device, instead authorizing bulk mail identical in size, weight, and content to be mailed without stamps. The Post Office had accepted Pitney's idea of a printed indicia taking the place of the adhesive stamp but had overlooked the essence of Pitney's invention: the recording, self-locking postage meter that would allow the posting of first-class mail without licking stamps.

In 1912 Pitney devised a way to detach the printing and registering mechanism from the mail-handling machinery so that it could be brought to the Post Office for resetting instead of having the postal representative travel to it. The new system was tested from January to May 1914 at the Addressograph Company of Chicago and other firms with sensational results. But war in Europe delayed the legislation Pitney needed to move forward with his invention.

In 1919 a pessimistic Arthur Pitney brought his invention to Walter Bowes (1882–1957), a manufacturer of post-office canceling machines. Pitney's patent rights were about to expire, and he had invested some $90,000 in the machine with little to show for it. Pitney and Bowes decided to join forces. Pitney, the inventor, would work to improve his brainchild, while Bowes, the promoter, concentrated his energy in Washington, lobbying for passage of the necessary legislation to allow first-class mail to be carried without postage stamps affixed.

Congress finally approved such a bill in March 1920. The Pitney-Bowes Postage Meter Company was formed the following month. In August, Walter H. Wheeler, Jr., Bowes's twenty-three-year-old stepson, successfully demonstrated the Model "M" postage meter. Post Office authorization came the following month. The Pitney-Bowes postage meter was put into commercial use for the first time at the Stamford, Connecticut, Post Office on November 16, 1920.

By 1924, Pitney-Bowes had installed over a thousand postage meters in 112 U.S. cities. By 1940, one of every five letters was metered mail. A desk-model postage meter, introduced in 1949, allowed even the smallest companies to enjoy the advantages of metered mail.

Today, with over a million postal meters in use, metered mail accounts for the largest single source of postal revenue. And, because it eliminates the culling, facing, and canceling of 47 billion pieces of adhesive-stamped mail, metered mail represents enormous cost savings to the postal service. Today Pitney-Bowes is a multinational business equipment, supplies, and financial services company.

Location/Access

A reconstruction of the Model "M" postage meter is on display in the main lobby of Pitney-Bowes, One Elmcroft Road, Stamford, CT 06926.

FURTHER READING

William Cahn, *The Story of Pitney-Bowes* (New York: Harper and Brothers, 1961).

Xerography

Columbus, Ohio

The story of the development of Xerography is really three stories: one, of a solitary wizard who recognized a need, then devoted his life to seeking technical solutions and financial support to fulfill it; another, of an innovative research organization with the foresight to invest in that idea and the technical talent to engineer it; and finally, of a small, entrepreneurial company with the courage to defy convention and risk its assets to bring a new concept and pioneering technology to the marketplace.

Xerox Model D copier, one of the first production units, at Battelle's Columbus laboratories.
Photograph by Jet Lowe, Library of Congress Collections.

Chester F. Carlson (1906–), a New York patent attorney with a degree in physics, conceived the idea of a "dry" office copier in 1935. In the course of his patent work, Carlson frequently needed copies of patent specifications and drawings and was dissatisfied with the available methods of photography or tracing. Carlson's starting point was the New York Public Library, where he spent evenings and weekends reading everything he could about photoconductivity and electrostatics. Carlson's concept was to provide an electrical charge on a metal plate coated with melted sulfur.

Carlson at first experimented in the kitchen of his home in Astoria, Queens. He soon hired Otto Kornei, a young Austrian physicist, to help him, and the two set up shop above a bar-and-grill in Astoria. There, in 1938, they succeeded in producing the world's first dry, electrostatic copy. Carlson inked a terse message —"10.-22.-38 ASTORIA" (the date and place of the experiment)—onto a glass slide, rubbed a sulfur-coated plate vigorously with a handkerchief to stimulate an electrical charge on its surface, put the slide over it, and exposed it by aiming the light of a flood lamp through the slide. Next, he dusted the plate with lycopodium powder (from the spores of a creeping evergreen plant) and placed a piece of waxed paper on the plate. The image was instantly transferred to the paper.

For the next several years, Carlson tried unsuccessfully to develop his invention by getting outside help in perfecting and marketing it. Finally, in 1944, Battelle Memorial Institute, a leading independent research-and-development organization, signed a contract with Carlson. Battelle agreed to develop Carlson's

The Xerographic Process

The surface of the Xerographic drum consists of a base of aluminum over which is laid a thin layer of aluminum oxide. On top of this is a layer of a selenium alloy. Selenium is a photoconductor; i.e., it will conduct only when exposed to light.

1. *The drum is charged electrostatically in the dark by rotating it under a corotron, a bare wire to which a high positive voltage is applied.*
2. *The image of the original to be copied is projected onto the drum by a series of lenses and mirrors. The white areas of the original reflect the light, destroying the charge on the drum, while the black areas do not reflect and therefore leave the charge intact, forming an image in static electricity on the drum.*
3. *A developer, or a mixture of a carrier and toner powder, is poured over the drum. The carrier consists of tiny plastic-coated glass beads, about 0.0098 inch (0.25 mm) in diameter; the toner is a fine black powder composed of a thermoplastic resin and carbon. Both substances are triboelectric; that is, they generate static electricity when rubbed together. The carrier receives a positive charge; the toner, a negative charge. Thus, the carrier beads become covered with a layer of toner, and there is now an image in toner on the drum.*
4. *The toner is held on the drum by the positive charge. To remove the toner from the drum onto the paper, the paper needs a higher positive charge. This is accomplished by another corotron, which charges the paper as it is pressed against the drum; the toner now clings to the paper.*
5. *The copy now passes under a radiant heater or through heated pressure rollers. The toner melts into the fiber of the paper to give, when cool, a permanent dry copy.*
6. *The copy is now finished, but the drum must be cleaned before making the next copy. The drum is discharged by means of a negative or AC corotron, followed by wiping and exposure to light, which neutralizes any charge left on it.*

copying process—he called it "electrophotography"—and to give Carlson 25 percent of all profits or royalties.

The Battelle scientists materially advanced Carlson's work. A major step forward was the discovery that the nonmetallic element selenium worked better than the sulfur. Just as important, the Battelle researchers made important improvements in the development of the powder image. The powder (toner) became a pigmented resin that could be fused to the paper; its combination with coarser particles made it possible to develop a clean, sharp image (see sidebar).

Battelle now sought a partner to support further research and eventually to produce and market the product. Nobody wanted it, including such giants as Addressograph-Multigraph and Lockheed. But Haloid Company, a small Rochester, New York-based manufacturer of photographic paper and supplies, was looking for something to lift the firm above its neighbor and rival, Eastman Kodak, and entered the first of several agreements with Battelle in 1946. Following consultation with a Greek scholar at Ohio State University, the cumbersome "electrophotography" was changed to Xerography, from the Greek *xeros*, "dry," and *graphein*, "to write." The new word was clearly descriptive of Carlson's revolutionary dry-graphics invention.

Xerography was publicly demonstrated for the first time on October 22, 1948, exactly ten years after Carlson had reproduced his first image, at a meeting of the Optical Society of America in Detroit. Haloid introduced its first commercial copier, the Model A, the following year, but the machine was crude and complicated to use. "Copyflo," the first completely automated Xerographic machine, was unveiled in 1955. It was the first copier to use a rotating drum instead of a plate as the photoconductive surface.

Haloid changed its name to Haloid-Xerox in 1958 and, to move the product to market, tried to find new capital. The firm offered to share the project with other, larger companies, including Bell & Howell and IBM, but was rebuffed for lack of interest. Haloid-Xerox took a gamble, risking all of its assets to develop the "Xerox 914" copier (so called because of its ability to copy sheets up to 9 by 14 inches, or 228 by 335 mm). The world's first dry office copier was a huge success despite its bulk (the machine was as large as an office desk) and cost ($29,500).

In 1961 Haloid-Xerox became Xerox Corporation. Revenues surged as the company tried to keep pace with phenomenal demand for its product. Xerox Corporation was, for now at least, the unrivaled master of a new industry. And Chester Carlson was a multimillionaire.

Location/Access

Battelle's Columbus Laboratories (505 King Avenue, Columbus, OH 43201-2693) exhibits a model of one of Chester Carlson's first Xerographic devices and the first commercial copier, the Model A, in its lobby. Phone (614) 424-6424.

FURTHER READING

John H. Dessauer, My *Years With Xerox: The Billions Nobody Wanted* (Garden City, N.Y.: Doubleday & Company, Inc., 1971).

J. Mort, *The Anatomy of Xerography: Its Invention and Evolution* (Jefferson, N.C.: McFarland & Company, Inc., 1989).

IBM 350 RAMAC Disk File

San Jose, California

An IBM 305 RAMAC computer delivered to the Zellerbach Paper Company in San Francisco in 1956 ushered in a new era of magnetic disk storage and interactive computer applications, inaugurating what would come to be called the Information Age. The magnetic disk was widely and rapidly accepted—for airline reservation systems, inventory control, banking, and word processing, to name only a few applications—triggering a huge demand for disk-storage devices and launching a new industry.

The storage component of the 305 RAMAC (for Random Access Method of Accounting and Control) was the IBM 350 disk file, developed in 1955 by a team of IBM engineers working in a small research laboratory in San Jose. (In its early development, the magnetic disk file was called the 305. It became the 350 when the 305 system, which included a central processor, a card reader, and a printer, was announced.) Consisting of fifty magnetically coded metal platters stacked one atop the other and rotated by a common drive shaft, the IBM 350 disk file held an astonishing 5 million bytes of data.

Early in 1952, IBM asked Reynold B. Johnson, a former high-school science teacher who had been hired by IBM to develop a test scoring machine, to set up

The IBM 350 RAMAC disk file ushered in the era of magnetic-disk storage.

a small research laboratory on the West Coast. The new laboratory was to research nonimpact printing and data reduction. In short order, Johnson leased and renovated a small concrete-block building in San Jose, advertised for engineers to staff the facility, and began interviewing applicants. By July 1952, IBM's new San Jose Research and Development Laboratory comprised a staff of thirty.

In 1952 there were only three ways of storing information for use by data-processing equipment: punched cards, magnetic tape, and, to a lesser extent, magnetic drums. The new laboratory's first assignment was to automate the punched-card "tub files"—large, rectangular trays containing master cards arranged in sequence by customer number, item number, size, color, etc.—to make the information they contained more readily accessible.

Despite the skepticism of his colleagues, Johnson decided to concentrate on magnetic disks. After a period of trial and error, engineers settled on aluminum laminates clamped under pressure and heated above the annealing temperature. A magnetic coating, made from a paint base, was applied onto the inner surface of the rapidly rotating disks, then spread evenly. An air bearing would "float" a reading and writing arm above the surface of the disks. Meanwhile, researchers worked on an electrical-servo drive system to provide disk-to-disk and track-to-track accessing motions. On February 10, 1954, the San Jose team achieved the first successful transfer of information from cards to disks and back. The unwieldy Rube-Goldberg contraption they had built was soon reevaluated and redesigned—the shaft holding the disks was made upright, for example, rather than horizontal—and in November the team was authorized to develop the 305 RAMAC system to use the new magnetic disk technology.

Since the introduction of the RAMAC 350 file, of course, magnetic disks have been vastly improved. Today's microminiature read-write heads use thin-film technology, permitting vastly increased storage density and much faster access time. A comparison between the 350 file and a modem disk file, the IBM 3380, tells the story: The 350's fifty 24-inch (609-mm) disks revolved at a speed of 1,200 rpm, with a resulting data rate of 100,000 bits per second; the 3380, using nine 14-inch (355-mm) disks, revolves at 3,600 rpm, with a data rate of 24 million bits per second. The 350 disk file stored 2,000 bits of information per square inch (25.4 mm); the 3380 packs 12 million bits in the same amount of space.

Location/Access

A historical display is located in the lobby of IBM's Building 12, 5600 Cottle Road, San Jose, CA 95193; phone (408) 256-9450.

Biomedical Engineering

INTRODUCTION by Euan F. C. Somerscales

Engineers are currently applying their skills and knowledge to solve some of the many problems associated with living systems. These activities have been described by a number of titles, including biophysics, bionics, bioengineering, biomedical engineering and clinical engineering. In general, this terminology is used rather loosely and indiscriminately, but no doubt, practitioners of these various subdisciplines would prefer to differentiate among the various terms. It is not appropriate here to go into a lengthy discussion on the differences, real or otherwise, among the titles, so we have chosen to lump together all these activities in which engineers are assisting in the solution of problems associated with living systems.

Although biomedical engineering is an area that has really blossomed since the close of World War II, its roots, in fact, go back to the eighteenth century or even earlier. Admittedly, the earliest examples of the practice of biomedical engineering were by those who had been trained as physicians rather than engineers, but, nevertheless, they were applying engineering principles to living systems.

Luigi Galvani (1737–98), an Italian physician, for example, was the discoverer of bioelectricity, which has had a profound effect on clarifying our understanding of the nature of the nervous system, and on the diagnosis and treatment of diseases of the heart and brain, organs that are both the center of intense electrical activity. J. L. Poiseuille (1799–1869), when he was a medical student, made the first attempt to measure blood pressure using the engineer's traditional mercury-filled U-tube. Hermann von Helmholtz (1821–94) was a physician who turned himself into a physicist by his studies of the law of conservation of energy, of acoustics, and of electricity.

At the present time, engineers are directly involved in fundamental research on the properties of bone and soft tissue and joints. They have contributed to the development of artificial organs, prostheses such as hip replacements, instrumentation such as the CAT scanner, assistance devices such as cardiac pacemakers, and computer aids to persons with communication and mobility limitations. The blood heat exchanger is a perfect example of the application of well-developed engineering techniques to the solution of an important medical need, namely, how to temporarily replace the pumping and oxygenating mechanism of the human body while the heart is replaced or defects corrected.

Engineers believe strongly that their activities are intended to be of benefit to society, and certainly this intention appears to be most directly attained in the area of biomedical engineering.

Blood Heat Exchanger

Amherst, New York

The blood heat exchanger marked a significant advance in the treatment of patients who undergo open-heart surgery by shortening the time required to cool patients before surgery and to rewarm them afterward. It was developed jointly in 1957 by research engineers at the Harrison Radiator Division of General Motors Corporation, Lockport, New York, and physicians in the Department of Surgery, Duke University School of Medicine, Durham, North Carolina.

Lowering the patient's temperature slows metabolic activity and reduces the body's oxygen needs, permitting lower blood-flow rates through the artificial heart-lung machine, protecting the brain and other vital organs and lessening the damage to delicate blood cells. Formerly, patients were cooled with ice packs or refrigerated blankets after the induction of anesthesia, which required an hour or more and was difficult to control; following surgery, it took four to five hours to rewarm them. With the blood heat exchanger, the body temperature could be safely lowered in five minutes, carefully controlled during surgery, then raised in just ten to fifteen minutes. Faster and more accurately controlled cooling and warming allowed notable improvements in surgical management, including the elimination of prolonged anesthesia.

In the mid-1950s, a team of physicians at Duke led by Dr. Ivan W. Brown, Jr., conducted research on hypothermia, i.e., chilling the body to subnormal temperatures. Brown thought there should be a way to cool or rewarm the blood while it was being pumped through the extracorporeal circuit; the blood vessels themselves would then rapidly cool the patient before surgery and rewarm the patient afterward. He asked engineers at Harrison Radiator for help in the design and development of a blood heat exchanger. A ten-month cooperative research program between the Duke physicians and Harrison research engineers, led by GM's W. O. Emmons, resulted in a marvel of ingenuity.

Blood is a suspension of living cells in a solution—plasma—comprised of sixty-odd proteins, complex molecules, and enzymes. The delicate nature of blood

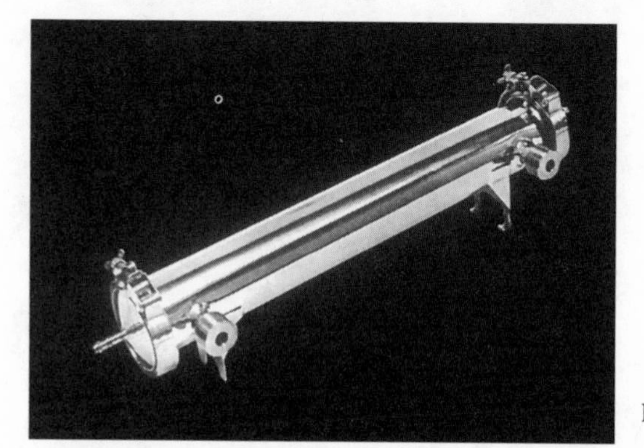

Blood Heat Exchanger.

and the intended function of the heat exchanger dictated an array of stringent specifications. The device had to be capable of thorough cleaning and inspection, free of bubble-trapping crevices, simple to assemble, and safe from intrusion of foreign matter into the bloodstream. All these requirements posed unique problems in the blood heat exchanger's design and development.

The blood heat exchanger was designed by determining the amount of heat to be dissipated, establishing a water-flow rate to do the job, and, finally, determining the material and the shape of the device. Made of stainless steel throughout, it consisted of an outer cylindrical jacket that was 15¼ inches (382 mm) long and 2¼ inches (57 mm) in diameter, through which 24 straight, thin-walled (0.035 inch, or 0.889 mm) tubes 0.18 inch (4.6 mm) in inside diameter ran longitudinally. The ends of the tubes were welded into a header plate at each end, and the surfaces in contact with blood were highly polished.

On each end of the exchanger was a stainless-steel cap fashioned with a beaded tube to which the plastic blood inflow and outflow tubing was attached. The inside of each cap was conically shaped to provide sloping surfaces at each end of the exchanger to avoid trapping any gas bubbles. The caps were tightened against Silastic O-rings by a special clamp closure to minimize the possibility of leaks and to facilitate cleaning.

The heat exchanger was mounted in a vertical position on a floor stand. Near the bottom of the exchanger jacket was a water inlet through which cold or warm water entered and flowed upward to a similar outlet at the top of the exchanger. Baffle plates inside the jacket of the exchanger insured thorough circulation of the cooling or warming water around the thin tubes carrying the blood. The water intake was connected by a short length of hose to an automatic thermoregulated water-mixing valve, into which hot and cold tap water was admitted by conventional rubber garden hoses.

The use of the blood heat exchanger for open-heart surgery quickly became standard practice. Today, these machines are used on heart-lung and artificial kidney circuits worldwide, although the original Brown-Harrison exchanger has been superseded by cheaper, disposable models.

Location/Access

The first commercial blood heat exchanger is on display at Capen Hall on the Amherst Campus of the State University of New York at Buffalo, Amherst, NY 14260; phone (716) 645-2000.

FURTHER READING

Ivan W. Brown, Jr., M.D.; Wirt W. Smith, M.D.; and W. O. Emmons, "An Efficient Blood Heat Exchanger for Use with Extracorporeal Circulation," *Surgery* 44 (August 1958): 372–77.

Williard O. Emmons and Demetrio B. Sacca, "The Design and Development of a Blood Heat Exchanger for Open Heart Surgery," *General Motors Engineering Journal* 5 (July–September 1958): 38–41.

FURTHER READING

The following is a list of books for readers who would like to extend their general knowledge of the history of mechanical engineering beyond the topics covered in this volume:

A. F. Burstall, *A History of Mechanical Engineering* (London: Faber and Faber, 1963; reprint, Cambridge, Mass.: The MIT Press, 1965).

T. K. Derry and T. I. Williams, *A Short History of Technology* (New York and Oxford: Oxford University Press, 1961; reprint, New York: Dover Publications, 1993).

Engineering Heritage, 2 vols. (London: Heinemann [for the Institution of Mechanical Engineers], 1963 and 1966).

R. B. Gordon and P. M. Malone, *The Texture of Industry* (New York: Oxford University Press, 1994).

L. C. Hunter, *A History of Industrial Power in the United States*, 1780–1930, vols. 1 and 2 (Charlottesville, Va.: The University Press of Virginia [for the Eleutherian Mills–Hagley Foundation], 1979 and 1985).

L. C. Hunter and L. Bryant, *A History of Industrial Power in the United States, 1780–1930*, vol. 3 (Cambridge, Mass.: The MIT Press, 1991).

R. S. Kirby, S. Withington, A. B. Darling, and F.-G. Kilgour, *Engineering in History* (New York: McGraw-Hill, 1956; reprint, New York: Dover Publications, 1988).

L. McNeil, ed., *An Encyclopedia of the History of Technology* (London: Routledge, 1990).

C. Singer, E. J. Holmyard, and A. R. Hall, eds., *A History of Technology*, 5 vols. (Oxford: Clarendon Press, 1954–58).

INDEX

Note: Introductory essays set each landmark in context with others within the same chapter. Although additional information may be available in these introductions, they were not indexed in detail in order to avoid duplication.